Garden of Egypt

NEW TEXTS FROM ANCIENT CULTURES

Founding Editor
Traianos Gagos†

Edited by
Francesca Schironi, The University of Michigan
Luigi Battezzato, Scuola Normale Superiore

Garden of Egypt: Irrigation, Society, and the State in the Premodern Fayyūm
by Brendan Haug

Settling a Dispute: Toward a Legal Anthropology of Late Antique Egypt
by Traianos Gagos and Peter van Minnen

Women of Jeme: Lives in a Coptic Town in Late Antique Egypt
by T. G. Wilfong

Wine, Wealth, and the State in Late Antique Egypt:
The House of Apion at Oxyrhynchus
by T. M. Hickey

New Literary Papyri from the Michigan Collection:
Mythographic Lyric and a Catalogue of Poetic First Lines
by Cassandra Borges and C. Michael Sampson

Materia Magica:
The Archaeology of Magic in Roman Egypt, Cyprus, and Spain
by Andrew T. Wilburn

Honor Among Thieves:
Craftsmen, Merchants, and Associations in Roman and Late Roman Egypt
by Philip F. Venticinque

Getting Rich in Late Antique Egypt
by Ryan E. McConnell

Recording Village Life: A Coptic Scribe in Early Islamic Egypt
by Jennifer A. Cromwell

Ancient Latin Poetry Books: Materiality and Context
by Gabriel Nocchi Macedo

Confiscation or Coexistence: Egyptian Temples in the Age of Augustus
by Andrew Connor

Garden of Egypt

IRRIGATION, SOCIETY, AND THE STATE IN THE PREMODERN FAYYŪM

Brendan Haug

UNIVERSITY OF MICHIGAN PRESS
ANN ARBOR

Published in the United States of America by the
University of Michigan Press
Printed and bound by CPI Group (UK) Ltd, Croydon, CR0 4YY

First published May 2024

A CIP catalog record for this book is available from the British Library.

Library of Congress Cataloging-in-Publication data has been applied for.

ISBN: 978-0-472-13352-9 (hardcover : alk. paper)
ISBN: 978-0-472-90440-2 (open access ebook)

DOI: https://doi.org/10.3998/mpub.11736090

The University of Michigan Press's open access publishing program is made possible thanks to additional funding from the University of Michigan Office of the Provost and the generous support of contributing libraries.

Down the great aqueduct, by freight of politics and gravity, |
came the excess waters

—Mark Arax, *The Dreamt Land*

Preface

This project was long in gestation. The 2012 Berkeley dissertation that inspired it was researched and written in haste during a difficult period. That it was completed at all is owed largely to my PhD adviser and mentor Todd Hickey, who gave me the time I needed to find both my footing and a research topic that would take me beyond the boundaries of traditional papyrology. The topic eventually emerged from the accidental but fortunate intersection of two unrelated (or so I thought) streams of research. The first was my hobbyist's interest in the Columbia Valley Project, the government-directed water reclamation project that transformed my home in arid southcentral Washington State into an agricultural and hydroelectric powerhouse. The second—Arabic and the history of early Islamic Egypt—was sparked by Jim Keenan's seminal work on Abū ʿUthmān al-Nābulusī's fiscal survey of the Fayyūm. Jim was an indispensable guide to this text and helped me begin to tease out further connections between al-Nābulusī's Fayyūm and the landscape more dimly revealed in the papyri. More important still, Jim put me in touch with Yossef Rapoport and Ido Shahar, who in 2011 graciously shared a preliminary draft of their English translation of al-Nābulusī. This greatly expedited my reading of the text and did much to improve my then rudimentary Arabic to boot. I also received assistance from Alan Mikhail, who fielded many emails about his work on irrigation in Ottoman Egypt, and from Cornelia Römer and Willy Clarysse, who introduced me to the Fayyūm's archaeological landscape during a November 2011 visit. Back at Berkeley, the other members of my dissertation committee—Susanna Elm, Carlos Noreña, and Maria Mavroudi—offered constant advice and encouragement. Maria's discomfort with the overly materialist framework of the dissertation was also instrumental in precipitating my eventual turn away from an obsessive and vaguely determinist emphasis on the material agency of water, soil, and salt, and toward the more humane approach reflected here. Finally, Ari Bryen's kindness, generosity, and much-needed companionship during these years will never be forgotten and cannot adequately be repaid.

I wrote this book entirely at Michigan, and I thank my library supervisors, chairs of Classics, and faculty mentor—James Hilton, Bryan Skib, Ruth Scodel, Sara Forsdyke, Artemis Leontis, Celia Schultz, and Arthur Verhoogt—for their unstinting support. Thanks also to my students in environmental history, graduate and undergraduate alike, for many stimulating discussions. Henry Upton provided valuable research assistance, and Eli Weaverdyck and Julian Thibeau produced the maps and drawings. Yossef Rapoport and Ido Shahar (again), Noha Abu Khatwa, Jessica Barnes, Aileen Das, Usama Gad, Khadija El-Hazimy, and Ibrahim Khalaylih helped me with Arabic, and Lance Jenott and Andreas Winkler with Egyptian. Todd Hickey and Lajos Berkes read the entire manuscript and offered valuable feedback. So too did an "anonymous" reader, whose lonely last stand against the split infinitive will long be remembered in song. Preliminary versions of some of my arguments were presented in New Orleans, Oxford, Leuven, Edinburgh, Tokyo, and Newcastle. My thanks to Ryan Boehm, Tyler Franconi, Sofie Waebens, Katelijn Vandorpe, Kimberley Czajkowski, Andreas Gavrielatos, Ryosuke Takahashi, Wakako Kumakura, Guillemette Crouzet, Jane Rowling, and Rebecca Wright for organizing these events. An earlier version of chapter 3 appeared in Japanese in the *Journal of Social Sciences and Humanities (Jinbun Gakuhō)* 515–19, and *History and Archaeology* 47, translated by Ryosuke Takahashi. An earlier version of chapter 4 was submitted in 2017 for inclusion in a forthcoming volume of *Studia Hellenistica* edited by Sofie Waebens and Katelijn Vandorpe.

Working with Ellen Bauerle at the University of Michigan Press, both on this project and others, has been a pleasure and I thank her for her professionalism and her patience. The readers whose commentary she solicited also offered many helpful critiques that substantially improved my arguments and organization. That I did not take all their advice is not a judgment on its merits but a reflection of my own idiosyncrasies—a passion for seeking out ever more (and ever more obscure) sources, and a preference for the journey over the destination. Whether my resulting tendency to err on the side of "more is more" is an asset or a liability I leave to the reader to decide.

Finally, none of this would have been possible without my wife, Amanda, and our daughters Evelyn and Lydia, who have had to endure my too many absences over the last few years. They are my life and my joy, and it is to them that this work, such as it is, is dedicated. Words cannot express my love and gratitude.

أود لو كان فمي كنيسة
وأحرفي أجراس

Contents

Digital materials related to this title can be found on the Fulcrum platform via the following citable URL: https://doi.org/10.3998/mpub.11736090

Maps

Figures

Tables

Aids for the Reader

1. Calendar

The Egyptian calendar consists of twelve months of thirty days, along with five intercalary days (six in leap years). The calendar below lists the names of each month (Greek, Bohairic Coptic, and al-Masaʿūdī's Arabic) along with non–leap year Julian equivalencies.

Θώθ (Thōth)	Ⲑⲱⲟⲩⲧ (Thōout)	توت (Tūt)	29 Aug.–27 Sept.
Φαῶφι (Phaōphi)	Ⲡⲁⲟⲡⲓ (Paōpi)	بابه (Bābah)	28 Sept.–Oct. 27
Ἀθύρ (Hathyr)	Ⲁⲑⲱⲣ (Hathōr)	هاتور (Hātūr)	28 Oct.–26 Nov.
Χοίακ (Choiak)	Ⲭⲟⲓⲁⲕ (Koiak)	كيهك (Kīhak)	27 Nov.–26 Dec.
Τῦβι (Tybi)	Ⲧⲱⲃⲓ (Tōbi)	طوبه (Ṭūbah)	27 Dec.–25 Jan.
Μεχείρ (Mecheir)	Ⲙⲉϣⲓⲣ (Meshir)	أمشير (Amshīr)	26 Jan.–24 Feb.
Φαμενώθ (Phamenōth)	Ⲡⲁⲣⲉⲙϩⲁⲧ (Paremhāt)	برمهات (Baramhāt)	25 Feb.–26 Mar.
Φαρμούθι (Pharmouthi)	Ⲫⲁⲣⲙⲟⲩⲑⲓ (Pharmouthi)	برموده (Baramūdah)	27 Mar.–25 Apr.
Παχών(Pachōn)	Ⲡⲁϣⲟⲛⲥ (Pashons)	بشنس (Bashans)	26 Apr.–25 May
Παῦνι (Payni)	Ⲡⲁⲱⲛⲓ (Paōni)	بونه (Baunah)	26 May–24 Jun.
Επείφ (Epeiph)	Ⲉⲡⲓⲡ (Epip)	أبيب (Abīb)	25 Jun.–24 Jul.
Μεσορή (Mesorē)	Ⲙⲉⲥⲱⲣⲓ (Mesōri)	مُسرى (Musrā)	25 Jul.–23 Aug.
Epagomenal days: 24–28 Aug.			

2. Units of measurement

- *aroura*: basic Graeco-Roman unit of land measurement = 2,756.5 m^2.
- *artaba*: basic Graeco-Roman unit for dry goods = 38 dry liters.
- cubit: Greek *pēchys* (πῆχυς) = 52.5 cm; Arabic *dhirā*ʿ (65.6 = (ذراع cm.

- *naubion* (ναύβιον): Egyptian measure of volume equaling 3 cubits in each direction = 3.9 m^3.
- *qabḍa* (قبضة) "fist width" (ca. 10.9 cm), the unit in which al-Nābulusī reports Fayyūm water rights.

3. Administrative terms

- *meris* (μερίς): administrative subdivision of the Graeco-Roman Fayyūm, of which there were three (Herakleidēs, Themistos, and Polemon).
- nome (νομός): administrative district in Graeco-Roman Egypt, of which there more than forty.
- *stratēgos* (στρατηγός): the Graeco-Roman governor of an Egyptian nome.

4. Irrigation and agricultural terms

- *ʾayyām al-Nīl* (أيام النيل): lit. "the days of the Nile," i.e., the Nile flood.
- *baḥr* (بحر): lit. sea or river, designating a large public canal, including the Baḥr Yūsuf.
- *chōma*, pl. *chōmata* (χῶμα, χώματα): earthwork embankments built along a canal to reinforce it in advance of the flood.
- command area: the total amount of land irrigated by a canal.
- *diōryx* (διῶρυξ): large public canal.
- *drymos*, pl. *drymoi* (δρυμός, δρυμοί): marshes along the outskirts of the ancient Fayyūm, artificially irrigated for the cultivation of brushwood.
- *emblēma* (ἔμβλημα): transverse dike or weir that slows canal flow in order to create a small reservoir or to divert water into adjacent land.
- *faddān* (فدان): Egyptian unit of land measurement = 6,368 m^2.
- *hydreuma* (ὕδρευμα): well through which the water of reservoirs was accessed.
- *khalīj* (خليج): term for a large public canal in Abū Isḥāq's survey of the Fayyūm.
- *masqa* (مسقة): contemporary term for a tertiary canal shared by a small number of irrigators.
- *nasba* (نسبة): contemporary term for a weir-cluster on a branch canal that governs the apportionment of water to multiple smaller *masqa*s.
- *pedion* (πεδίον): lit. "plain," the agricultural area of a village.
- *perichōma*, pl. *perichōmata* (περίχωμα, περιχώματα): square or rectangular ring-dike surrounding a field to contain water for flood-recession irrigation.
- *phrontis* (φροντίς): lit. "concern," referring to a division of a large estate in the third century CE.

- *sāqiya* (ساقية): animal-powered waterwheel. Greek *mēchanē* (μηχανή).
- *ṣayfī* (صيفي): the canal-irrigated summer crop unique to the Fayyūm.
- *shādūf* (شادوف): counterpoise lift for drawing water from a well.
- *shatawī* (شتوي): standard flood-irrigated winter crop in the Nile Valley, Delta, and Fayyūm.
- sluice gate: Greek *thyra* (θύρα), Arabic *bāb* (باب), movable gate for modulating the flow of water through a dam or weir.

Transliteration and Translation

Arabic is transliterated in accordance with standards of the *International Journal of Middle East Studies*. For Greek, long vowels *ēta* (η) and *ōmega* (ω) are indicated by a macron except in proper names. All translations are the author's own except where indicated.

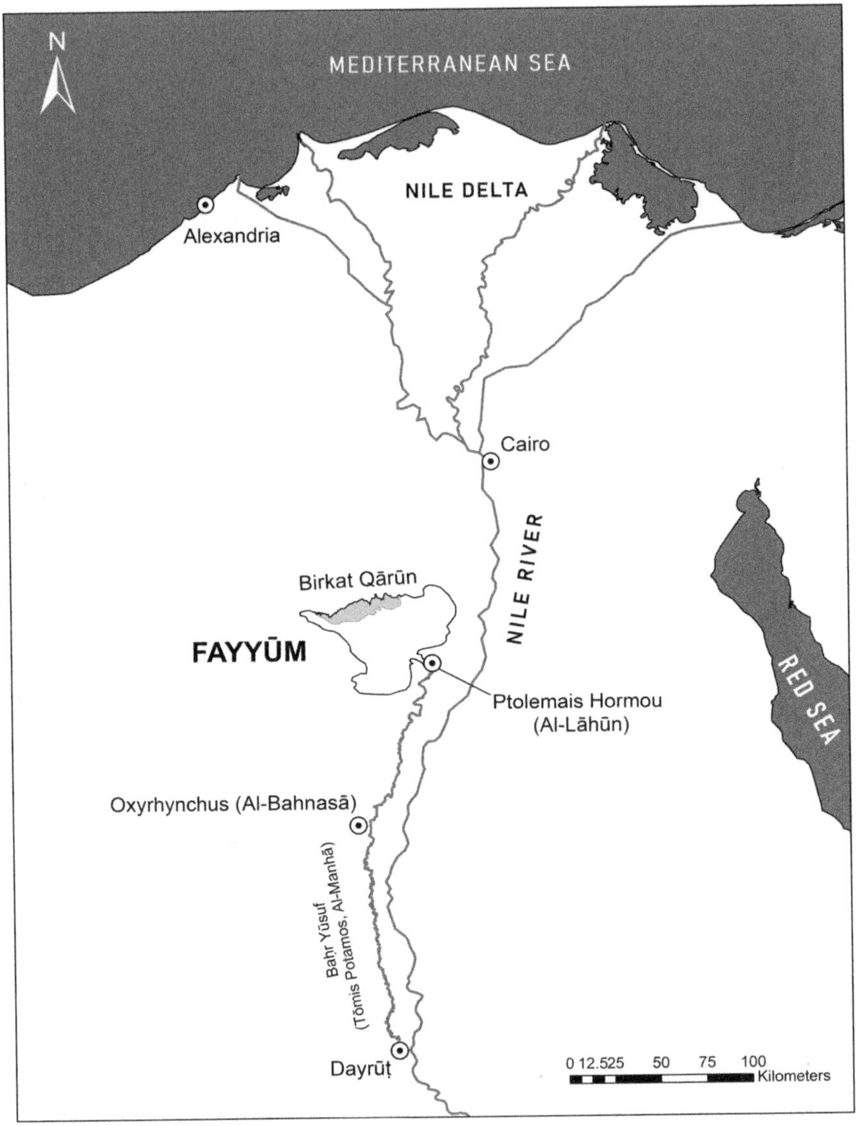

Map 1. Egypt and the Fayyūm (Cartographer: Julian Thibeau).

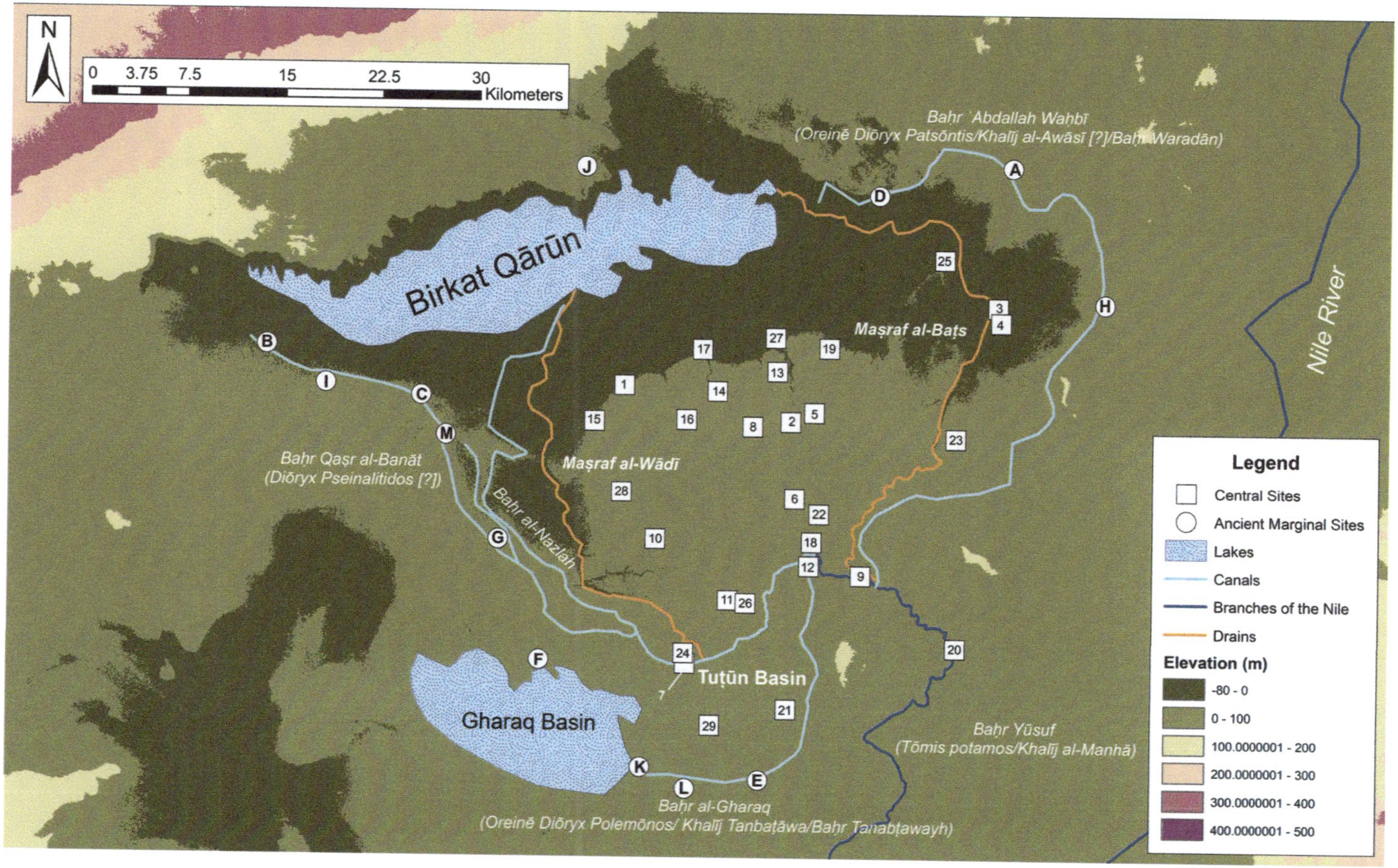

Map 2. The premodern Fayyūm (Cartographer: Julian Thibeau).

Key to Map 2

Ancient Marginal Sites

A. Bakchias (Umm al-Athl)
B. Dionysias (Qaṣr Qārūn)
C. Euhemeria (Qaṣr al-Banāt)
D. Karanis (Kaum Aushīm)
E. Kerkesoucha Orous (Buljusūq)
F. Narmouthis (Narmūda/Madīnat Māḍī)
G. Perkethaut/Philagris (Ḥāmūlī)
H. Philadelphia (Al-Kharaba al-Kabīr)
I. Philoteris (Waṭfā)
J. Soknopaiou Nesos (Dīma al-Sibāʿ)
K. Talithis/Talei (Ṭalīt)
L. Tebtynis (Umm al-Burayjāt)
M. Theadelphia (Baṭn Ihrīt)

Central Sites

1. Abū Ksā
2. Akhṣāṣ Abū ʿUṣayya (Ḥarfūsh)
3. Al-Rawḍa
4. Al-Rūbiyyāt
5. Andriantōn Kōmē/Piamouei (Biyahmū)
6. Arsinoitōn Polis (Al-Fayyūm)
7. ʿIzbat Abū al-Nūr (Arsinoe on the Lock?)
8. Banū Majnūn
9. Hauēris (Hawwāra)
10. Ihrīt
11. Itṣā
12. Mouchis (Dumūshiyya)
13. Naqalīfa
14. Phentemin (Fidimīn)
15. Pisais (Ibshāyat al-Rummān)
16. Psenaryo (Sīnarū)
17. Psenhyris (Sanhūr)
18. Pseonnophris (Ṣunūfar)
19. Psineuris (Sinnūris)
20. Ptolemais Hormou (al-Lāhūn)
21. Qalamshāh
22. Qushūsh
23. Selē (Sīla)
24. Shidmū
25. Tamauis (Ṭāmiyya)
26. Tebetny (Difidnū)
27. Tirsā
28. Ṭubhār
29. Tuṭūn

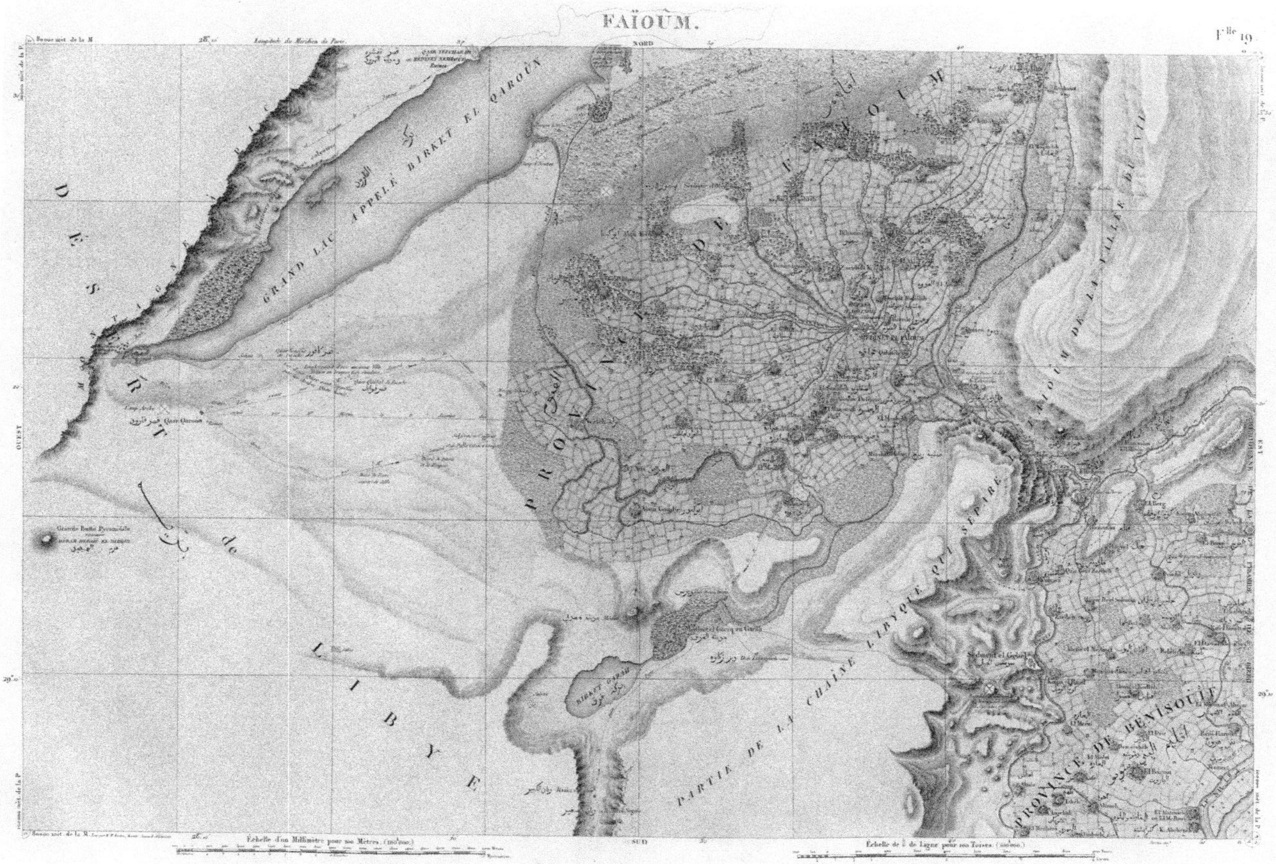

Map 3. Fayyūm in the *Description de l'Égypte*. Source: Pierre Jacotin, *Carte topographique*, *Carte* 19, *Faïoùm (1818)*. Image courtesy of David Rumsey Map Collection, David Rumsey Map Center, Stanford Libraries.

Map 4. The Fayyūm in 1843. Source: Linant de Bellefonds, *Mémoire sur le lac Moeris*. Image courtesy of gallica.bnf.fr / Bibliothèque nationale de France.

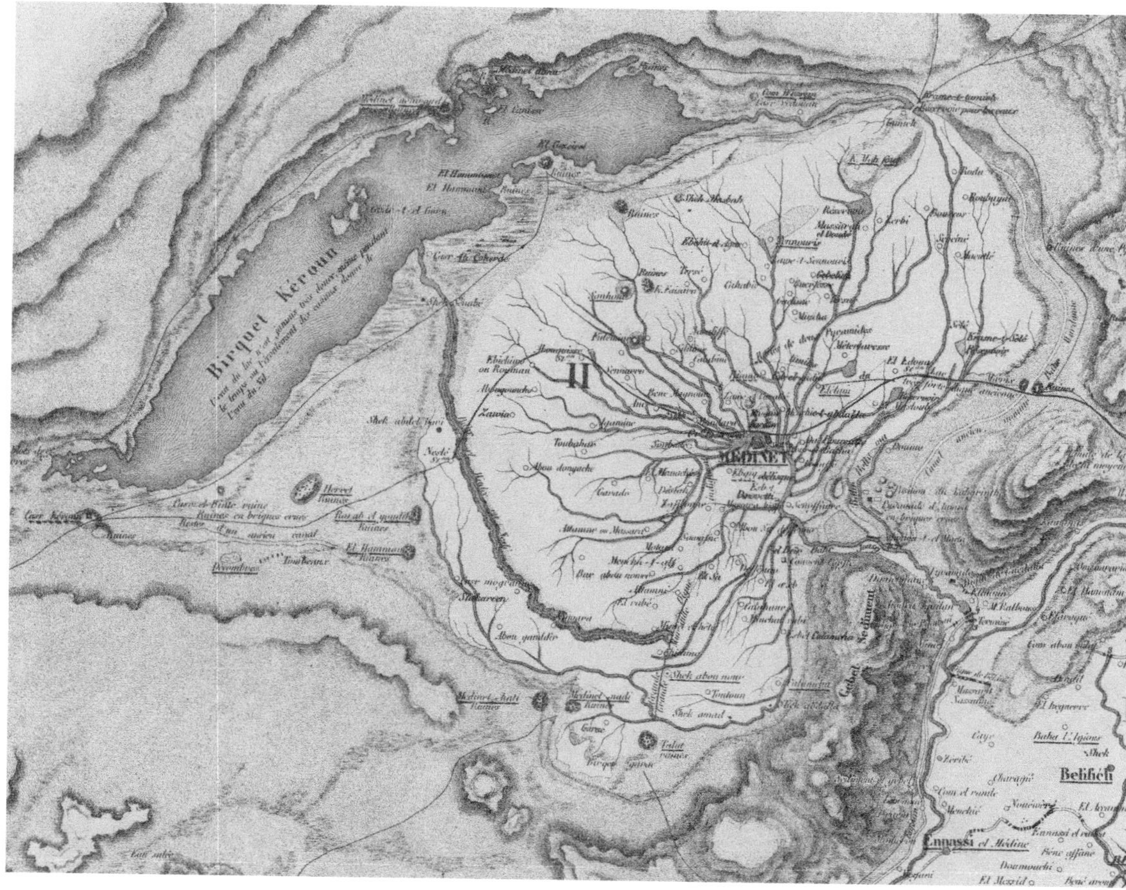

Map 5. The Fayyūm in 1854. Source: Detail from Linant de Bellefonds, *Carte hydrographique de la moyenne Égypte*. Image courtesy of gallica.bnf.fr / Bibliothèque nationale de France.

Introduction

From Water, Everything[1]

This low-lying place has come to resemble the heavenly realm
—Al-Nābulusī, Villages of the Fayyūm (1244 CE)

an emerald isle of verdure in the all-surrounding wastes
—John Ward, "The Oasis of Roses" (1902 CE)

A Strange and Unexpected Beauty

Egypt was not always so green. Before modern irrigation infrastructure began to impound the Nile's annual flood, thereby enabling irrigation and cultivation year-round, the river's annual rise and fall progressively altered the color of the countryside. Drawing on a tradition ascribed to Egypt's Arab conqueror ʿAmr ibn al-ʿĀṣ, the tenth-century historian and geographer al-Masʿūdī evocatively described this cycle in his *Murūj al-dhahab wa-maʿādin al-jawhar* ("Meadows of Gold and Mines of Gems").[2] During the Egyptian months of Abīb, Musrā, and Tūt when the floodplain was submerged beneath the Nile's inundation, the surface of the waters glistened like a white pearl (*luʾluʾa bayḍāʾ*) and the villages atop their hillocks and tells, inaccessible save by boat, stood out like the stars in the sky. In Bābah, Hātūr, and Kīhak, the season of black musk (*miska*

1. *Qurʾān* (hereafter *Q.*) 21:30: أَوَلَمْ يَرَ الَّذِينَ كَفَرُوا أَنَّ السَّمَاوَاتِ وَالْأَرْضَ كَانَتَا رَتْقًا فَفَتَقْنَاهُمَا ۖ وَجَعَلْنَا مِنَ الْمَاءِ كُلَّ شَيْءٍ حَيٍّ ۖ أَفَلَا يُؤْمِنُونَ ("Have those who disbelieved not considered that the heavens and the earth were a joined entity, and We separated them and made from water every living thing? Then will they not believe?").
2. Al-Masʿūdī, *Murūj*, 2:356–58; Butler, *The Arab Conquest*, 433–34.

saudāʾ), the waters retreated, leaving the dark and silty earth, moist and exhaling a musk-like fragrance, ready to receive seed. Throughout the following months of growth, Ṭūbah, Amshīr, and Baramhāt, the country's flourishing pastures and vegetation shone green like an emerald (*khaḍrāʾ k-al-zumurrada*). Finally, at the approach of the harvest in Baramūdah, Bashans, and Baunah these green fields first blanched (*yabyaḍ*) and then reddened (*yatawarrad*), at which point the landscape took on both the appearance and the value of an ingot of gold (*sabīkat al-dhahab*).[3]

But even after the harvest when Egypt's fields were dry and bare, and the Nile had shrunk to an unpalatable trickle—"heated in its bed, becoming greenish, fetid, and filled with worms" in one French orientalist's acerbic account[4]—premodern observers consistently remarked that the Fayyūm, an oasis-like depression west of the Nile Valley in Middle Egypt, remained verdant, its canals full and flowing. Marveling at its uniqueness, the first-century BCE Greek geographer Strabo wrote that "this nome is the most remarkable of all in its appearance, prosperity, and design; it is the only one planted with large and full-grown olive trees, which produce a fine crop."[5] Centuries later, Arab writers would offer still more fulsome appreciations. In his history the *Futūḥ Miṣr* ("The Conquest of Egypt"), the ninth-century Egyptian author Ibn ʿAbd al-Ḥakam credited the Hebrew patriarch Joseph (*Yūsuf*), pharaoh's vizier, with the Fayyūm's design and construction. At pharaoh's command the divinely guided prophet (*nabī*) drained the marsh from this wasteland pit (*jawba*) and excavated its unique canal system, thereby transforming the Fayyūm into a garden whose canals always delivered the right amount of water at the right time. "Those who look upon what Joseph brought to life from the Fayyūm," Ibn ʿAbd al-Ḥakam averred, "know nothing like it in all of Egypt."[6] In the following century al-Masʿūdī likewise extolled the virtues of Fayyūm irrigation, particularly the dam near the village of al-Lāhūn that regulated the head of its canal system, which he loftily described as "among the most wondrous objects

3. Cf. Volney's description of the floodplain *Voyage*, 1:234: *inondé pendant trois mois, fangeux et verdoyant pendant trois autres, poudreux et gercé le reste de l'année* ("inundated for three months, muddy and green for three others, dusty and cracked the rest of the year").
4. Volney, *Voyage*, 1:18: *elle s'échauffe dans son lit, devient verdâtre, fétide, et remplie de vers*. Green vegetal matter washed down from Sudan in this late season presaged the coming of the flood. Derr, *The Lived Nile*, 28–29 with refs.
5. Strabo, *Geog*. 17.1.35: ἔστι δ᾽ ὁ νομὸς οὗτος ἀξιολογώτατος τῶν ἁπάντων κατά τε τὴν ὄψιν καὶ τὴν ἀρετὴν καὶ τὴν κατασκευήν: ἐλαιόφυτός τε γὰρ μόνος ἐστὶ μεγάλοις καὶ τελείοις δένδρεσι καὶ καλλικάρποις.
6. Ibn ʿAbd al-Ḥakam, *Futūḥ Miṣr*, 15–16: فنظروا فاذا الذى أحياه يوسف من الفيوم لا يعلمون له بمصر كلها مثلا.

and most ingenious structures, an edifice that has long endured upon the face of the earth." The dam, he claimed, was designed to admit only such water as the Fayyūm needed and no more. Moreover, every ruler who had ever conquered Egypt traveled to admire it, drawn by the renown of its construction and workmanship.[7] Yet even this breathless hyperbole was surpassed by the late-Ayyūbid functionary Abū ʿUthmān al-Nābulusī, author of a detailed fiscal survey of the Fayyūm entitled *Bilād al-Fayyūm* ("Villages of the Fayyūm," 1244 CE). In al-Nābulusī's account, the Fayyūm's ceaselessly flowing waters and perennial bloom were nothing less than an earthly reflection of the Qurʾānic Garden of Paradise (*Janna*) with its "orchards and trees and 'gardens beneath which rivers flow.'"[8] A local Fayyūmī couplet that al-Nābulusī transcribed put it similarly: "How wonderful is the land of the Fayyūm among countries / like the Garden of Eternity (*Jannat al-khuld*) in its rivers and trees."[9]

Nor has this unusual landscape, the so-called "Garden of Egypt," lacked for modern admirers.[10] The German theologian, traveler, and manuscript collector Johann Michel Wansleben, the first modern European to record an extensive tour of the province (1672–73), exclaimed that its central canal was "remarkable, because it keeps fresh water all the year,"[11] irrigating a flourishing landscape of crops unseen elsewhere in Egypt, including vineyards, orchards of diverse fruit-bearing trees, and roses. "All that grows here," he asserted, "is of a better taste than in the other provinces."[12] Although the Ottoman Turkish traveler Evliya Çelebi (1685) was annoyed by the ceaseless twittering of birds in the Fayyūm's orchards, the "mournful creak of thousands of waterwheels and the frightful gurgle of myriads of gushing fountains," he still celebrated the many parks and rose gardens that carpeted the environs of its capital in flowers: "at dawn," he claimed, "one can even smell the roses in Tamiya, a town two hours away from Faiyum [City]." The produce of the Fayyūm's many orchards—grapes, apples, pears, peaches, plums, pomegranates, dates, and quinces—also filled the markets of Cairo for a full five months out of the year, he claimed, complemented by other characteristically Fayyūm produce like

7. Al-Masʿūdī, *Murūj al-dhahab*, 2:386: حجر الهون من أعجب الأمور، ومن أحكم البنيان، ومن البناء الذي يبقى على وجه الأرض.
8. Al-Nābulusī, *Villages of the Fayyūm* (hereafter *VF*), 31, quoting *Q*. 3:15: مِّن ذَٰلِكُمْ لِلَّذِينَ اتَّقَوْا عِندَ رَبِّهِمْ جَنَّاتٌ تَجْرِي مِن تَحْتِهَا الْأَنْهَارُ ("For those who fear God there are, with their Lord, gardens beneath which rivers flow").
9. Al-Nābulusī, *VF*, 36: يا حبذا بلد الفيوم من بلد \ كجنة الخلد انهارا واشجارا.
10. So dubbed by Karl Lepsius, *Discoveries in Egypt* (London: Richard Bentley, 1852), 85.
11. So also Lucas, *Voyage*, 2:46; and Lane, *Description of Egypt*, 231 and 236.
12. Wansleben (Vansleb), *The Present State of Egypt*, 153–54.

white bread, rose water, and linen.[13] Çelebi's account is echoed by that of the French traveler and antiquities collector Paul Lucas, who visited Fayyūm in September 1707: "one encounters entire fields filled with roses, small woods composed of fig trees, and gardens everywhere planted with the most beautiful fruit trees, such as apples, oranges, lemons, peaches, prunes, apricots, palms, and others, whose fruits the Phioumites travel or send to Cairo to sell, the dates above all, of which there is always a very great quantity."[14] Entranced by roses, the eighteenth-century French orientalist Claude Étienne Savary lapsed into heady romance in his *Lettres sur l'Égypte* (1786), writing that "the surrounding air is perfumed and when the weather is warm under a clear sky one feels still more intensely the pleasure of breathing in the fragrance of the rose mingled with the sweet scent of orange blossom."[15] This floral landscape, Egypt's principal source of attar of roses in the early modern period, was still visible at the beginning of the twentieth century. Indeed, in addition to dubbing it an "emerald isle of verdure," the Irish artist and popular author John Ward christened the Fayyūm an "oasis of roses." After his trip across the barren desert the weary traveler found a welcome respite in this "land of running waters, amid the noise of streams, the plash of rapid brooks of clear crystal, and the music of countless water-wheels."[16]

Yet of all of these admirers it is perhaps the Irish classicist and historian John Pentland Mahaffy whose gaze best encompasses the peculiar otherness of the Fayyūm—an oasis of cultivation bounded by desert, at once intimately linked to yet somehow distinct from the Nile Valley:

> When the train leaving Wasta on the Nile has passed a long cutting in the desert, through the saddle of high ground separating the oasis of Arsinoe from the Nile valley, the traveler suddenly looks down upon a band of the richest green—orchards, gardens, farms—which extends north and south as far as the eye can reach; from its east border he looks downward about five or six miles, till the gradual slope reaches a long very blue lake, stretched out as the western boundary of the oasis, and beyond it the amber mountains of the Libyan desert, rising abruptly from its shores. The scene is one of strange and unexpected beauty, and probably the most fascinating in all of Egypt.[17]

13. Dankoff, Tezcan, and Sheridan, *Ottoman Explorations*, 358.
14. Lucas, *Voyage*, 2:47–48.
15. Savary, *Lettres sur l'Égypte*, 2:44.
16. Ward, *Pyramids and Progress*, 64; cf. Ward, "The Oasis of Roses."
17. Mahaffy, *Empire of the Ptolemies*, 173.

Figure 1. View of the Fayyūm from Kaum Aushīm (Karanis). Photo by author, November 2011.

Today Mahaffey's strange and unexpected beauty is readily accessible from Cairo via a desert road that runs southwest from the capital for some 60 km, eventually passing by the Graeco-Roman archaeological site of Kaum Aushīm (ancient Karanis) on the Fayyūm's northeastern outskirts. Yet thanks to modern perennial irrigation, much of rural Egypt is now blanketed in green for the majority of the year, thereby diminishing the Fayyūm's immediate visual distinctiveness and making it difficult to recapture the emotional and psychological impact of these historical encounters. But even if the Fayyūm's perennial green now blends seamlessly into that of the Nile Valley, a glance to the south from the desertified mound of Karanis still presents the striking vista of a vast garden ringed by desert. After crossing the Baḥr ʿAbdallah Wahbī, a large irrigation canal that runs past Kaum Aushīm and marks the northern border of the Fayyūm's agricultural area, the descent into the cultivated portions of the depression is rapid and the greens quickly grow darker and denser as the field crops and scattered palms of the margins give way to orchards irrigated by a complex network of gravity-driven canals. Fed by the water of the Baḥr Yūsuf, a side channel of the Nile that enters the depression in its southeast, the gravity-driven canals flow swiftly along the steep slopes of the central plain, powering the undershot waterwheels that are so unique to the Fayyūm that they emblazon its provincial crest.[18] This easily watered and verdant plain—the "green belt" of orchards, vineyards, and field crops so admired by travelers over the centuries[19]—continues its sharp decline toward the north until it reaches some forty-five meters below sea level (mbsl) along the shores of the Birkat Qārūn, a shallow, brackish lake that occupies the deepest northern portions of the depression and that receives much of the region's agricultural runoff. As one climbs up and away from the lake toward either the western or southern margins of the depression, the fields begin to thin once again as water, so plentiful in the center near the head of the canal system, grows scarcer. And as in the north, the landscape suddenly reverts to desert beyond the border canals that flow along the depression's western and southern rims. Here the desert hills that enclose the Fayyūm—Mahaffey's amber mountains—begin to rise abruptly just beyond the cultivated area, sometimes only meters from the edge of the last field, prohibiting any major extensions of the gravity-canal network.

Despite these topographical obstacles, farmers on the margins continue to capture water and bring new land under cultivation. Wherever the local terrain allows, long narrow fields reach out into the higher desert ground beyond the border canals, irrigated by reused drainage or water mechanically pumped, at times illegally, from farther up the canal system. In some places long slender mounds of earth are simply heaped directly atop the sand and fitted with drip-irrigation tubing that carries a steady trickle to struggling tomato plantings, their bases encrusted with the salt and minerals left behind by the rapid evaporation of reused and heavily polluted drainage water. If this precarious water supply gives out, farmers will either attempt to restore it or find another source. Failing that, they may abandon their marginal plantings altogether and move inward in search of lands higher up the canals where water is more abundant. But if they are later able to restore their water supply, they will return to the margins, their fields stretching out as far as the water can be made to flow.[20]

• • •

18. On the ubiquity of waterwheels see already Shafei, "Fayoum Irrigation," 290: "The drone of their oilless shaft bearings is a peculiarity of the Fayoum."
19. So characterized in Price, "The Evolution of Irrigation," 299; cf. the remarks in Girard, "Mémoire sur les irrigations," 3:334, on the ease of gravity-powered water distribution in the immediate environs of the capital city: "comme ils sont dominés par les canaux qui viennent de Médine, il est aise d'y faire descendre les eaux et de les diriger à volonté."
20. For agricultural life and mobility at the tail end of the canal system, see Barnes, *Cultivating the Nile*, esp. 69–70, 121–22, 163–64, and 176–77.

This picture of a perpetually fluid and flourishing landscape is a recent introduction to the scholarship on the premodern Fayyūm, a region often thought to have suffered a sudden collapse in late antiquity, heralding centuries of stagnation and obscurity to come. The common narrative holds that the Fayyūm prospered from the Ptolemaic period until the later Roman era, becoming on one reading the most fertile and productive landscape in Rome's Mediterranean empire.[21] Sometime in the 300s CE, however, a sudden "hydraulic and/or administrative crisis"[22] struck and many villages were subsequently abandoned. The phenomenon has commonly been attributed to the later Roman state's failure to maintain the canal system—a portrait in miniature, some have claimed, of broader Roman decline at the end of antiquity.[23] With the retreat of the state, water ceased to flow and many formerly vibrant villages were returned to the desert, there to await the revival of the irrigation system under British colonial rule.[24] Yet even though this spate of village abandonments is an incontrovertible historical fact (though as we will see, some apparent abandonments were only shifts), it is increasingly recognized that the perceived rupture between antiquity and the early Islamic period is more apparent than real, largely the product of longstanding disciplinary divides between scholars of Graeco-Roman Egypt and Arabists.[25] This book thus represents the first study of the premodern Fayyūm to both collate and systematically exploit an evidentiary corpus composed of ancient papyri, medieval Arabic historical, geographical, and administrative texts, as well as modern comparative evidence. It simultaneously brings the tools of environmental history to bear on the Fayyūm's unique canal irrigation system in order to elucidate the human relationships with water that have sustained the region since antiquity. Indeed, as an object of study water encourages such diachronism. Existing both inside and outside of historical time, it transcends the various temporal, linguistic, cultural, and material categories by which scholars periodize Egypt's past and

21. Bagnall, "Agricultural Productivity and Taxation," 297.
22. Davoli, "The Archaeology of the Fayum," 155.
23. Bowman, "Egypt," 324. Classic accounts in Boak, "Irrigation and Population"; and Westermann, "The Development of the Irrigation System." Recent restatements include Alston, "Urban Population in Late Roman Egypt"; Kelly, *Petitions, Litigation, and Social Control*, 23; Rathbone, "Surface Survey," 18. On Theadelphia as an example of wider Roman decline see Bell, "The Byzantine Servile State" 96; and Ermatiger, "The Economic Death of Theadelphia."
24. So Boak, "Irrigation and Population," 364. Cf. Scott-Moncrieff, prefatory note to *The Fayûm and Lake Moeris*, by Brown, v.
25. The seminal statement is Keenan, "Deserted Villages."

apportion its study among various heavily siloed disciplines.[26] Eschewing such disciplinary particularism, this study draws on archaeologist Matthew Edgeworth's characterization of rivers as "entanglements of nature and culture" to argue that the Fayyūm's irrigated landscape should be regarded in all periods, past and present alike, as an unstable and dynamic entanglement between flowing water and its human dependents.[27] Within this dense network of shared canals, every reorientation of prevailing water flows, whether purposeful or not, compels downstream irrigators to adapt their cultivation habits and, at extremes, even their settlement patterns to the new water regime. Yet such adaptations necessarily alter canal flows once again, in turn stimulating still further human adjustments to the surrounding land- and waterscapes, a perpetual dialectic between flowing water, canals, and the rural communities they bind and sustain. Productive and disruptive in equal measure, I argue that this liquid mutability is the defining feature of the Fayyūm's landscape, a place where water, space, and society are deeply entangled, mutually constitutive, and in constant flux.

From Reclamation to Retrenchment: Historical Overview

The Fayyūm occupies the deepest portions of a homonymous geological depression in Egypt's Western (Libyan) Desert. The origins of the depression are uncertain but it was likely formed over millions of years either (or both) by the erosive action of wind (aeolian deflation) or the dissolving of underground channels and caverns by rainwaters acidified by their passage through the upper limestone strata (solution weathering), a process that gradually caused the earth to subside.[28] Although often referred to as an oasis, unlike true oases the depression does not overlie groundwater sufficient to support agriculture on a wide scale. It instead receives its water via the Nile's largest distributary, the Baḥr Yūsuf ("Joseph's River"), so named for its creator in Islamic myth.[29] After branching from the Nile near the southern city of Dayrūṭ, the Baḥr Yūsuf follows a sinuous course northward for some 190 km before bending sharply to the west and debouching into the Fayyūm depression through a small inlet

26. Cf. Terje Oestigaard, *Water, Christianity and the Rise of Capitalism* (London: I.B. Tauris, 2013), 1.
27. Edgeworth, *Fluid Pasts*.
28. Sampsell, *The Geology of Egypt*, 90.
29. Whitehouse, "Le Bahr Youssouf."

straddled by the villages of Hawwārat ʿAdlān and al-Lāhūn. This connection to the Nile was intermittent throughout much of prehistory, becoming consistent only during the early Holocene some ten thousand years before present. Thereafter, the Baḥr Yūsuf carried a portion of the Nile's annual flood into the Fayyūm each year.[30] Yet apart from this single inlet, the Fayyūm is an enclosed (endorheic) basin from which water can escape only through evaporation. As a result, the annual influx of floodwater created and sustained a large lake that once filled most of the depression.

The Birkat Qārūn in the northwest of the Fayyūm is all that remains of this ancient lake, which was reduced during two phases of land reclamation between the pharaonic and Hellenistic periods.[31] The first reclamation is not well understood due to a lack of hard evidence and may be more legend than fact.[32] It is nonetheless often repeated in scholarly literature that during the Middle Kingdom the pharaohs Sesostris II (ca. 1842–1837 BCE) and Amenemhat III (ca. 1818–1770 BCE) successfully reclaimed from the ancient lake roughly 450 km^2 of land on the Fayyūm's central plateau, an area ca. 23–27 meters above sea level (masl).[33] That the Fayyūm was important to pharaohs of this period is undeniable as the presence of their pyramids and other monuments amply attests. Yet it remains impossible at present to determine the extent of Middle Kingdom hydraulic works in the region.[34] Regardless, the shallow, marshy lake that continued to occupy the depression—dubbed *Šy* ("Lake") or *Šy-wr* ("Great Lake") during the Middle Kingdom—continued to be a fishing and fowling destination throughout the New Kingdom (ca. 1550–1077 BCE), a "water-meadow" frequented by those keen "to snare birds and catch many fish."[35] It was during this latter period that the lake began to be called *pa-yom* ("The Lake") by Egyptians. The toponym survived for centuries and was variously pronounced *Piam*, *Peiōm*, and *Phiōm* in later Coptic Egyptian dialects. Arabized as *al-Fayyūm* after the Islamic conquest, it remains in use to this day.

---

30. Marks et al., "Holocene Lake Sediments."
31. Drawing on language in Thonemann, *The Maeander Valley*, 18. For a pictorial overview of the landscape history of the depression see now the series of maps by Matthias Rebentisch entitled "Fayyum Blue Moeris"; and Rebentisch, "Notes on the Maps 'Fayyum Blue Moeris,'" both in Römer, *News from Texts and Archaeology*, 187–93.
32. Malleson, *The Fayum Landscape*, 49.
33. Manning, *Land and Power*, 99–102.
34. Malleson, *The Fayum Landscape*, 56.
35. So characterized in a New Kingdom poem entitled "The Account of the Pleasure of Fishing and Fowling." Cited from Blouin, *Triangular Landscapes*, 220.

A massive reclamation of this lake—known to Greeks since the fifth century BCE as the *limnē Moirios* ("Lake of Moeris")[36]—was undertaken by Egypt's first two Hellenistic kings Ptolemy I (323–283 BCE) and his son Ptolemy II Philadelphos (283–246 BCE). Although no written record of their method(s) survives,[37] it appears that much of the Baḥr Yūsuf's annual influx was diverted from the al-Lāhūn inlet, which deprived the lake of annual recharge and allowed the high rate of evaporation—ca. 2 m per year for standing water—to reduce its surface area.[38] After the lake had retreated, Ptolemaic engineers excavated a radial canal system to distribute the waters of the Baḥr Yūsuf throughout the newly reclaimed land. By exploiting the natural slope of the depression's terrain, these canals could deliver water by gravity, a method of irrigation largely absent from the Nile Valley where the level terrain and the dramatic fall of the river's water level after the flood rendered radial canalization all but impossible. The head of this innovative system at the al-Lāhūn inlet was regulated by a dam that was opened to admit the waters of the flood and later closed again when the inundation began to ebb. Once within the depression, these floodwaters were then distributed by the radial canal network. Along the eastern, western, and southern rims of the depression, three large, long border canals irrigated settlements located along the depression's near-desert margins, while the interior was watered by a dense network of smaller canals that filled the Fayyūm's central alluvial plain. A system of drains subsequently channeled wastewater into two massive natural ravines—known today as the Maṣraf al-Baṭs and the Maṣraf al-Wādī—that discharged into the much-reduced lake in the north, now repurposed as the Fayyūm's primary drainage sump.[39] This virgin terrain was then dotted with dozens of new villages and settled by Graeco-Macedonian klerouchs (*klērouchoi*), military reservists given plots of land (*klēroi*) on long-term loan in return for their service to the crown. Voluntary Egyptian immigration along with the influx of native agricultural laborers soon contributed to the cultural makeup of the new region, its emerging Graeco-Egyptian village toponymy indicating the arrival of settlers

36. So Herodotus, *Hist.* 2.149–50.
37. Papyrological evidence of the original reclamation is limited to a single text describing works in a late phase of the project, *SB* 20.14624 (258–256 BCE). It preserves a proposal for work on 70,000 *arourai* to be undertaken by fully 15,000 laborers. See also Van Beek, *The Archive of the Architektones*, 14–15.
38. See most recently Römer, "The Nile in the Fayum"; and Willems et al. "The Analysis of Historical Maps," 302–4. Evaporation rate: Wolters et al., "Managing the Water Balance," 106.
39. Brief overview of the new canal system in Morini, "La nuova gestione indrica"; Thompson, "Fayum Irrigation in the Third Century BC"; and Thompson, "Irrigation and Drainage."

from various parts of the Delta and Middle Egypt.[40] In 257 BCE the burgeoning nome at last received an official designation. Hitherto referred to by Greeks as simply "the lake" (*hē limnē*), the Fayyūm was rechristened the Arsinoite Nome (*Arsinoitēs Nomos*) in honor of Ptolemy Philadelphos' deceased sister-wife Arsinoe II (d. ca. 270).[41] By century's end its population had swelled dramatically, reaching upwards of 85,000 or 95,000 persons.[42]

Absent contemporary documentation, the speed and scale of the Ptolemaic reclamation remain unknown and scholars vary in their estimates of the amount of cultivable land it produced. The standard estimate of between 1200 and 1600 km^2 is perhaps too high, however, since it is now known that the lake was not immediately reduced to its present size and level (43 mbsl) during the reigns of Ptolemy I and II.[43] The clustering of some of the earliest Ptolemaic village foundations at several meters above or below sea level likely indicates the approximate extent of the lake at the time of their founding in the mid- to late-third century BCE.[44] A lake at roughly sea-level will therefore still have occupied some 960 km^2, leaving just under 900 km^2 available for settlement and cultivation.[45] The lake nonetheless continued to retract during antiquity. By the first centuries CE it reached some 20 mbsl, exposing ca. 1350 km^2 of cultivable terrain.[46] Newly discovered Roman-era village sites to the north of the lake also prove that it dropped as low as 40 mbsl at certain points during Roman rule, potentially opening up as much as 1700 km^2 in some periods.[47] Yet due to regular variations in the size of the Nile's annual flood, the volume of water that entered the Fayyūm fluctuated from year to year. Inhabitation and cultivation along the lake's shifting shoreline were consequently sporadic and unpredictable throughout Graeco-Roman antiquity.[48] Notwithstanding this relative instability, the reclamation of the Fayyūm was the single greatest expansion of

40. Clarysse and Thompson, *Counting the People*, 90-92.Clarysse, "Toponymy of Fayyum Villages"; and Clarysse, "Graeco-Roman Oxyrhyncha." Egyptian labor, see e.g., *P.Lond.* 7.1954 and 1955 (257 BCE).
41. It is still so called in the Ptolemaic Revenue Laws of 259/8 BCE (*P.Rev.*) and other near-contemporaneous documents, e.g., *P.Zen.Pestm.* 29 (257 BCE).
42. Clarysse and Thompson, *Counting the People*, 101.
43. 1200–1600 km^2 is the standard range. See Manning, *Land and Power*, 107n49 for references.
44. Cf. Römer, "The Nile in the Fayum," 180.
45. Lake size in Shafei, "Fayoum Irrigation." My figure assumes the contemporary agricultural area of 1,850 km^2 as the maximum extent of the irrigable space.
46. On historical lake levels see Hassan and Tassie, "Modelling Environmental and Settlement Change"; Marks et al., "Holocene Lake Sediments."
47. Cappers et al., "The Fayyûm Desert."
48. Hobson, "Agricultural Land and Economic Life."

Egypt's agricultural area during premodernity. Hovering around 1500 km^2 or more by the early Roman period and boasting ca. 170,000–200,000 inhabitants, it represented some 7.5 percent of Egypt's agricultural area and roughly 3 percent of the country's population.[49] Rigorous, state-coordinated maintenance of the canal system simultaneously ensured that water flowed smoothly throughout the region, irrigating fields that supplied the Roman state with plentiful grain taxes in kind.

Troubles are generally thought to have begun in the latter half of the second century CE in the form of a population shock following a disease event that afflicted the Roman Empire between 165 and 180 CE—traditionally called the Antonine Plague.[50] As we will later see, the effects of this disease event remain highly debated. But whatever its long-term effects, if any, on Egyptian demographics, the Fayyūm's agricultural landscape was radically changed by the third century CE due to the emergence of large private estates in various parts of the nome, which enclosed both former smallholdings as well as public lands that had once been worked by lessees.[51] By the middle of this century the settled area also began to contract, in some cases precipitously. The Egyptian temple-town of Soknopaiou Nēsos (Arabic Dīmah al-Sibāʿ) is commonly regarded as the first ancient village site to be abandoned since the last secure reference to the site dates to the year CE 239.[52] Sacred to the Fayyūm's tutelary crocodile god Sobek (Greek Souchos) and located atop a low mound in the desert north of the lake, the village lacked both a fresh water source and reliable access to an agricultural territory of its own, making it unusually precarious.[53] But by the early fourth century even some formerly prosperous Ptolemaic foundations in the northwestern Fayyūm were teetering on the edge of abandonment.[54] The village of Dionysias (Qaṣr Qārūn), the farthest-flung agricultural settlement in the western reaches of the nome, clung to a half-life as a

49. Assuming a maximum cultivable area during the Roman period of ca. 20,000 km^2. For the latter figure see Scheidel, *Death on the Nile*, 220–23; and Haug, "Agriculture in Roman Egypt," 524–25. Population estimates in Monson, *From the Ptolemies to the Romans*, 40. Percentages assume a population of six to seven million persons, for which see Haug, "Agriculture in Roman Egypt."

50. Bruun, "The Antonine Plague." See also Andorlini, "Considerazioni sulla 'peste antonina' in Egitto"; Elliott, "The Antonine Plague"; and Harris, "The Great Pestilence." Environmental-historical perspective in Blouin, *Triangular Landscapes*, 243–97.

51. Rathbone, *Economic Rationalism*; and Schubert, *Philadelphie* are the principal studies of the phenomenon.

52. *SB* 14.11907 (239 CE). The two possible later references in Trismegistos are, as noted there, dubious at best: *SB* 26.16540 (250–325 CE), *P.Oslo.* 3.162 (ca. 360 CE).

53. Davoli, "Archaeological Research" 54; Hobson, "Agricultural Land and Economic Life."

54. For a recent reassessment see Römer, "Why Did the Villages in the Themistou Meris Die?"

Roman military encampment into the second half of the fourth century and may even have endured into the sixth although the evidence for this is a single attestation of the toponym in a late papyrus.[55] Neighboring Euhemeria (Qaṣr al-Banāt) 13 km to the east is last mentioned in 347 CE.[56] The fate of Theadelphia (Baṭn Ihrīt), 2 km east of Euhemeria, is better known thanks to the survival of the papyrus archive of Aurelius Sakaon, one of early fourth-century Theadelphia's leading residents. As described in his papers, the village suffered from a minuscule population, persistent water shortages, and an unsustainable tax burden in the first decades of the 300s CE. Its problems proved irresoluble and the last few residents were probably gone by the middle of the century.[57] Similar waves of abandonment affected the east and northeast of the Fayyūm in the following centuries. Here, the long-established Ptolemaic villages of Philadelphia (Kaum al-Kharāba al-Kabīr), Bakchias (Umm al-Athl) and Karanis appear to have been in decline by the 400s or 500s CE although there is no evidence for systemic conflicts over their water supply as in Theadelphia.[58] Archaeology nonetheless suggests that Karanis held on well into the sixth century CE, far beyond the latest securely dated papyri from the site.[59] It may even have persisted into the seventh century but, like Dionysias, the evidence amounts to a single papyrological attestation of the toponym.[60]

Although the Fayyūm's southern margins were likewise eventually abandoned, settlements here endured for centuries longer. The southwestern village of Narmouthis (Madīnat Māḍī), another old Egyptian temple-town, even remained in use as an administrative center well in the eighth century CE. Arabized as Narmūda during this latter period, it served as the base of the Muslim Arab administrator in charge of the south/southwest portion of Fayyūm province.[61] Along the Fayyūm's southern border canal, several settlements also remained active on their original sites into the eleventh century CE: Tebtynis, Talithis, and Kerkesoucha Orous, known in this period by Arabized versions of

55. *P.Laur.* 3.93 (sixth century CE) seems to refer to the elders (*meizones*) of Dionysias but provides no additional evidence.
56. *SB* 22.15728.
57. Bagnall, "The Population of Theadelphia."
58. *Pace* Van Minnen, "Deserted Villages," 50–51.
59. Pollard, "The Chronology and Economic Condition."
60. *SPP* 10.291 (seventh century CE) at l. 19: [δ(ιὰ) Ἰ]ωάννου Καρανείτου.
61. Sijpesteijn, *Shaping a Muslim State*. Cf. briefly Berkes and Haug, "Villages, Requisitions, and Tax Districts." The village is attested in Greek in the eighth century: Keenan, "Fayyum Villages in *SPP* XX 229," 489.

their native Egyptian toponyms as Ṭuṭūn, Ṭalīt, and Buljusūq.[62] By the thirteenth century, however, only Tebtynis/Ṭuṭūn still survived, having abandoned its ancient site (Umm al-Burayjāt) and migrated north toward the Fayyūm's interior. Now Taṭūn, the village is still extant some 4.5 km north of its original location.[63]

This much of the Fayyūm's history is known in part through archaeology but primarily through everyday handwritten texts on papyri, ostraka (potsherds), and other media preserved in the dry sands that covered the region's marginal villages following their abandonments. Of these, documents in Greek overwhelmingly predominate, amounting to more than 16,400 texts published to date, though smaller *corpora* of documents composed in various forms of the Egyptian language, Latin, and Arabic survive as well. All told, the papyri provide historians with minute insights into the administration and agricultural society of the Fayyūm's margins between the early Ptolemaic period and the initial centuries of Muslim Arab rule. By contrast, the Fayyūm's wet and continuously inhabited interior has preserved vanishingly few ancient documents from country villages, ensuring that this portion of the depression's landscape remains largely obscured throughout the Graeco-Roman millennium. The interior first appears, albeit dimly, in the early Arab period thanks to a cache of administrative documentation recovered in 1877–78 from the ruins of the Fayyūm's ancient capital. Among these is a large number of village lists dating to the seventh and eighth centuries CE, records compiled by the central administration in order to track incomes and arrears from the countryside. Mostly fragmentary, often alphabetical, and devoid of accompanying detail, the lists nonetheless offer a tantalizing administrative bird's-eye view of an agricultural landscape that continued to thrive even in the face of the progressive desertion of its outer margins.[64]

This robust interior emerges more clearly in Arab literature, though the earliest discussion is Ibn ʿAbd al-Ḥakam's telling of the Fayyūm's Islamic creation myth.[65] In his account, several of pharaoh's advisers had become jeal-

62. Respectively *Tȝ-Tn*, *Tȝ-ʿlṭ* (*Tȝ-ȝlyt*), and *Pȝ-r-glg-Sbk*. On Kerkesoucha/Buljusūq see the references in Berkes and Haug, "Villages, Requisitions, and Tax Districts," 215. Medieval attestations in Gaubert and Mouton, *Hommes et villages*, esp. 167–68.

63. The toponym's initial *ṭāʾ* (ط) in the Arabic papyri has been exchanged for the *tāʾ* (ت) found in literary texts like Nābulusī's *VF*. Gaubert and Mouton, *Hommes et villages*, 167n4.

64. Banaji, *Agrarian Change*, 241–50; Morelli, "I χωρία in α"; Keenan, "Fayyum Villages in *SPP* XX 229"; Keenan, "Deserted Villages"; Berkes and Haug, "Villages, Requisitions, and Tax Districts."

65. The following three paragraphs depend on Rapoport, *Rural Economy*, 32–51.

ous of the now-aged vizier Joseph and spitefully convinced the king to put the old man's powers to the test with an all but impossible task. Pharaoh accordingly ordered Joseph to reclaim the vast and nigh-unreachable desert depression known as "the Pit" (*al-Jawba*), which at that time served as an outlet (*muṣāla*) for the excess waters of the flood. Joseph accepted the task and set his workmen to excavate a massive canal called al-Manhā (now the Baḥr Yūsuf), which took off from the Nile in the far south at al-Ashmūn (al-Ashmūnayn) and ran north to al-Lāhūn. He then excavated the Fayyūm's canal system and cleared the depression of reeds and brushwood. Thereupon the water of the Nile poured into the al-Manhā and the new canal system, transforming the pit into a gulf (*lujja*) of the Nile and a vast expanse of cultivable land watered by precisely regulated perennial canals. Completed in a mere seventy days, Joseph's accomplishment so impressed the pharaoh that he dubbed it the work of "one thousand days" (*alf-yaum*), whence the popular Arabic folk etymology of *al-Fayyūm*.[66] Repeated and elaborated by later authors, Ibn ʿAbd al-Ḥakam's pious fiction marks the beginning of a long fascination in Islamic literature with the Fayyūm's canal system and the productivity it enabled. Later writers also often repeated the myth that Joseph had founded 360 villages in the Fayyūm, one for each day of the Egyptian lunar year. Gifted by Joseph with abundant perennial water, each village was sufficiently productive to feed the whole of Egypt for a single day, thus ensuring that the Fayyūm could provision all of Egypt in years of poor flood.[67] While such folktales may preserve cultural memory of its actual ancient reclamation(s), they also reimagined the Fayyūm's landscape as a distinctly Islamic-Egyptian space by linking its establishment both to native Egyptian pharaohs and to the biblical Joseph, who is revered as a prophet (*nabī*) in the Islamic tradition.[68]

Later historians and geographers indicate that the Fayyūm remained highly productive throughout the first Fatimid century (969–1068 CE). In addition to traditional grain and garden produce, it was known in this period for producing water-intensive crops not found in the Nile Valley such as sugarcane and rice, both of which had been introduced after antiquity. According to the eleventh-century Andalusian geographer al-Bakrī (ca. 1068 CE), the ability of the

66. Ibn ʿAbd al-Ḥakam, *Futūḥ Miṣr*, 14–16. See also Malleson, *The Fayum Landscape*, 150–53.
67. Al-Maqrīzī, *Khiṭaṭ*, 1.668–69; Ibn al-Kindī, *Faḍā'il Miṣr*, 33.
68. On early Islamic Egyptian identity in Ibn ʿAbd al-Ḥakam see Omar, "'The Crinkly-Haired People of the Black Earth.'" For Joseph/Yūsuf in Islam see John Kaltner and Younus Mirza, *The Bible and the Qur'an: Biblical Figures in the Islamic Tradition* (London: Bloomsbury T&T Clark, 2018), 99–104.

Fayyūm's unique canal system to capture and store water also allowed it to produce a second, summer (*ṣayfī*) harvest in addition to the single flood-irrigated winter crop harvested in the rest of Egypt.[69] As late as the mid-eleventh century, this canal system was closely regulated by Egypt's central administration. The detailed knowledge possessed by the state emerges in an account of regulations (*dustūr*) concerning the operation of major canals composed in 1031 CE by the otherwise unknown Egyptian functionary Abū Isḥāq. Although the original text is lost, the *dustūr* survives in a much-abridged twelfth-century recension penned by the Fatimid tax official al-Makhzūmī (ca. 1189 CE), whose text was transmitted by the fifteenth-century Egyptian historian and geographer al-Maqrīzī (see appendix below).[70] Even in this abbreviated form, Abū Isḥāq's *dustūr* is the only synoptic premodern account of the flow and functionality of the Fayyūm's canal system. While it reveals a thriving and well-administered system, it nonetheless simultaneously demonstrates how far the Fayyūm had retracted since the Graeco-Roman period. Of the three large border canals excavated along the east, west, and southern margins under the Ptolemies, only the southern border canal of Tebtynis (Greek *oreinē diōryx Tebtyneōs*) seems to have been in full operation under the name Khalīj (canal) Tanabṭawa. The eastern canal may also have been in use in this period, but this is far from certain (see below in chapter 2).

Beginning in 1069 CE, the Fayyūm's second Fatimid century is more poorly documented. It is nonetheless during this period that multiple ancient sites along the southern border canal were finally abandoned, completing the long retrenchment that had begun at Soknopaiou Nesos some eight centuries earlier.[71] This last phase of retraction was instigated in part by an economic crisis under the caliph al-Mustanṣir (1068–1174 CE) and allegedly compounded by several low Nile floods and resultant famine. While the southern border canal continued to function and support some agricultural settlement in the area, the remainder of the depression's cultivated area would be confined to the Fayyūm's central alluvial plain, an area bounded on east and west, respectively, by the al-Baṭs and al-Wādī ravines. Memory of these abandonments persisted and Ayyūbid authors of the twelfth and thirteenth centuries accord-

69. Al-Bakrī, *Al-Masālik wa'l-mamālik*, 1:515 (no. 866). Al-Bakrī's account is excerpted by the fifteenth-century Andalusian geographer al-Ḥimyarī in his *Kitāb al-Rawd al-Mi'tār*, Textos medievales 10, ed. María Pilar Maestro Gonzalez (Valencia: Gráficas Bautista, 1963), 445.

70. Al-Maqrīzī, *Khiṭaṭ*, 1:669–75.

71. Gaubert and Mouton, *Hommes et villages*, 172–73.

ingly describe the Fayyūm in their period as a shadow of its former self. Indeed, al-Makhzūmī's abridgement of Abū Isḥāq's *dustūr* was prompted by the intervening abandonment of many of the villages mentioned in the original eleventh-century text. The world geography of the encyclopedist Yāqūt al-Rūmī (1179–1229 CE) entitled *Mu'jam al-buldān* ("Dictionary of Countries," 1228 CE) sums up this perception, noting that the 360 villages of Joseph's Fayyūm were a thing of the past, this once-miraculous landscape now reduced to a mere tenth of its former size.[72]

Despite this final spate of abandonments, the Fayyūm described by al-Nābulusī in his thirteenth-century *Villages of the Fayyūm* (hereafter *VF)* was in the process of rehabilitation. Long-neglected maintenance was underway on both the Baḥr Yūsuf's inlet and the al-Lāhūn dam and multiple abandoned villages were being reestablished. According to al-Nābulusī, the recent recovery had achieved all that could reasonably be accomplished without undue state coercion of the native population. Yet all of this resurgence was still confined to the central Fayyūm. The ancient margins remained entirely unreclaimed and the ruins of long-abandoned villages were still visible along the silt-clogged and disused border canals. Al-Nābulusī catalogues these desolate sites in full by their local Egyptian-Arabic toponyms, remarking that this great swath of derelict land could never be reclaimed without an amount of labor and expense so immense that it would fill the inhabitants' hearts with fear and risk putting them to flight.[73] Yet in contrast to this evocative desolation, the more than one hundred active villages in the central plain were highly productive, cultivating a wide array of field crops—for example, wheat, barley, rice, vetch, sesame—as well as water-thirsty flax, cotton, sugarcane, dates, olives, and grapes, along with livestock. Confirming the earlier assertion of al-Bakrī in exhaustive detail, al-Nābulusī's village-by-village survey documents that most of the Fayyūm's settlements produced both a flood-irrigated winter crop (*shatawī*) and a summer crop (*ṣayfī*) watered by the perennial canals (*saqī*) that were unique to the region. Impressed by this singularly productive irrigation system, al-Nābulusī prefaced his survey with a detailed autopsy of the Fayyūm's water supply and canal system, stressing its distinctiveness throughout. He nonetheless complains at various points that this abundance of water produced a swampy, stinking, and miasmic environment that impaired both the health and moral charac-

72. Yāqūt, *Mu'jam al-buldān*, 4:288.
73. Al-Nābulusī, *VF*, 47.

ter of its inhabitants (as well as that of any visitor unlucky enough to remain too long in this miserable environment). Together with his more sober analyses of the Fayyūm's peculiar hydrology, these caustic asides—the grumblings of an urban sophisticate contemptuous of the pungent smells of the rural countryside—open rare windows onto mundane human entanglements with water, a subject about which the Graeco-Roman papyri are less forthcoming.

Reading for Water: Proximity, Distance, and Perspective

Thanks to this abundant diachronic evidence there is no other rural Egyptian landscape whose history can be traced in such minute detail between Graeco-Roman antiquity and the early Islamic period. Yet evidence for human-water relationships is preserved unevenly throughout this diverse corpus. We must therefore read closely and carefully, alert to the various ways in which the region has been viewed, interpreted, and represented over time, between cultures, and across a multiplicity of textual genres. Our resulting picture will therefore be something of a patchwork stitched together from the testimony of a wide array of authors, each of whom observed the landscape from a different vantage point and, as a result, saw something distinct.

The documentary papyri, the greatest source of everyday writing from antiquity, bring us closest to life at ground level and might thus be expected to contain the most abundant and detailed information on human interactions with water in rural Egypt. Yet while they are at times revelatory, they more often disappoint. This is largely because much of the documentary corpus was generated by the demands of the state, taxation above all. The papyri thus tend to view the countryside through the prism of the fisc and to document only those hydrological or hydraulic events that threatened to reduce agricultural productivity and subsequent tax revenues, such as lands flooded either insufficiently[74] or to excess,[75] as well as conflicts over water between neighboring farmers and neighboring villages.[76] Such episodes might prompt the affected irrigators to petition the state for tax relief or other sorts of assistance, thereby

74. *Abrochia* declarations: Habermann, "Aspekte des Bewässerungswesens."
75. E.g., *BGU* 2.571 (151–52 CE), referring to waterlogged hollows (*koilōmata embrocha*) exempt from tax.
76. E.g., *P.Lond.* 7.1967 (225 BCE); *P.Tebt.* 1.61B, ll. 350–80 (118–117 BCE); *P.Tebt.* 1.50 (112/111 BCE); *P.Sakaon* 35 (ca. 332 CE).

generating written evidence of human relationships with water at a particular place and time. By contrast, unproblematic, everyday encounters with water required no such documentation and are therefore referenced only briefly and sporadically in the papyri, and even then only obliquely. Of course, this tendency to overlook the ordinary is itself unremarkable. As papyrologist Andrea Jördens has remarked, "the smooth running of affairs generally has, as one of its chief characteristics, that it leaves no obvious traces in the papyrological record."[77] This habit persists in the present day. In a recent study of contemporary Fayyūm irrigation practices, geographer Jessica Barnes notes that "as a substance that runs through all aspects of daily life, water is so apparent and obvious it does not need constant discussion." Though the lives of modern Fayyūmīs are literally and metaphorically entwined with the flow of Nile water through the canal network, these flows are taken for granted until they are contested, interrupted, or fail altogether.[78] Yet even during such periods of contestation and conflict, contemporary Fayyūm irrigators prefer to settle disputes locally instead of seeking the intervention of external authorities through petition—this, in order to avoid prolonging and intensifying the conflict and because outside authorities "are not aware of the organizational details of irrigation" at the local level and therefore lack the ability to adjudicate any conflict fairly and impartially.[79] So too in antiquity, where only in periods of disruption and dispute did water generate heated debate and, much rarer still, documentation.

Notwithstanding the occasional revealing exception, the papyrological record therefore tends to document only small-scale disturbances to prevailing patterns of water flow that, for whatever reason(s), eluded local resolution and hence generated a written record designed to solicit state intervention on the complainants' behalf. Regardless, these occasional glimpses can help us to establish how both the physical and the social infrastructure of Fayyūm irrigation was supposed to function at specific times and places—or at least how complainants claimed that they ought to function. Yet a second methodological hurdle extends from these very documents, namely their questionable representativeness. This is a critical consideration since the ancient village papyri derive only from the abandoned settlements that lined the Fayyūm's outer rim. Sitting at or near the tail ends of a long and complex canal network with their

77. Jördens, "Communicating with Tablets and Papyri," 237.
78. Barnes, *Cultivating the Nile*, 33; Barnes, "Governing Flow," 239.
79. Gouda, *Social Capital*, 168–69.

backs to the desert, these villages were marginal in every sense of the word and thus particularly vulnerable to water shortages and other related issues such as soil salinization, sand encroachment, and desertification. By contrast, the Fayyūm's interior has preserved vanishingly few village papyri, consequently obscuring agroenvironmental conditions within this most robust and enduring portion of the depression. The pattern is repeated throughout Egypt, a "survival of the least fit" in papyrologist Roger Bagnall's incisive phrasing. The papyrological corpus thus reliably illuminates peripheral or desert settlements while leaving "the most fertile and continuously occupied lands of Egypt . . . documentarily blank."[80] We must accordingly be wary of generalizing from the evidence of these marginal settlements to the rest of the Fayyūm, let alone all of rural Egypt where conditions were far different.

A third and related concern is the perspective afforded by the village papyri. This paradoxically stems from the greatest strength of the corpus, namely, the proximity it grants to everyday life in antiquity. Such documentary immediacy draws papyrologists directly into the world of Graeco-Roman Fayyūm villages, allowing us to observe rural society in often intimate detail. Yet like rural communities elsewhere in time and place, the world of a Fayyūm village was small and its horizons narrow, often scarcely extending beyond the borders of the village itself or the small constellation of its nearest neighbors. The resulting parochialism of the village papyri accordingly restricts the historian's field of view, making it difficult to see any farther afield than ancient villagers themselves. Nowhere is this constraint more keenly felt than in our attempts to contextualize and interpret the village desertions that began in late antiquity. Indeed, the only robust contemporary testimony to the phenomenon is the aforementioned early fourth-century CE archive of Aurelius Sakaon of Theadelphia, whose protagonists portray the collapse of their community as an existential crisis. In various petitions and hearings before state officials Sakaon and his companions vividly describe persistent water shortages, an unsustainable tax burden, and the near-total desolation of their village. In one uniquely evocative courtroom *narratio*, Sakaon and two other men claim through their advocate that they were Theadelphia's only remaining inhabitants. "We sold ourselves!" they plead, yet were still unable to make good their fiscal dues on lands permanently desiccated by allegedly water-thieving villages upstream.

80. Bagnall, *Reading Papyri*, 11. The most significant exception is the carbonized municipal rolls from Roman Thmouis in the Nile Delta. See Blouin, *Triangular Landscapes*.

Perched on high ground (*en hypsēlois topois*) at the end of a shared canal, they had come to inhabit a deserted village (*erēmon kōmēn*).[81]

Courtroom speech is designed to persuade, of course, but even when steeled against its rhetoric it is difficult to read such desperate accounts without being struck by a pang of sympathy—the "certain affective tremor" described by archival historian Arlette Farge.[82] Such sympathy then all too easily shades into credence, subtly encouraging us to see the world through our subjects' eyes alone. Yet our understanding of the evolution of this irrigated landscape will remain incomplete if we interpret it only through the narrow, emotive, and deeply personal frameworks of a few individuals living through the death of their village. We must therefore approach such testimony with historian and anthropologist Harvey E. Goldberg's admonition borne firmly in mind: "There may be ethical value in recording the views of those whose voices are rarely heard, but that, in itself, does not make good history. On the other hand, history in which one cannot hear people speaking faces the danger of being condemned to irrelevance."[83] Viewed in this light, the voices of fourth-century Theadelphians are essential but insufficient to the interpretation of landscape change in the Fayyūm. They humanize the region's transition between late antiquity and the Islamic period by laying bare both the hardships that accompanied village decline and the various strategies by which farmers adapted to a changing landscape. As we will see later on, the Theadelphians' problems also highlight the crucial links between internal communal integrity and successful water management. All the same, I read Theadelphia's story as one deeply rooted in time and place. Absent confirmatory evidence, these local events cannot be regarded as exemplary of a widespread contemporary collapse of the irrigation system, let alone uncritically shoehorned into larger metanarratives of late Roman decline and fall.[84] Indeed, what may reasonably have been perceived by a handful of farmers at ground level and in real time as an agroenvironmental crisis might appear otherwise from the perspective of distance.[85]

Since the village papyri cannot offer such distance and the broader perspective it affords, it is here that the utility of the later Arabic evidence becomes clear, the autopsies of al-Masʿūdī, al-Nābulusī, and al-Maqrīzī in particular. In

81. *P.Sakaon* 35 (ca. 332 CE), l. 18: κατεπωλήσαμεν δὲ ἑαυτοὺς.
82. Arlette Farge, *The Allure of the Archives* (New Haven: Yale University Press, 2013), 31.
83. Goldberg, *Sephardi and Middle Eastern Jewries*, 49.
84. For this critique see already Cook, "Landscapes of Irrigation," 79.
85. On environmental crises in perspective see Fisher, Hill, and Feinman, "Environmental Studies," 8–9.

contrast to the hyperlocalism of the papyri, these authors treat the Fayyūm panoptically and dwell on issues that are fundamental to an environmental history of irrigation in the region yet are largely invisible in the Graeco-Roman documentation, such as the hydrological regime of the Baḥr Yūsuf, the form and function of the al-Lāhūn dam, the administration of canal flow, and the agroenvironment of the Fayyūm's interior. Such evidence enables us to view the canal system holistically and deepens our understanding of the wider land- and waterscapes in which the marginal villages of antiquity were embedded. Of course, this broader view necessarily comes at the expense of papyrological granularity, which even al-Nābulusī's detailed village survey cannot duplicate; his narrow remit was to record the agricultural produce and prescribed fiscal value of each settlement, not to describe the mundane human relationships with water that sustained them.[86] His record of village water rights, for instance—the amount of water to which a village was entitled and from what canal—is prescriptive rather than descriptive and consequently obscures the ways in which water flow was produced, sustained, negotiated, and contested within and between villages. These limitations notwithstanding, his survey can still be mined for revealing insights on the water regimes of single settlements, the links between crop choice, canal flow, and landscape change, and the often-unpleasant lived experience of perennial water.

This is not to say that Arabic sources are otherwise unproblematic, since these authors often situate their accounts within the mythographic tradition established by Ibn ʿAbd al-Ḥakam. Indeed, both effusive praise of the Fayyūm's productivity and laments for its decline were already *topoi* by the tenth century CE and must be treated with caution. For example, al-Nābulusī's extravagant depiction of the region's plentiful and perennial water supply (*VF* 41) is at times belied by the testimony of his own survey, which records variable water availability and even water shortages in certain parts of the depression, especially at the tail ends of canals during the Nile's low-water season. Though surely influenced by observations of the many deserted ancient villages that limned the Fayyūm's desert margins, assertions of the region's precipitous decline must also be balanced against the clear and consistent evidence of robust continuity in the central alluvial plain. These caveats aside, the general reliability of early-Islamic-era descriptions of the Fayyūm's landscape and hydrology is confirmed by comparison to the works of modern European

86. On al-Nābulusī's "blind spots" see Rapoport, *Rural Economy*, xxvii.

authors, who observed and described the region before its conversion to modern perennial irrigation at the end of the nineteenth century. As we will see in chapter 1, the frequent agreement between these two independent *corpora* overwhelmingly demonstrates that Arabic sources provide a firm foundation for the reconstruction of premodern environmental conditions.[87]

• • •

Documenting the Fayyūm's irrigated landscape from above and below, from the center and the margins, and across more than sixteen centuries, this body of evidence unsettles the traditional historiography of the region in two significant ways. First, by revealing the continuity of irrigated agriculture, it challenges the persistent assumption that the village abandonments of late antiquity and the early Islamic period represented a systemic crisis. Painful though these events surely were to those affected, when viewed diachronically they instead appear not as a short, sharp shock to an otherwise stable equilibrium but as a long, transformative phase in the evolution of a landscape that was, in papyrologist James Keenan's seminal phrasing, "not static but shifting, and not always for the worse."[88] Second, by illuminating a complex array of changing human entanglements with the Fayyūm's hydrology, this corpus will enable us to see and to represent the flow of water through its canal system as a socioenvironmental phenomenon, that is, an entangled coproduction of nature and human agency. Thus construed, it becomes both conceptually and methodologically dubious to ascribe changes in the landscape to single exogenous variables such as climate change,[89] infrastructure collapse,[90] or the administrative neglect of the canal system.[91] While these were perhaps part of the story, so too were the myriad human interventions that continually reproduced the flow of water through the system. It is ultimately this complex interplay between flowing water and everyday life with which this environmental history is concerned.

87. *Pace* Römer, *The Fayoum Survey Project*, 17, who is less sanguine about al-Nābulusī's accuracy based on his confusion about the original routes of the ancient border canals.
88. Keenan, "Deserted Villages," 138.
89. Elliott, "The Antonine Plague"; Huebner, "Climate Change"; and McCormick et al., "Climate Change."
90. Römer, "Why Did the Villages?"
91. Bonneau, "Fiscalité et irrigation," 65.

Water, Flow, and Premodern Environmental History

Born of the nascent green movement of the 1960s and 1970s and initially concerned with the consequences of industrial modernity, environmental history is a relative newcomer to the study of Mediterranean premodernity and it remains methodologically underdeveloped as a result.[92] While classical philologists have become adept at elucidating environmental representations and mentalités in Greek and Latin literature,[93] progress on the historical side has been slower largely because these same texts preserve little in the way of hard evidence for everyday human interactions with nature. Archaeology can fill some of this gap by providing compelling material evidence for human adaptation to local or regional ecologies and for the long-term evolution of physical landscapes but yet material remains rarely reveal the voices of humans in nature that environmental historians so prize.[94] The few existing studies of ancient environmental history are therefore either broad ecological surveys,[95] cursory treatments of major environmental phenomena,[96] or works marred by the attempt to impose a single interpretive paradigm on the whole of Mediterranean antiquity.[97]

In the attempt to push past these roadblocks, some have turned to innovative scientific methodologies such as paleoclimatology and paleoepidemiology in the hope of modeling the impact of exogenous environmental phenomena on ancient society as a whole, particularly during periods of political upheaval or

92. For a general introduction to the field see J. Donald Hughes, *What Is Environmental History?* (Malden, MA: Polity, 2016).
93. E.g., C. Bosak-Schroeder, *Other Natures: Environmental Encounters with Ancient Greek Ethnography* (Berkeley: University of California Press, 2020); R. F. Kennedy and M. Jones-Lewis, eds., *The Routledge Handbook of Identity and the Environment in the Classical and Medieval Worlds* (London: Routledge, 2016); A. Merrills, *Roman Geographies of the Nile: From the Late Republic to the Early Empire* (Cambridge: Cambridge University Press, 2017); and Patricia Cox Miller, *In the Eye of the Animal: Zoological Imagination in Ancient Christianity* (Philadelphia: University of Pennsylvania Press, 2018).
94. For an innovative approach see Andrea L. Brock, "Rome at Its Core: Reconstructing the Environment and Topography of the Forum Boarium" (PhD diss., University of Michigan, 2017).
95. Giangiacomo Panessa, *Fonti greche e latine per la storia dell'ambiente e del clima nel mondo greco* (Pisa: Scuola Normale Superiore, 1991); Robert Sallares, *The Ecology of the Ancient Greek World* (Ithaca: Cornell University Press, 1991); Horden and Purcell, *The Corrupting Sea*. See also Alfred Thomas Grove and Oliver Rackham, *The Nature of Mediterranean Europe: An Ecological History* (New Haven: Yale University Press, 2003).
96. Lukas Thommen, *An Environmental History of Ancient Greece and Rome*. (Cambridge: Cambridge University Press, 2012).
97. Günther E. Thüry, *Die Wurzeln unserer Umweltkrise und die griechisch-römische Antike* (Salzburg: Otto Müller Verlag, 1995); J. Donald Hughes, *Environmental Problems of the Greeks and Romans: Ecology in the Ancient Mediterranean* (Baltimore: Johns Hopkins University Press, 2014).

major structural change. The methods and interpretive goals of this new scientism are on clearest display in three recent, high-profile contributions.[98] In a widely publicized 2017 study, an interdisciplinary team of scientists and humanists led by Ptolemaic historian Joseph Manning exploited ice-core records of historical volcanic events to argue that volcanic suppression of sub-Saharan rainfall depressed Nile flood levels several times during the Hellenistic period. These flood shocks, in turn, correlate with periods of significant internal revolt against the Ptolemaic state.[99] In a far more broad-reaching contribution published in the *Journal of Interdisciplinary History*, a similar team led by early medievalist Michael McCormick compiled a wide array of climate proxy data ranging from tree rings to ice cores to argue that climate change was a critical component in the destabilization of the later Roman empire. McCormick's team concluded that climate conditions in the late Roman West were less favorable than those in the East, thereby contributing to the failure of the Western half of the empire and the continued vitality of the East.[100] The most accessible yet by far the boldest contribution, however, is the 2017 monograph *The Fate of Rome: Climate, Disease, and the End of an Empire* by Kyle Harper, a member of McCormick's team. Synthesizing both original scholarship and copious recent publications on ancient climate and disease, Harper's study identifies the end of the so-called "Roman Climate Optimum" in ca. 150 CE and a subsequent rise in epidemic disease as the primary drivers of Rome's progressive decline from its second-century height. Other proximate causes, in Harper's analysis, were drought on the Eurasian steppe, which drove the Huns into the Roman Empire, decimating an already weakened army and splitting the empire in two at the end of the fourth century CE.[101] Climate and disease, in other words, were Rome's ruin.

While this new scientism's reliance on unorthodox evidence and methodologies is controversial,[102] its robust interdisciplinarity offers a riposte to

98. Humanist Russell Blackford has criticized the deprecatory term "scientism" because of its roots in eighteenth- and nineteenth-century theological objections to scientific thought. This critique notwithstanding, I retain the term here as a shorthand for a mode of ancient environmental history in which empirical data derived from the hard sciences drive the narrative, while traditional sources are employed only as support. R. Blackford, "The Sciences and Humanities in a Unity of Knowledge," in *Science Unlimited? The Challenges of Scientism*, ed. Maarten Boudry and Massimo Pigliucci (Chicago: University of Chicago Press, 2017), 11–29.

99. Manning et al., "Volcanic Suppression."

100. McCormick et al., "Climate Change." See also McCormick, "What Climate Science."

101. Harper, *The Fate of Rome.*

102. See, e.g., the three-part critical review of Harper, *The Fate of Rome* in Haldon et al.; as well as Marx, Haunschild, and Bornmann, "Climate and the Decline and Fall."

Roman historian William Harris' concern that "no historian knows enough science and no scientist knows enough history" to write the environmental history of antiquity successfully.[103] Yet Harris' remark simultaneously elides a fundamental if seemingly unasked methodological question: must premodern environmental history be so dependent on scientific forms of knowledge production? While scholars of antiquity—forever short on evidence—should be receptive to new methodologies and new datasets whatever their disciplinary origins, there are compelling reasons for caution. Practically speaking, elevating the approach of Manning, McCormick et al., and Harper as the standard model for ancient environmental history risks creating a ghettoized and self-referential subfield walled off from the rest of the discipline of ancient history by terminology and techniques with which traditional historians are ill-equipped to engage critically. More importantly, the conviction that complex historiographical problems can be resolved or at least made more tractable by appeal to data-driven empiricism is dubious. In the first place, fields like paleoclimatology are obviously dominated by scientists with little historical training and historians must approach their conclusions warily. The science of ancient environments is also rapidly evolving, and its datasets are continuously, indeed exponentially, expanding. Any conclusions based on so protean a foundation are accordingly both premature and insecure.[104] Yet all the more troubling from a methodological perspective is the failure of the new scientism to fully integrate social, political, economic, and cultural factors into its grand narratives.[105] By effacing human agency in favor of empirical data, studies like Harper's rest entirely on the alleged temporal coincidence of environmental perturbations and sociopolitical change. Not only is this a simple fallacy of correlation implying causation, by treating human communities as the passive subjects of a tyrannical nature it also lapses into neodeterminism, even if authors claim to reject such deterministic monocausality.[106] But despite the perverse appeal of environmental determinism in this era of existentialist climate alarmism, it cannot withstand close scrutiny.[107] As Romanist Paul Erdkamp argues in an

---

103. Harris, "What Kind of Environmental History for Antiquity?" 1.
104. So Marx, Haunschild, and Bornmann, "Climate and the Decline and Fall."
105. So Haldon et al., "Plagues, Climate Change, and the End of an Empire (3)," 7; and Sessa, "The New Environmental Fall of Rome."
106. Sessa, "The New Environmental Fall of Rome," 219–20. On contemporary neodeterminism see William B. Meyer and Dylan M. T. Guss, eds., *Neo-Environmental Determinism: Geographical Critiques* (Basingstoke, Hampshire: Palgrave Macmillan, 2018), esp. 39–96.
107. See the favorable, activist receptions of Harper's monograph in Ellie Robins, "How the Environment Topples Empires: On Kyle Harper's *The Fate of Rome*," *LA Review of Books*, 7 December

essential statement on the current debate, the effects of emergent natural phenomena on premodern societies were neither as predictable nor as preordained as the new scientism assumes.[108] Indeed, a plethora of regional case studies from various parts of the Mediterranean demonstrate that ostensibly adverse environmental factors like climate change rarely correlated with the sudden simplification or deterioration of social structures during the Holocene. Rather, such transformations were regularly stimulated by political, economic, or institutional change.[109] In Erdkamp's formulation, exogenous natural phenomena were always mediated through a dense array of social, economic, and political institutions, all of which provided a greater or lesser degree of resilience to environmental pressures. For instance, although climate change *may* have led to more frequent harvest failures in the late Roman West, such events were a regular feature of life throughout Roman history. It was therefore not climate shocks per se that destabilized the western half of the late Roman Empire, Erdkamp concludes, but rather the collapse of social and administrative infrastructure that reduced overall societal resiliency to such events.

Given the evident interpretive limitations of scientistic approaches, environmental historians of the premodern Mediterranean must therefore begin to refine both our understandings and representations of natural-environmental causality. As historian Kristina Sessa counsels, this means moving beyond the current first phase of scholarship in which "science (and the work of scientists) ultimately drives the inquiry, while historical evidence (and the work of historians) serves largely to validate their hypotheses."[110] Sessa's proposed second phase would represent true interdisciplinarity by better integrating human experience, social organization, and culture into scientifically informed studies of the material environment. In my own formulation this means eschewing the binary and deterministic apposition between humans *and* nature that characterizes the new scientism and refocusing on humans *in*

2017, accessed 28 June 2019, https://lareviewofbooks.org/article/how-the-environment-topples-empires-on-kyle-harpers-the-fate-of-rome; Madeline Ostrander, "When Rome Fell, the Chief Culprits Were Climate and Disease. Sound Familiar?" Undark, 16 March 2018, accessed 28 June 2019, https://undark.org/article/book-review-harper-fate-of-rome; Williamson Murray, "Reviewing the Fate of Rome," Real Clear Defense, 30 August 2018, accessed 28 June 2019, https://www.realcleardefense.com/articles/2018/08/30/reviewing_the_fate_of_rome_113755.html.

108. Erdkamp, "War, Food, Climate Change."

109. See the special issue of *Quaternary Science Reviews* 136 (2016), ed. Alexandra Gogou, Adam Izdebski, and Karin Holmgren. See also Xoplaki et al., "Modelling Climate and Societal Resilience"; and the methodological review of Haldon et al., "History Meets Palaeoscience."

110. Sessa, "The New Environmental Fall of Rome," 230.

nature, thereby highlighting dynamic, mutually constitutive entanglements between ancient peoples and the environments within which they were embedded.

Narrower case studies are surely the best starting point for this study of human-nature embeddedness. Firmly grounded in all the evidence from specific times and places, the higher granularity of such work permits historians to elucidate more clearly and convincingly the complexity and contingency that have always characterized human entanglements with nature. There has already been some important work in this vein, most notably Peter Thonemann's historical geography of the Maeander River Valley between antiquity and the Byzantine period.[111] Magisterial though it is, the scope of Thonemann's study is nonetheless expansive and its arguments diffuse, reflecting the impossibility of describing the totality of human relationships with nature even within a tightly circumscribed geography. As the water historian Terje Tvedt has already trenchantly observed, "the terms 'nature' and 'environment' cover such a myriad of variables and aspects of importance to societies that meaningful, empirically oriented research or precise discussions become virtually impossible."[112] In consequence, attempts to analyze a given society's innumerable and ever-evolving relationships with "nature" must either expand to unwieldy length[113] or else sacrifice analytical depth in favor of broad but shallow coverage.[114] A more rigorous and revelatory approach, Tvedt argues, is to probe human interactions with particular elements or subsystems of the natural world. At first blush, this methodological exhortation might seem redundant, since environmental historians have already been producing such close analyses for decades.[115] Even the few outstanding studies of Graeco-Roman environmental history thus far produced are narrow in compass.[116] Tvedt nonetheless proposes a novel approach to long-term environmental history that foregrounds human relationships with water. Why water? Essential to biological life and thus of paramount importance to all societies in all periods, water is indisputably universal. Yet owing to environmental, cultural, socioinstitutional, and

---

111. Thonemann, *The Maeander Valley*.
112. Tvedt, "'Water Systems,'" 144–45.
113. Horden and Purcell, *The Corrupting Sea*.
114. Thommen, *The Maeander Valley*, .
115. https://aseh.org/past-recipients, accessed 3 November 2023.
116. Robert Sallares, *Malaria and Rome: A History of Malaria in Ancient Italy* (Oxford: Oxford University Press, 2002); Gregory S. Aldrete, *Floods of the Tiber in Ancient Rome* (Baltimore: Johns Hopkins University Press, 2007); Scheidel, *Death on the Nile*; Blouin, *Triangular Landscapes*.

technological diversity, this universality manifests itself in myriad ways across space and time—a "simultaneous embodiment of the universal and the particular."[117] To embrace such diversity, Tvedt urges the interrogation of human-water relationships at three interlocking analytical levels: (1) the natural-environmental characteristics of particular hydrological systems, (2) evolving human exploitation and manipulation of these systems, and (3) the ideologies and institutions that structure human-water interactions within specific temporal and cultural contexts. Ultimately, this approach is founded on the unabashedly materialist assertion that "water exists independently from cultural ways of knowing it."[118] In other words, its innumerable and shifting cultural resonances notwithstanding, water possesses intrinsic physical properties that remain constant over time and that necessarily inform but do not determine human interactions with it. Although outwardly commonsensical, these observations in fact push back against the increasingly vigorous critique of nature/culture dualism in contemporary water studies. In this literature "nature" and "culture" elide into "hybrid socionatures,"[119] while the "hydrological cycle" is transformed into the "hydrosocial cycle,"[120] thereby collapsing the human and the nonhuman into a dense, ever-shifting web of relations. From this relational-dialectical perspective, no material substance possesses physical properties inherent to itself; rather, such "properties" are better characterized as "relations" since they are manifest only at moments of interface between two substances. Water thus becomes little more than a "process of engagement," a substance whose properties are emergent and are realized only through its relationships with other substances or human beings.[121] On this reading, water's capacity to flow, for example, does not exist per se, since it is expressed only in relation either to another material surface—for example, the bed of a canal, stream, or river—or to human perceptions.[122] As geographer Bruce Braun writes, such monism abolishes every distinction between nature and culture and represents "all ecological, social, cultural, and political forms as historical and relational effects"[123] Thus the influential water theorist Erik

117. Tvedt, "'Water Systems,'" 146.

118. Tvedt, "'Water Systems,'" 148. See also Tvedt and Coopey, "A 'Water Systems' Perspective on History," 5.

119. E.g., Karpouzoglou and Vij, "Waterscape." See esp. Swyngedouw, *Social Power*; and Swyngedouw, *Liquid Power*.

120. Linton and Budds, "The Hydrosocial Cycle."

121. Linton, *What Is Water?*, 30.

122. Alberti, "How Does Water Mean?"

123. Braun, *The Intemperate Rainforest*, 266.

Swyngedouw, who has programmatically declared that "nature outside the social and the political does not exist."[124]

Such literature is often inscrutable to the uninitiated, its arguments cloaked in a "veil of verbal mist" that obscures more than it reveals.[125] Still more critically, the total elision of nature and society raises a fundamental methodological obstacle to environmental history as it is commonly practiced, for how can we study humans in nature if the two spheres cannot be distinguished? Yet one need not adopt a relational-dialectical mode wholesale to agree with the central (and sensible) proposition that water ought to be viewed as existing and operating within both spheres simultaneously: "neither purely 'natural' nor purely 'social,' but rather simultaneously and inseparably both"[126] Indeed, Tvedt's proposed water systems methodology merely urges us not to ignore or discount the materiality of water, since its social manifestations are significantly informed by the unique and unchanging physical and chemical properties of H_2O, along with its absolute indispensability to biological life. As Jessica Barnes describes it, taking the materiality of water seriously simply means "attending closely to water's physical characteristics—its ability to flow, reflect, absorb, collect, divide, dissolve, disperse, transform, and erode."[127] As we will see, this attention to materiality is hardly deterministic, since water's physical characteristics inform rather than dictate the myriad ways in which it is put to use by its human dependents.

Argument and Organization

Inspired by Tvedt's water systems approach, as well as by geographer Jessica Barnes' and archaeologist Matthew Edgeworth's scholarship on water flow and its social production, this study interrogates human entanglements with flowing water in the Fayyūm, the socioenvironmental relationship central to the evolution of its irrigated landscape.[128] Its arguments are thus firmly

---

124. Swyngedouw, *Liquid Power*, 229.
125. John McNeill, review of *Liquid Power,* 160–61.
126. Perreault, "Introduction to Part II," 118. Cf. Strang, *The Meaning of Water*, 5: "As the substance that is literally essential to all living organisms, water is experienced and embodied both physically and culturally . . . Engagement with water is the perfect example of a recursive relationship in which nature and culture literally flow into each other."
127. Barnes, *Cultivating the Nile*, 29.
128. Barnes, *Cultivating the Nile*; and Edgeworth, *Fluid Pasts*.

grounded in the material, specifically the hydrology and the hydraulics of Fayyūm canal flow. While such a perspective necessitates some engagement with the science of the Fayyūm's physical environment, I introduce such information sparingly and demonstrate throughout that natural-environmental phenomena were never the primary drivers of historical change. In further contrast with the new scientism, this study draws its evidence primarily from textual sources, ancient, early Islamic, and modern alike. It may therefore be read, at least in part, as a programmatic assertion of the continued importance of text-based local environmental histories of the premodern Mediterranean world, which will lend much-needed texture and detail to the macroscopic pictures sketched by authors working at the level of the Mediterranean basin. Indeed, only written sources can shed light on the issues at the heart of this study: the lived experience of human-water entanglement and the human interventions that continuously reshaped its flow. This approach, moreover, will demonstrate to scholars of Graeco-Roman Egypt the value of later sources for the study of Egyptian landscapes. Moreover, by consistently highlighting both the crucial role of human agency and the significance of (micro)regional variation,[129] this study cautions future environmental historians of Mediterranean premodernity not to reduce the complexity and contingency of human-nature relationships to simple narratives of mechanistic environmental causality.

In its attempt to represent water flow as an entangled coproduction of environmental phenomena and human agency, this study moves from the macro- to the microscopic, following the flow of the Fayyūm's water from its origins in southern Egypt, northward down the Baḥr Yūsuf, into the mouth of the depression, through its canal system, and finally to the moribund village of Theadelphia near the northwestern terminus of the system in the fourth century CE. The goals of this approach are threefold. First and most broadly, I hope to clarify the nature and function of the premodern Fayyūm's canal system, a subject about which there is some confusion in the scholarship.[130] A more refined and holistic understanding of the environmental parameters of Fayyūm irrigation and the various methods by which the region's water supply was governed, distributed, and utilized over time will make a significant contribution to the landscape, agricultural, administrative, and socioeconomic history of premodern Egypt. Second and more specifically, this environmental-

129. Drawing on the arguments of Horden and Purcell, *The Corrupting Sea*, though the focus of this study is far narrower.

130. E.g., Römer, "The Nile in the Fayum," esp. 172–73.

historical interpretation of Fayyūm water flow will serve as the context for a reexamination of the troubles that afflicted the village of Theadelphia in its final years, thereby enabling us to see more clearly than before the complex yet highly localized socioenvironmental circumstances that precipitated the village's decline. Third, by consistently accentuating environmental particularity—the particularity of the Fayyūm as a whole as well as the distinctly local struggles of late Roman Theadelphia—this work also endeavors to explore premodern human-nature relationships where they are most meaningful and deeply felt: on the ground, wet and dirtied with the water and soil of home.

• • •

The first two chapters lay the materialist and technical foundations of the rest of the book by establishing the hydrological and hydraulic conditions that informed the agroenvironmental and socioinstitutional features of Fayyūm irrigation discussed later on. Chapter 1 reconstructs the hydrological regime of the premodern Baḥr Yūsuf, arguing that although the channel rose and fell in time with the Nile's annual cycle, it could still provide some water to the Fayyūm year-round, albeit in much reduced quantities during the dry season. The peculiar hydrology of the Baḥr Yūsuf notwithstanding, the form and function of the Fayyūm's physical irrigation infrastructure, particularly the large reservoirs that dotted the landscape in all periods, indicate that perennial cultivation in the region was largely enabled by its unique ability to capture, store, and later distribute floodwater for the irrigation of a second annual crop. Perennial irrigation in the premodern Fayyūm therefore did not require the constant, high-level influx from the Nile that characterizes the modern irrigation system. The agroenvironments produced by this unique irrigation system are the subject of chapter 2, especially the apparent discrepancy between the double-cropped landscape described by later Arabic and early modern European authors and the primarily (though not exclusively) single-cropped environment documented in Graeco-Roman papyri. Rather than regarding these contrasting images as mutually exclusive, I argue that differential productivity was a feature both of natural-environmental phenomena and human decision-making. That is, the Baḥr Yūsuf's hydrology, the topography of the depression, and the functionality of the canal system all combined to produce two distinct agroenvironments in the Fayyūm: a robust, resilient, and well-watered central plain, and the drier single-cropped margins.

Having sketched the material-environmental parameters of Fayyūm irrigation, the remaining chapters more closely investigate the social production of water flow, beginning in chapter 3 with the role(s) of Egyptian state governments. This focus on central state agency departs from most contemporary Egyptological work on irrigation, which instead emphasizes local water control as a means of combatting the lingering influence of Karl Wittfogel's theory of hydraulic despotism. While such work has done much to reveal the enduring contributions of local actors to the making of Egypt's irrigated landscape, it has too readily dismissed the possibility of significant and sustained yet nondespotic state intervention. As Ottomanist Alan Mikhail has convincingly demonstrated, a premodern state could be deeply involved in encouraging, coordinating, and supporting the annual local maintenance of public irrigation infrastructure—a system of "coordinated localism" that nonetheless never amounted to totalizing control over rural irrigation practices.[131] This chapter accordingly urges a more thoroughgoing reintegration of the state into the study of premodern Egyptian water management, albeit in a way that respects the often significant differences in state interests, capacity, and configuration between (and even within) discrete historical periods. With the Fayyūm as case study, I trace the evolution of state intervention between the early Ptolemaic, Roman, and late ʾAyyūbid periods, arguing that government oversight was essential to the maintenance of the original shape, extent, and productive capabilities of the region's unique canal system. This intervention nonetheless took markedly different forms between the Ptolemaic and Roman periods, evolving from a centralized administration that emerged from the state-directed Ptolemaic reclamation of the depression to a form of coordinated localism under Rome, whose extractive interests in Egyptian agriculture incentivized the formalization of local rituals of annual dike- and canal-maintenance as a closely supervised and rigorously documented corvée. By contrast, the late Ayyūbid state had no significant fiscal stake in Fayyūm agriculture and had consequently retreated from the countryside, leaving the maintenance of the canal system almost entirely in the hands of local actors. Although locals proved more than capable of operating the much smaller Ayyūbid-era canal system, the advanced degradation of key components of its physical infrastructure testify to the significant practical and material contributions of earlier, more interventionist governments.

But beyond these broad programmatic aims, this chapter draws particular

131. Mikhail, *Nature and Empire*; and Mikhail, *Under Osman's Tree*, 19–110.

attention to the abundant and richly textured written record generated by Roman coordinated localism, arguing that it was central to the production of rural subjectivity in Roman Egypt.[132] Here at ground level, state authority was keenly felt through the administrative and documentary practices that bound cultivators to their villages of record (*idia*) and compelled them each year to contribute corvée labor for the maintenance of its public irrigation infrastructure (and at times critical infrastructure elsewhere in the depression). Constituted at the site of these compulsory entanglements with earth and water, rural subjectivity in Roman Egypt was a phenomenon at once embodied and environmental—the product of the work of annually reproducing the Fayyūm's irrigated landscape for the benefit of an extractive imperial state. Yet this individualized and spatially circumscribed vison of the rural subject—single bodies, fixed (theoretically) in place, and there compelled to recursively reproduce existing patterns of water flow—was ultimately in conflict with the internal constitution of Fayyūm irrigation communities. While this disjuncture was largely obscured and unproblematic during periods of communal prosperity, it became increasingly apparent and exacerbated during phases of communal dissolution.

Chapter 4 accordingly pivots toward these rural irrigation communities and the socioenvironmental production of water flow at the local level. Building upon the discussion of the first two chapters and drawing upon comparative historical and anthropological work on small-scale, internally self-organized irrigation communities, I foreground the issue of water rights—the means and methods by which public waters were apportioned among irrigators—arguing that the contrasting watersharing regimes attested in Greek papyri and medieval Arabic evidence represent context-specific human adaptations to the hybrid hydrology and hydraulics of the Fayyūm's canal system. In the center of the depression where water flows were more plentiful and predictable, a system of strict water allotments obtained (at least in the Islamic period, when this portion of the Fayyūm becomes visible). By contrast, along the flood-irrigated Graeco-Roman margins where the water supply was unpredictable in both volume and the time of its arrival, water was instead treated as a common resource to which all irrigators had a right of access. Maintaining such access nonetheless required significant investments of social capital, ranging from monitoring the water use of canal-sharing neighbors, to contesting the obstruction of public waterways, and even to aggressive, potentially violent, collective action in defense of communal rights to the commons.

132. Drawing on language in Derr, *The Lived Nile*, esp. 3.

Yet irrespective of the methods by which they were apportioned, the production of rural water flows was entirely contingent upon annual communal maintenance of local public irrigation infrastructure, the informal social practice formalized as a corvée during the Roman period. Moreover, it was through these annually recurring labors—the shared work of producing a shared water supply—that Fayyūm irrigation communities were internally constituted. Seen in this light, participation in the yearly maintenance of village irrigation infrastructure can be regarded not as an individual but rather a communal obligation, while the water flows such labor produced likewise appear as material-environmental expressions of communal integrity annually reinscribed on the landscape. Here, then, the disjuncture between Roman constructions of rural subjectivity and the internal constitution of Fayyūm irrigation communities becomes evident. By reconfiguring these internally generated collective social obligations to *community* into externally imposed individual corvée obligations to *place*—both *idia* and nome—Roman coordinated localism was sustainable only in the context of healthy and internally cohesive irrigation communities collectively invested in the maintenance of the particular stretch of the Fayyūm's public water infrastructure upon which their livelihoods depended. Yet if an irrigation community were to destabilize and its population start to trickle away, the social relationships that had once enmeshed its members in bonds of mutual obligation to one another and to their shared irrigation infrastructure would begin to fray, a process that would only intensify as it grew harder and harder to keep the water flowing.

The final chapter brings all of these arguments to bear on the case of fourth-century CE Theadelphia, whose terminal struggles over water are a microcosm of the themes explored in the rest of the book. As recounted in petitions and related documentation preserved in the archive of Aurelius Sakaon, the sparsely populated settlement lay on high ground toward the tail end of the canal system and suffered persistent water shortages allegedly produced by the illicit or malicious actions of irrigators located upstream along shared public waterways. The result, the Theadelphians claim, was the virtual desolation of their village and their attendant inability to meet their fiscal responsibilities to the state. Although these events have traditionally been viewed as evidence of Roman administrative neglect of the Fayyūm's canal system,[133] the most recent scholarship instead posits epidemic disease or climate change as the principal

133. See above nn. 21–22.

drivers of decline.[134] After a critique of these arguments, I argue that Theadelphia's fourth-century struggles were internal and socioenvironmental in origin: the end result of a long population decline that had progressively undermined the coherence of this marginal irrigation community and left it unable to produce the water flows it needed. Readings from Sakaon's archive indicate that the Theadelphians understood irrigation in much the same way, for they consistently described their problems as a unique concatenation of local socioenvironmental phenomena—field height, a disadvantageous position in the canal network, population depletion and their subsequent weakness in the face of upstream competition—that had together resulted in their complete inability to irrigate by traditional means. Narrow and particular, the narrative that emerges from Sakaon's archive reflects a deeply contextualized local knowledge born of the work (increasingly futile) of moving water to their own fields. The Roman state, however, viewed these conflicts through the prism of its own constructions of rural subjectivity, which prioritized the perpetual reproduction of the Fayyūm's irrigated landscape. In consequence, its attempts to resolve these conflicts only reinforced the bonds that tied the remaining Theadelphians to their *idia*—the territory and relict water infrastructure of an irrigation community already defunct in all but name. This, if anything, was the real tragedy of late antique Theadelphia: the failure of the Roman state to perceive how the Fayyūm's water was made to flow and thus to realize that here at Theadelphia it no longer could.

Note on Irrigation in the Nile Valley and the Fayyūm[135]

Egyptian irrigation practices have been familiar to modern Western readers since at least Shakespeare's day if not earlier.[136] Broadly speaking, the agricultural year was divided into three principal segments: the inundation (*Akhet*), the growing season (*Peret*, "emergence"), and the harvest season (*Shemu*, prob. "low water"). During the inundation season in the Nile Valley, the flood

134. Elliott, "The Antonine Plague"; and Huebner, "Climate Change."

135. For a technical overview of irrigation from sources of the twelfth through the fourteenth centuries see Sato, "Irrigation in Rural Egypt."

136. .William Shakespeare, *Antony and Cleopatra*, act 2, scene 7: "Thus do they, sir: they take the flow o' the Nile / By certain scales i' the pyramid; they know / By the height, the lowness, or the mean, if dearth / Or foison follow: the higher Nilus swells / The more it promises; as it ebbs, the seedsman / Upon the slime and ooze scatters his grain, / And shortly comes to harvest."

was captured and contained for forty days or more within dense networks of earthwork basins that covered the cultivable land of the floodplain. This first cleansed the soil by leaching away accumulated salts and other minerals deleterious to plant life and, second, fertilized the ground by depositing on the floodplain a thin layer of silt that had been carried down from the highlands of Ethiopia by the Blue Nile during the inundation. As the flood ebbed, these basins were breached and wastewaters returned to the Nile, leaving the wet, fertile soil ready for the sowing of a single annual cereal crop (the winter or *shatawī* harvest of later Arab authors like al-Nābulusī). Commonly dubbed "basin irrigation," the practice is more properly known as flood recession irrigation or recessional agriculture.[137] Yet the strict seasonality of Nile Valley irrigation must be distinguished from irrigation practices in the Fayyūm. The twelfth-century CE Egyptian bureaucrat Ibn Mammātī, chief of all ministries under Salāḥ al-Dīn (r. 1174–93) and his successor al-ʿAzīz ʿUthmān (r. 1193–98), makes this distinction clear in his administrative survey of Egypt *Qawānīn al-Dawāwīn* ("Regulations of the Ministries"). In his chapter on Egypt's canals and dikes, Ibn Mammātī surveys the annual breaching of dikes and feeder canals in the Delta provinces of Buḥayra, Hauf Ramsīs, and Damanhūr, which principally occurred at the arrival of floodwater early in the month of Tūt.[138] The Fayyūm, however, was radically different, home to "many renowned canals, filled to the brim and overflowing, as well as sluicegates, waterways, and divisors; every village has a fixed water-right at a known time."[139] Open to the Nile when the flood was at its height, these canals were then alternatingly closed and opened on a repeating schedule during the next six months of the year (Hātūr–Baramūdah). As discussed in chapter 2 below, this regulatory schedule governed the peculiarly abundant water supply of the province and enabled the cultivation of two annual harvests—the standard *shatawī* as well as a summer or *ṣayfī* crop—in portions of the depression. A more detailed account of the schedule also appears in Abū Isḥāq's redacted *dustūr* of 1031 CE, which is translated in the appendix.

---

137. "Basin irrigation" technically refers to the practice of flooding a field at a low level and allowing the water to infiltrate, rather than draining it as in flood recession irrigation.

138. Ibn Mammātī, *Qawānīn*, 205–33. Translation in Cooper, "Ibn Mammātī's Rules," 60–77. For a study of this system as described by the contemporary bureaucrat al-Makhzūmī see Borsch and Sabraa, "Qānūn al-Riyy."

139. Ibn Mammātī, *Qawānīn*, 229: فيه خلج كثيرة مشهورة عامرة وغامرة، وأبواب وتراع ومقاسم ولكل ناحية شرب معلوم، في وقت مفهوم

Chapter 1

Capturing the Flood

And the river of the Fayyūm emerges from the Nile of Egypt upon its rise.

—Abū-l-Fidā', *Taqwīm al-buldān*

The Baḥr *Yūsuf and the Water Supply of the Fayyūm*

The Fayyūm receives water not directly from the Nile but through a natural side-channel of the river known as the Baḥr Yūsuf ("Joseph's River"), which is either an ancient subsidiary branch of the Nile or the partial remains of its prehistoric course.[1] The Baḥr Yūsuf presently takes off from the western bank of the Nile near the city of Dayrūṭ and travels some 190 km north before bending sharply to the west and emptying into the Fayyūm at the village of al-Lāhūn. The location of its headwaters has nonetheless been unstable over time. Nineteenth-century British engineers claimed to have identified an early, albeit undated, mouth at Mafalūṭ some 35 km south of Dayrūṭ, while certain early Islamic geographers variously located its origin at Ashmūnayn or farther to the south at Sōhāg—a 165 km stretch of the southern Nile Valley in all.[2] Beginning with Ibn ʿAbd al-Ḥakam in the ninth century, however, Islamic authors most frequently linked the head of the canal to a region called al-Manhā located somewhere in the vicinity of Dayrūṭ (then Darwat Sarabām, later known as Darwat al-Sharīf). Arab geographers subsequently applied this obscure toponym metonymously to the entire length of the Baḥr Yūsuf between Dayrūṭ

1. Sampsell, *Geology*, 90.
2. Cooper, *The Medieval Nile*, 101–2. See also Said, *The River Nile*, 63; and Sampsell, *Geology*, 90.

and al-Lāhūn, which was called the *Khalīj al-Manhā* ("al-Manhā Canal") and attributed to the prophet Joseph.[3] The thirteenth-century geographer Yaqūt accordingly describes the region of al-Manhā near Dayrūṭ as "the name of the mouth (*famm*) of the stream (*nahr*) excavated by the Righteous Joseph, which brings to the Fayyūm what it receives from the Nile."[4] According to al-Nābulusī, the al-Manhā canal branched from the Nile "at the southern part of the district of al-Ahmūnayn above a village known as Darwat Sarabām."[5] After flowing past al-Bahnasā (ancient Oxyrhynchus), he continues, the channel entered the Fayyūm at al-Lāhūn. Al-Nābulusī then gives several names to the final section of the channel within the Fayyūm including the *Baḥr al-Fayyūm al-Aʿẓam* (Grand Canal of the Fayyūm) and the *Baḥr al-Aʿẓam al-Yūsufī* (Grand Canal of Joseph). In the fifteenth century, al-Maqrīzī continued to distinguish between the channel's southern arm (*Khalīj al-Manhā*) and its terminal stretch inside the Fayyūm (*Khalīj al-Fayyūm*), the latter then colloquially known as the Baḥr Yūsuf.[6] In 1672 the German Wansleben still refers to the channel as both the *Calitz il Menhi* and the *Bahr Jusef*, though the Ottoman traveler Evliya Çelebi in 1685 calls it only the Baḥr Yūsuf. Both nonetheless agree that its head lay at Dayrūṭ.[7]

In order to ascertain the water supply of the premodern Fayyūm, we must first remember that the perennial flow of the contemporary Baḥr Yūsuf is a relatively recent development, the product of nearly a century of modern hydraulic engineering in southern Egypt. This current perennial link between the Nile and the Baḥr Yūsuf was first established only in 1873 via the opening of the Ibrāhīmiyya canal, which branches from the river at Asyūṭ and discharges into

3. Ibn ʿAbd al-Ḥakam, *Futūḥ*, 6 and 15. Early modern European authors variously transliterated منهى as *Mounha*, *Manhi*, or *Menhi*. The pronunciation مَنهى is established by Yāqūt and Ibn ʿAbd al-Ḥakam and is so rendered in Rapoport and Shahar's translation of al-Nābulusī. For further references see Maspero and Wiet, *Matériaux*, 83; and Timm, *Das christlich-koptische Ägypten*, 1562. According to Carsten Peust, the Egyptian name of the channel, *Tm.t*, is likely derived from the root *tm*, "to be over/at an end," while *manhā* stems from the synonymous √*nhy*, suggesting that the name is to be interpreted as "last canal," i.e., not leading to any other waterway (Peust, *Die Toponyme*, 103). An attractive but probably fanciful Coptic etymology is ⲙⲁ ⲛ̄ϩⲁⲉ, "place at the end," i.e., of the canal.

4. Yāqūt, *Muʿjam al-buldān*, 5:217. Cf. the entry for Darwat Sarabām in Yāqūt, *Muʿjam al-buldān*, 2:453.

5. Al-Nābulusī, *VF*, 34: بحرا يسمي المنهى فوهته المتصلة بالنيل في الطرف القبلي من اعمال الاشمونين فوق قرية تعرف بذروة [دروة] سربام

6. Al-Maqrīzī, *Khiṭaṭ*, 1:190. The portion of the channel within the Fayyūm is today officially known as the Baḥr Yūsuf behind (*khalf*) al-Lāhūn. Barnes, *Cultivating the Nile*, 22.

7. Wansleben, *The Present State of Egypt*, 153; and Dankoff, Tezcan, and Sheridan, *Ottoman Explorations*, 204 and 354–55. Çelebi also refers to the stretch of the channel within the Fayyūm as the "Canal (ترعة) of Joseph."

the head of the Baḥr Yūsuf some 60 km north alongside Dayrūṭ.[8] Yet before the completion of the Asyūṭ Barrage (1903), which diverts water into the head of the Ibrāhīmiyya, this new canal had no headworks whatsoever and its water level consequently rose and fell in time with the annual cycle of the Nile.[9] The flow of the newly perennial Baḥr Yūsuf was therefore subject to considerable interseasonal variability in its early decades. Writing in 1889, the French engineer Julien Barois claims that the discharge of the Ibrāhīmiyya into the Baḥr Yūsuf during the low-water season was insufficient for the irrigation of every district along the Baḥr Yūsuf's course—Asyūṭ, Minyā, Banī Sweif, and the Fayyūm. It was therefore necessary to curtail all cultivation along the southern stretches of the channel during these months in order to conserve the entirety of its limited flow for the Fayyūm. According to Barois' calculations, the channel's discharge at al-Lāhūn was some 85 m^3/second (7.3 million m^3/day) during the flood, falling to 17 m^3/second (1.47 million m^3/day) during the low-water season, a decrease of some 400 percent.[10] Figures supplied in 1892 by the British engineer Robert Hanbury Brown, Inspector-General of Irrigation for Upper Egypt, are roughly identical: a discharge at al-Lāhūn of 6.5–7 million m^3/day during the flood compared to 1–1.5 million m^3/day at low water, a decrease of between 300 percent and 600 percent.[11] Yet even after the completion of both the Aswān Low Dam (1902) and the Asyūṭ Barrage the following year, discharge at al-Lāhūn still fluctuated between 3 million m^3/day at low water and 8 million m^3/day during the flood.[12] Only the opening of the Aswān High Dam (1970) has stabilized the flow of the channel at a high level throughout the year, thereby completing its conversion into a truly perennial irrigation canal.[13]

In premodernity, the connection between the Nile and Baḥr Yūsuf was seasonal. Indeed, European colonial engineers noted that the mouth of the

---

8. Before the construction of the Ibrāhīmiyya, Linant de Bellefonds proposed to divert the tail of the natural Sohagiyya canal at Delgā, some 14 km northwest of Dayrūṭ, into the Baḥr Yūsuf to "maintenir les eaux d'étiage" in the latter channel. Linant de Bellefonds, *Mémoires*, 402. See also the drawing of the connection in Linant de Bellefonds, *Carte hydrographique de la partie septentrionale de la haute Égypte*.
9. Willcocks and Craig, *Egyptian Irrigation*, 1:434–35. See 438 for the annual discharge of the Ibrāhīmiyaa in the early twentieth century.
10. Barois, *Irrigation*, 41.
11. Brown, *The Fayûm*, 12.
12. Willcocks and Craig, *Egyptian Irrigation*, 1:444. Cf. the similar 1904 figures in Barois, *Les irrigations*, 184: 15 m^3/second (ca. 1.3 million m^3/day) during *étiage* and 95 m^3/second (8.2 million m^3/day) during the flood, a 533 percent increase.
13. Wolters et al., "Managing the Water Balance." See the graph in Barnes, *Cultivating the Nile*, 11, for Nile discharge at Aswān before and after 1970.

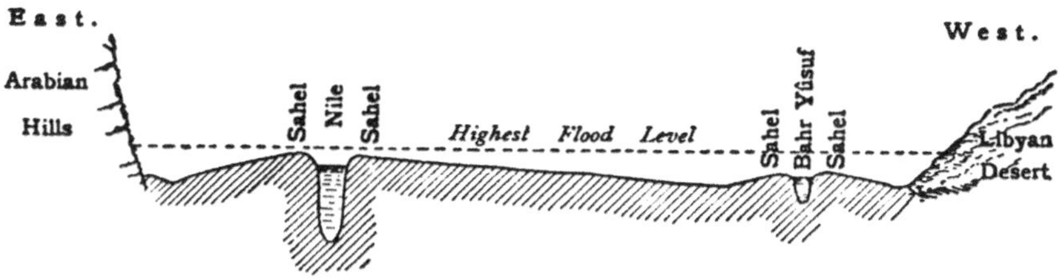

Figure 2. Cross section of the Nile Valley. Brown, *The Fayûm and Lake Moeris* (1892).

Baḥr Yūsuf at Dayrūṭ lay at a higher level than the low-water level of the river. Consequently, before the opening of the Ibrāhīmiyya in 1873 the channel was linked to the Nile's surface flow only during the inundation. After the flood receded, this connection was again severed until the following year.[14] This cycle surely resulted in far greater seasonal fluctuations of the Baḥr Yūsuf's flow than those recorded in the early years of perennial irrigation. Unfortunately, the thoroughgoing restructuring of Nile hydraulics in the nineteenth and twentieth centuries has made it impossible to observe and document the channel's natural cycle, and we must instead rely on scattered ancient, early Islamic, and modern observations. Such testimony nonetheless consistently describes a channel whose rhythms were intimately (and unsurprisingly) entangled with the annual increase and decrease of the Nile. Yet several observers also remark that the Baḥr Yūsuf alone of the Nile's major canals and branches continued to carry a reduced amount of water even after the flood had receded and its visible connection to the river had been broken. This peculiar semiperenniality notwithstanding, the remainder of this chapter argues that the channel delivered the overwhelming majority of the Fayyūm's water supply during the inundation. This annual influx thus influenced the form and function of the Fayyūm's premodern irrigation infrastructure, which was designed not to regulate a perennial flow of water through the al-Lāhūn gap but to capture and contain the flood.

Arabic sources are consistent in their description of the seasonality of the Baḥr Yūsuf. In a passing reference to the hydrology of Egypt, the fourteenth-century geographer Abū-l-Fidāʾ remarks that "the river of the Fayyūm (*nahr*

14. See the cross section in Brown, *The Fayûm*, 10. Cf. Barois, *Irrigation in Egypt*, 28; Buckley, *Irrigation Works in India and Egypt*, 13; Brown, "The Bahr Yusuf"; Willcocks, *Egyptian Irrigation*, 60; Maury, "Irrigation et agriculture," 82.

Figure 3. The seasonal mouth of the Baḥr Yūsuf at Dayrūṭ. Detail from Linant de Bellefonds, *Carte hydrographique de la partie septentrionale de la haute Égypte* (1855). Image courtesy of Princeton University Library.

al-Fayyūm) emerges from the Nile of Egypt upon its rise."[15] That is, direct communication between the Baḥr Yūsuf and the Nile was annually reestablished upon the rising (*ziyāda*) of the Nile in flood. Ibn ʿAbd al-Ḥakam describes the situation succinctly in his mythographic narrative, writing that "[after] the water of the Nile rose, it then entered the head (*raʾs*) of the al-Manhā and flowed through it until it reached al-Lāhūn."[16] The precise point during the flood stage at which this linkage occurred cannot be determined with certainty and probably changed over time in response to shifts in the course of the Nile and the Baḥr Yūsuf, the migration of the channel's head, and the progressive siltation of the floodplain. Writing in the tenth century, al-Masʿūdī claims that "when the rise of the water has reached nine cubits (*tisʿat adhruʿ*, roughly 6 m), water enters the Canal of al-Manhā and the Canal of the Fayyūm."[17] Although this passage is quoted verbatim in later authors and seems to have become traditional,[18] it puts the point of communication at just past the halfway mark of a minimally optimal Nile Valley flood of 16 or 17

15. Abū-l-Fidāʾ, Taqwīm al-buldān, 46: يخرج من نيل مصر نهر الفيوم عند زيادته
16. Ibn ʿAbd al-Ḥakam, *Futūḥ Miṣr*, 15: وارتفع ماء النيل، فدخل في رأس المنهى، فجرى فيه حتى انتهى إلى اللاهون.
17. Al-Masʿūdī, *Murūj*, 2:368: وكان اذا بلغ في زيادته تسعة اذرع دخل خليج المنهى وخليج الفيوم. See also Sijpesteijn, *Shaping a Muslim State*, 28.
18. Al-Maqrīzī, *Khiṭaṭ* 1:158–59; Al-Maqrīzī, *al-Sulūk li-maʿrifat duwal al-mulūk* (Beirut: Dar al-Kutub al-ʿIlmiyya, 1997), 1:80; and Ibn Tagrībirdī, *Nujūm al-zāhira fī mulūk Miṣr wa-l-Qāhira* (Leiden: Brill, 1855), 1:59.

cubits (10.5–11.3 m).[19] Yet whenever the connection was established, the water level of the channel subsequently began to rise. As recounted by the engineer Pierre-Dominique Martin in the *Description de l'Égypte*, this increasing downstream flow soon enabled the irrigation of rural districts along the Baḥr Yūsuf's southern course: "during the inundation, the Baḥr Yūsuf communicates with the other parallel canals [i.e., the irrigation canals that take off from it] and through them covers the lands that lie between it and the river."[20] In heavy flood years, the waters carried by the channel could lie long on the countryside. The French engineer Edme-François Jomard remarks in the *Description* that a massive flood in the year 1800 still stood high at the end of December, covering all the lands between Banī Sweif and the Fayyūm and making foot travel between the two provinces impossible.[21]

While papyrological evidence is scant, the progressive downstream rise of the Baḥr Yūsuf is alluded to in the fragmentary record of a hearing held in 208 CE before the prefect of Egypt Subatianus Aquila in the city of Oxyrhynchus on the western bank of the channel, which was then known by the Hellenized Egyptian name *Tōmis potamos* (Tomis River).[22] The hearing concerned the organization of the shipping of tax-grain downstream from Oxyrhynchus in the direction of Alexandria. As described by one Ammonios, head of the Oxyrhynchite city council, it was customary to empty the public granaries that lay alongside channel and to transport their contents only during the flood, when the water level was high:

> ὁ ποταμὸς ὁ ἡμέτερος οὗτος ὁ παρακείμενος τῇ ἀναβάσι (*l.* ἀναβάσει) ἐπίροιαν ἴσχι (*l.* ἴσχει) καὶ δαψίλιαν ὕδατος. ἀξιοῦμεν τότε τὰ πλοῖα πέμπεσθαι καὶ τὰς παραποταμίους κώμας πρῶτον βαστάζεσθαι \μετὰ τούτου/ [[διὰ τοῦδε]] τοῦ ποταμοῦ, μετὰ τοῦτο τὸ κατὰ πάσσαλον τὸ σύνηθες εἶναι καὶ τὸ νενομισμένοννον.

> This river of ours [the Baḥr Yūsuf], which lies adjacent to the flood, has an influx and an abundance of water. So we ask that boats should be sent at that time and that the villages adjacent to the river be cleared first by means of this

19. Sixteen-cubit flood: Sijpesteijn, *Shaping a Muslim State*, 20 and note 29. Seventeen-cubit flood: al-Masʿūdī, *Murūj*, 2:362; and al-Maqrīzī, *Khiṭaṭ* 1:157.
20. Martin, "Description hydrographique," 12–13. Cf. Barois, *Les irrigations*, 180. See also Willems et al., "The Analysis of Historical Maps."
21. Martin, "Description hydrographique," 20.
22. From Egyptian *tꜣ-ʿm*, "the channel."

> river, and afterward the usual [system] according to the peg (*to kata passalon to synēthes*) be in effect beginning, as is customary, from the upper regions, and that each granary be emptied and [its contents] carried to the usual place.[23]

Although this local terminology was obscure to the prefect, Ammonios clarified that rather than leaping from village to village, under the customary "peg" system each district began in its upper, that is, southern, districts and emptied its granaries together in time with the flood, starting with those directly adjacent to the Baḥr Yūsuf. In other words, Oxyrhynchite granaries were cleared from south to north, moving downstream with the rising waters. The "peg" (*passalos*) of the system perhaps referred to Nilometers in miniature, local measuring devices that indicated when the channel had risen high enough to bear a laden vessel safely.[24] The seasonality of shipping on the Baḥr Yūsuf is likewise noted by the tenth-century geographer al-Muqaddisī, by al-Nābulusī's survey, and by the seventeenth- and early eighteenth-century travelers Wansleben and Lucas, all of whom write that only during the flood was the water level of the channel high enough to enable the safe passage of vessels.[25] In a separate treatise on Ayyūbid administration, al-Nābulusī also reports that in one recent year (i.e., sometime during the early 1240s CE), the level of the Nile fell before all of the tax grain could be transported by boat out of the Fayyūm through the dam at al-Lāhūn (on which see below), thus trapping some 80,000 *ardabbs* of grain (ca. 7.2 million dry liters) inside the depression.[26] A late ninth- or early tenth-century Arabic letter sent from the Fayyūm to the capital Fusṭāṭ likewise attests to the seasonal impassibility of the Lāhūn inlet. The sender writes that, shortly after his arrival in the depression, "no one is arriving from or leaving for Fusṭāṭ and there is no longer any water in the Fayyūm Canal (*khalīj al-Fayyūm*) and no boat enters or leaves."[27]

The recession of the flood thus dramatically reduced the water level of

23. *P.Oxy.* 22.2341 (208 CE), ll. 4–10.
24. So the ed. pr. 114.
25. Al-Nābulusī, *VF*, 42. Al-Muqaddasī, *Kitāb aḥsan al-taqāsīm*, 208 (cited from Rapoport and Shahar, "Irrigation," 7); Wansleben, *The Present State of Egypt*, 153; and Lucas, *Voyage* 2:46. See also Cooper, *The Medieval Nile*, 102n10 with refs.
26. Rapoport, *Rural Economy*, 80. See the partial translation of al-Nābulusī's *Luma*ʿ *al-qawānīn* in Owen and Torrey, "Scandal in the Egyptian Treasury," 74.
27. فلا احدا يقدم ولا يخرج الى الفسطاط وليس في خليج الفيوم ما فليس يدخل قارب ولا يخرج. (*P.Cambr.* UL Inv. Michael. A 1337, ll. 19–20). Published in Tillier and Vanthieghem, "Un voyageur." On the relatively poor navigability of the Baḥr Yūsuf in general see Martin, "Description hydrographique," 3, where he claims that the channel "*suffit à peine à la navigation de quelques légères barques*."

the Baḥr Yūsuf, eventually delinking it from the river's surface flow. As described by the English traveler James Augustus St. John in 1845, this caused portions of the channel's bed to dry out entirely, permitting foot traffic along its exposed bed:

> By [a local farmer's] aid we traversed the bed of the great arm of the Bahr Yusuf, by which, at the season of the inundation, the waters of the Nile are conducted into Lake Mœris, and diffused in innumerable smaller streams all over the province, which they fertilize and beautify. In several parts of the channel, now dry, we observed immense quantities of oyster-shells, bright and shining like mother of pearl.[28]

Yet the channel was not totally dry, for St. John elsewhere describes it as "reduced, during the hot season, to a chain of small shallow ponds, in many cases miles asunder."[29] The source of this moisture is not altogether clear since there has been no scientific study of the phenomenon. Some is surely attributable to residual floodwater that pooled in low-lying areas of the channel. Moreover, a recent analysis of eighteenth-century maps suggests that the Baḥr Yūsuf was the only viable drain for wastewaters released from the Middle Egyptian districts that the channel had itself irrigated during the flood.[30] Robert Hanbury Brown confirms this suspicion, describing the Baḥr Yūsuf as the "channel of discharge for the basins which it has filled, or helped to fill."[31] Devoid of fertilizing silt and contaminated with salt and other minerals leeched from fields during irrigation, this drainage water would nonetheless have supplemented the flow of the channel for a brief period during the low-water season.[32] Still, persistent infiltration from the Nile Aquifer probably accounted for much of the Baḥr Yūsuf's dry-season flow, especially late in the year when the period of drainage was long past. Underlying most of the floodplain and recharged by infiltration from the earthen beds of canals and fields irrigated by flooding, the aquifer's water can be exploited through both seepage and wells.[33] Indeed, aquifer seepage would account for al-

28. St. John, *Egypt and Nubia*, 182.
29. St. John, *Egypt and Nubia*, 192.
30. Willems et al., "The Analysis of Historical Maps," 295–300 and 309. See also Subias, Fiz, and Cuesta, "The Middle Nile Valley."
31. Brown, *The Fayûm*, 11.
32. The higher salinity of water entering the Fayyūm through the Baḥr Yūsuf was noted by William Willocks: Willcocks and Craig, *Egyptian Irrigation*, 2:854.
33. Omran, "Land and Groundwater Resources."

Nābulusī's description of the continual replenishment of the Baḥr Yūsuf after the flood season. In his words, the rural districts of al-Bahnasā and al-Ashmūnayn along the southern arm of the al-Manhā continued to irrigate from the channel after the flood, drawing so much water that it would have been depleted within in a month were it not for perpetual seepage (*nazaz*) from underground sources (*nab*ʿ). He claims that no matter how much water was withdrawn, "this seepage immediately substitutes for it, mixing with what was obtained from the Nile's [flood] water." Although the text is not entirely clear, it seems that seepage water within the final stretch of the Baḥr Yūsuf continued to supply gravity-driven lateral canals in the depression during the low-water season:

> أقول وبالله التوفيق ان بحر المنهى فيه خروق في ارضة متصلة بالنبع ينز منها الماء وكذلك في ذيول حافاته من مواضع فيه متعددة وذلك لنزول سمته عن سمت ماء النيل المبارك ومن الحكمة البالغة انه كلما انجر الماء منه في الفوهات المذكورة الى رأضي البلاد الفيومية ومزارعها وبساتينها واشجارها واقصابها وبالسواقي التي عليه حتى ينقص شيء اخلفه في الحال النزز المتصل به مع الحاصل الذي فيه من الماء النيل. هكذا دائما ابدا.

> I say—may God grant success—there are openings (*khurūq*) in the bed of the al-Manhā Canal and in the lower parts of its banks at numerous locations, which are fed by an underground source (*nab*ʿ) from which water seeps. This is because its level descends below the level of the water of the Blessed Nile. It is part of its ingenious design that whenever water is drawn through these openings toward the lands of the villages of the Fayyūm—its cultivated fields, gardens, trees, and sugar plantations—or by means of water wheels, thereby decreasing the water level [in the Baḥr Yūsuf], the seepage immediately substitutes for it, mixing with what was obtained from the water of the Nile [i.e., the flood]. This is always and forever the case.[34]

Modern observers described the Baḥr Yūsuf in similar terms, such as Wansleben's aforementioned remark that the channel "keeps fresh water all the year,"[35] Lucas' assertion that it "never lacks water,"[36] and Jomard's observation that "the canal that irrigates the Fayyūm carries water all year."[37] Yet this water was explicitly received from sources other than the surface flow of the Nile. As

34. Al-Nābulusī, *VF*, 41. Translation lightly modified from Rapoport and Shahar.
35. Wansleben, *The Present State of Egypt*, 153–54.
36. Lucas, *Voyage*, 2:46.
37. Jomard, "Mémoire sur le lac de Moeris," 160.

described in 1854 by the French engineer Linant de Bellefonds, chief of Egypt's Ministry Public Works from 1831–69, the Fayyūm, like the rest of rural Egypt, was inundated by water carried into the depression by the Baḥr Yūsuf during the annual flood. Unlike the rest of Egypt, however, "in the season of low water (*l'étiage*), this natural canal [i.e., the Baḥr Yūsuf] still provides waters for irrigation, but they no longer come directly from the river and are only waters that come from springs that emerge from the bed of the canal."[38] Linant says much the same in his 1871 memoirs, writing that that "the Joussoufi [i.e., Baḥr Yūsuf] is the only canal in Egypt that, despite not receiving water from the river during low stages (*étiages*), nevertheless retains [water], which serves for the irrigations of the Fayyum."[39] Elsewhere he clarifies the situation, remarking that during the flood season, the Baḥr Yūsuf "has as much water as is necessary for flood-recession irrigation (*inondation*) and canal irrigation (*arrosage*). But since its mouth (*prise d'eau*) at Dayrūṭ is not cut deeply enough to receive water from the Nile during the low season, it "bears only those waters that arise from the depths of its bed."[40] Later authors echoed these descriptions. As described in 1891 by the former chief irrigation engineer of the Egyptian Sudan Alfred Jacques Chélu, "at low water (*l'étiage*), the Bahr Youssef, nearly dry, brought [to the Fayyūm] only brackish infiltrations from the lands of Middle Egypt, along whose lowest parts it ran."[41] So too the British engineer William Willcocks, who in 1913 states succinctly that "the Bahr Yusuf used to obtain its summer supply from infiltrations only" before the opening of the Ibrāhīmiyya Canal.[42]

That the Baḥr Yūsuf was never completely desiccated is thus abundantly clear. Yet if St. John's account of a dry bed dotted by "small shallow ponds" is representative of typical conditions, this suggests that the channel did not carry a steady, if shallow, stream throughout its entire length during the dry season. Al-Maqrīzī suggests as much in his brief account of the al-Manhā-Fayyūm canal, claiming that the channel was merely wet in patches along its southern extremities, becoming a true stream only when it neared the Fayyūm:

---

38. Linant de Bellefonds, *Carte hydrographique de la moyenne Égypte*.
39. Linant de Bellefonds, *Mémoires*, 4.
40. Linant de Bellefonds, *Mémoires*, 18, cf. also 55: "Le Bahr Joussef, ou Bahr Youssef, est un cours d'eau naturel qui peut avoir existé de toute anti uité, mais qui n'a dû couler que pendant les inondations." So also Shafei, "Fayoum Irrigation," 298: "Its [i.e., the Baḥr Yūsuf's] bed was higher than the Nile summer level and only infiltration water reached Lahûn after the flood."
41. Chélu, *De l'équateur à la Méditerranée*, 384.
42. Willcocks and Craig, *Egyptian Irrigation*, 1:305.

خليج الفيوم والمنهى مما حفره نبي الله يوسف الصديق—عليه السلام—عندما عمر الفيوم كما هو مذكور في خبر الفيوم من هذا الكتاب. وهو مشتق من النيل لا ينقطع جريه أبدًا، وإذا قابل النيل ناحية دورة سريام التي تعرف اليوم بدورة الشريف. تشعبت منه في غربيِّه شعبة تسمى المنهى تستقل نهرًا يصل إلى الفيوم، وهو الآن يُعرف ببحر يوسف، وهو نهر لا ينقطع جريانه في جميع السنة، فيسقي الفيوم عامّة سقيًا دائمًا، ثم يبحر فاضل مائه في بحيرة هناك. ومن العجب أنه ينقطع ماؤه من فوهته، ثم يكون له بلل دون المكان المندَّى، ثم يجري جريًا ضعيفًا دون مكان البلل، ثم يستقل نهرًا جاريًا لا يُقطع إلا بالسفن، ويتشعب منه أنهار، وينقسم قسمًا يعم الفيوم يسقي قراه ومزارعه وبساتينه وعامّة مساكنه، والله أعلم.

Khalīj al-Fayyūm and *al-Manhā*, which Allah's prophet the Righteous Joseph—Peace be upon Him—excavated when he created the Fayyūm, as is recounted in the report on the Fayyūm in this book.[43] It derives from the Nile and its course is never interrupted. When the Nile reaches the region of Darwat Sarabām, which is today known as Darwat al-Sharīf . . . an arm branches from its western side called al-Manhā, which receives a stream that leads to the Fayyūm, now known as the Baḥr Yūsuf. The course of this stream is uninterrupted throughout the entire year and it irrigates the Fayyūm perennially, channeling its excess water to the lake that is there. Remarkably, its water [flow] is cut off at its mouth [at Dayrūṭ], but [further downstream] there is moisture and then a place covered in wetness. From this wet place, the channel flows weakly, thereafter becoming a flowing stream that cannot be crossed except by boats. Streams branch from it and they are distributed throughout the Fayyūm, irrigating its villages, farms, gardens, and all its dwelling places. Allah knows best.[44]

The premodern Baḥr Yūsuf was thus something of a hybrid: an inundation canal that received water directly from the Nile only during the flood, but also a semiperennial channel that carried drainage and infiltration during the rest of the year. This water regime scarcely resembles that of the present day, in which the now-perennial Baḥr Yūsuf carries a steady stream of water into the Fayyūm with only minimal interruption throughout the year.[45] It is therefore important to avoid retrojecting into premodernity the contemporary hydraulics of the channel.[46]

43. Multiple variants of the legend are recorded in Al-Maqrīzī, *Khiṭaṭ*, 1:655–68.

44. Al-Maqrīzī, *Khiṭaṭ*, 1:190. See the French translations of this passage in Silvestre de Sacy, *Chrestomathie arabe*, 2:308; Quatremère, *Mémoires géographiques*, 1:402–3; and Toussoun, *Mémoire*, 1:175–76.

45. Water is cut off for several weeks in January, a period known as the *gafaf* ("drying out"), during which cleaning and other maintenance work is performed. Price, "Evolution of Irrigation," 261–64.

46. *Pace* Römer, "The Nile in the Fayum."

Indeed, in marked contrast to the current situation, Arab observers often remarked that the primary distinguishing feature of the Fayyūm's irrigation system was its ability to store floodwater for use in double-cropping. Thus the Egyptian physician Ibn Riḍwān (988–1061 CE), who writes that "the Fayyūm stores within itself (*yakhzan fīhi*) the waters of the Nile and is cultivated multiple times per year."[47] This following section investigates this phenomenon, arguing that the Fayyūm's major irrigation infrastructure was designed not to regulate and distribute a perennial influx as it does today but to capture and contain the flood.

Capturing the Flood: The al-Lāhūn Dam

Arriving in Aswān in early June, the flood began to rise rapidly during July and August. According to Arabic historiographical tradition, the administrative *dustūr* of Abū Isḥāq, and al-Nābulusī's survey, Fayyūm irrigation required a rise in the Nile of only 12 cubits (ca. 7.8 m), compared to the 16 or 17 cubits deemed optimal in the Nile Valley. It was a marvel, al-Bakrī writes, for "there is not in the land of Egypt a place irrigated from these [twelve] cubits apart from the Fayyūm."[48] This was surely a function of topography. Whereas in the Nile Valley the flood had to reach a height sufficient both to overtop the river's embankments and expand to cover the entire agricultural landscape, water flowed directly into the Fayyūm depression through the Baḥr Yūsuf and was then swiftly transported across the sloping terrain of the depression by gravity, which may have enabled irrigation at lower flood-volumes. The moment at which the flood arrived nonetheless surely differed from year to year. Wansleben records that the inundation of 1672 arrived on 2 August and purified the province's waters, which had become scant, stinking, and so polluted by agricultural waste that they were undrinkable. It is at this time, he writes, that "the Cisterns of the Town [Madīnat al-Fayyūm] are fill'd with Water, which the Inhabitants drink all the year long, therefore this inundation makes all the people of the Country round about to rejoice."[49]

47. Ibn Riḍwān, *Kitāb daf' maḍār al-abdān*, 16.: الفيوم يخزن فيه ماء النيل، ويزرع عليه مرّات في السنة. Quoted by Al-Maqrīzī at *Khiṭaṭ*, 1:669 (see appendix below).

48. Al-Bakrī, *Al-Masālik wa'l-mamālik*, 1:515 (no. 866): وليس من ارض مصر موضع يشرب من هذا الذراع. غير الفيوم Abū Isḥāq in al-Maqrīzī, *Khiṭaṭ*, 1:670; and al-Nābulusī, *VF*, 44. Cf. Sijpesteijn, *Shaping a Muslim State*, 28 with note 82. For flood levels and the Nile Valley see Borsch, "Nile Floods and the Irrigation System," 133.

49. Wansleben, *The Present State of Egypt*, 161–64.

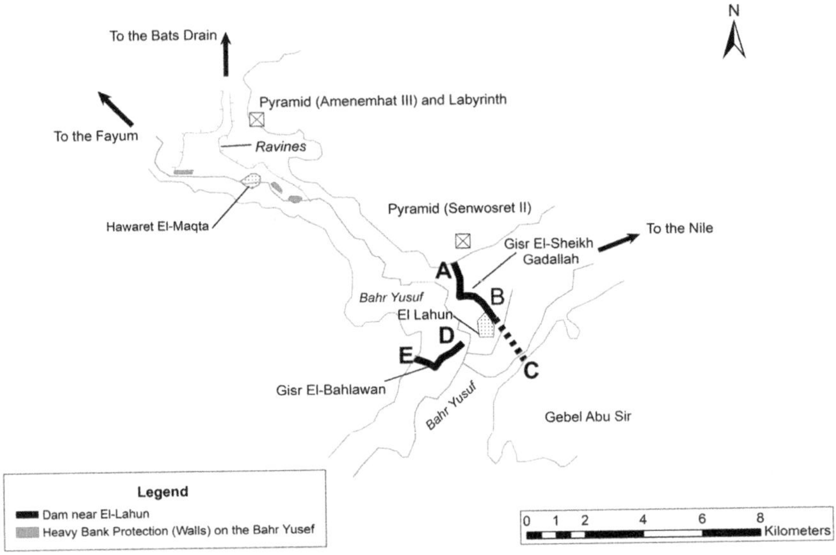

Figure 4. Control works at al-Lāhūn. Redrawn by Julian Thibeau from Garbrecht, "Historical Water Storage," 67, fig. 15.

Figure 5. The al-Lāhūn dam in the *Description de l'Égypte*. Detail from Jacotin Pierre, *Carte topographique, Carte 19, Faïoùm* (1818). Image courtesy of David Rumsey Map Collection, David Rumsey Map Center, Stanford Libraries.

The initial capture of the flood at al-Lāhūn was controlled by two large dikes, the stone-clad Jisr al-Shaykh Gādallah and the earthwork Jisr al-Bahlawān, the latter of Ptolemaic date and the former probably dating to the pharaonic period.[50] These two structures helped to divert floodwaters westward into the inlet. The inlet itself was then blocked by a masonry dam that spanned the breadth of the channel. According to Ibn ʿAbd al-Ḥakam the dam was constructed by Joseph during the reclamation of the depression, whence its common appellation the "Rock of Joseph" (*al-Ḥajar al-Yūsufī*).[51] The earliest extant literary description of the structure is that of the first-century BCE geographer Strabo, who simply mentions that "locks" (*kleithra*) on the canal allow engineers to store (*tamieuousin*) water within the Fayyūm.[52] There is little evidence in the papyri with which to compare this account. The only ancient references to control works at al-Lāhūn (Greek Ptolemais Hormou) derive from the third-century BCE archive of Kleon and Theodoros, two chief engineers (*architektones*) who successively administered the Fayyūm's irrigation system in the first decades after the Ptolemaic reclamation (see chapter 3 below). Yet it is not altogether clear whether the few references in this archive are to the dam itself or other downstream infrastructure that governed the initial apportionment of water into the Fayyūm's various canals.[53] The most probable notice comes from a fragmentary letter of 260–249 BCE that refers to water flowing from the "locks" (*katakleides*) at Ptolemais Hormou through the "great canal" (*megalē diōryx*), the latter probably a reference to the terminal stretch of the Baḥr Yūsuf within the depression.[54] The archive elsewhere preserves references to sluice gates (*thyrai*) at Ptolemais Hormou, which could be opened or closed to regulate the flow of water through the rest of the canal system. In a damaged letter of 256 BCE, the sender refers to the potential closing of one such gate (*mian thyran*) at Ptolemais Hormou in order to moderate water flow.[55] A contemporaneous letter from a writer based in Philadelphia on the Fayyūm's eastern margins requests that multiple sluice gates (*tas thyras*) be opened since the water level of an unnamed canal (perhaps the eastern desert

50. Hassan and Tassie, "Modelling Environmental and Settlement Change," 40.
51. Ibn ʿAbd al-Ḥakam, *Futūḥ Miṣr*, 15.
52. Strabo, *Geog.* 17.1.37.
53. Most notably the so-called "six-gated sluice" (ἑξάθυρος) mentioned in some two-dozen *penthēmeros* receipts, e.g., *BGU* 4.1075 (148 CE) and *P.Kron.* 68 (150 CE). See Pearl, ΕΞΑΘΥΡΟΣ.
54. *P.Petrie Kleon* 34 (260–249 BCE).
55. *P.Petrie Kleon* 18 (26 August 256 BCE).

canal) has not sufficiently risen.[56] Even if these are regarded as explicit references to the dam, they offer little beyond the indication that it was equipped with sluice gates.

Detailed accounts of the form and function of the dam begin to appear only in the early Islamic period.[57] Al-Masʿūdī provides the earliest notice, a passage that confirms the existence of sluice gates (*qanāṭir*) that allowed water to pass through the dam itself. His reference to the "days of its [the dam's] closing (*ayyām saddihi*)" nonetheless suggests that its primary function was to admit and then to contain the flood, just as Strabo had remarked centuries earlier:

> حائط الحجر فوّارات بعضها اليوم يخرج منه الماء، وبعض لا يرى، وفيما بين سطح الحجر . . . شاذروان . . . وإنما يدخل الماء الفيوم بوزن الحجر، وجعلت الاستقالة—وهي القناطر—ليخرج الماء منها، ولا يعلو الماء الحجر أيام سدّه، فبالتقدير بُني حجر اللاهون، وبقدر ما يكفي الفيوم من الماء يدخل إليها.
>
> In the wall of the rock are streams (*fawwārāt*), through some of which water flows today yet some cannot be seen. And at the upper part of the rock . . . is a conduit (*shādhrawān*) . . . and water enters the Fayyūm [in proportion to] the weight-bearing capacity of the Rock. And the *isqāla*—these are channels (*qanāṭir*)—were created for water to flow through and so that the dam might not be submerged during the days of its closing. And as for the proportions in the construction of the Rock of al-Lāhūn, they were calculated such that only the water that is sufficient for the Fayyūm passes through it.[58]

The later Andalusian geographer al-Bakrī appears to draw on al-Masʿūdī, likewise describing a dam with both a large central conduit (*shādhrawān*) and multiple "streams" (*fawwārat*). Basing himself on unknown sources in a later passage, he offers yet another description of the "day of the closing of the Rock of al-Lahūn" (*yaum sadd ḥajar al-Lahūn* [*sic*]) as the flood receded, an event attended by the local district *amīr*, juristic officials, engineers (*ahl al-handasa*) who oversaw the closing of the structure, and even the *amīr* of the Fayyūm, all accompanied by crowds and music.[59] Similar events sur-

56. *P.Petrie Kleon* 19 (13 October 256/55 BCE).
57. The following draws in part on Rapoport and Shahar, "Irrigation," 6–9.
58. Al-Masʿūdī, *Murūj*, 2:385–86.
59. Al-Bakrī, *Al-Masālik wa'l-mamālik*, 1:514–15 (no. 865).

rounded the breaching of dikes at Cairo during the flood—the *kasr* or *fatḥ al-khalīj*—festivities that celebrated the annual arrival of the waters that filled the city's canals and cisterns.[60]

The sluice gates first described in the tenth century were gone by the eleventh. The geographer al-Muqaddasī, who visited Egypt in this period, describes a simple spillway dam that could be passed over by boats when the Nile was high. Glass outlets at the base of the dam (*manāfis*) also allowed water to escape when the dam was shut and the Fayyūm had received all that it needed.[61] Writing in 1031 CE, Abū Isḥāq similarly suggests that the dam's form had simplified since the Hellenistic and early Islamic period, though the glass *manāfis* were still present at its base. In his account, the structure was composed of two sections, a long stretch running north-south and a shorter wall running east-west, which intersected the longer span at its southern terminus. This likely refers not to the dam itself but to the Jisr al-Sheikh Gādallah and the Jisr al-Bahlawān. The primary purpose of the dam proper—The Rock of Joseph (*al-Ḥajar al-Yūsufī*)—was to hold back floodwaters until they had reached the requisite height of 12 cubits, at which point the dam would be breached to allow water to flow toward the capital city. Since the ancient *qanāṭir* were no longer in use, water entered the Fayyūm in Abū Isḥāq's period through two openings in the dam, one in the south and one in the north, each of which was later closed via an embankment (*jisr*) made of local grass and vegetation (see the appendix below).

Al-Nābulusī later offers a detailed description of the annual opening and closing of the dam via its central spillway. Closed with a block (*qiṭʿa*) fashioned of a palm log wrapped in rags and rope to increase its girth, the spillway was opened during the inundation to admit floodwater and to allow the passage of vessels that might otherwise founder if they attempted to pass over the stone cap of the dam, which would eventually be submerged by the rising waters. Missing entirely from al-Nābulusī's description is any indication that the dam in this period could admit significant amounts of water after the flood.

فالقطعة المشار اليها عبارة عن نخلة طويلة يعمل عليها القش والجب تربط بحبال حتى يبقى لها
تما يعمل عليها غاظ كبير تكون الحبال القوية في طرفها واطراف الحبال في ايدي جمع كثير من

60. Echols and Nassar, "Canals and Lakes of Cairo," 207.
61. Al-Muqaddasī, *Kitāb aḥsan al-taqāsīm*, 208.

الرجال في البر المتصل بالضيعة المسماة اللاهون والبر المقابل لها يرخون الحبال قليلا قليلا مع حمل الماء القطعة المذكورة وجذبه لها الى الفوهة المذكورة وثم في البناء المسمى باللاهون المقدم ذكره بين البنيان المذكور فوهة يخرج منها الماء أيام النيل وتدخل فيها المراكب وتخرج خوفا من ان تمر على الحجر فيكسرها لقوة اضطراب الماء عليه. فاذا نزل النيل وظهر الحجر بقي الماء يخرج من هذه الفوهة وقد جمع جمع كثير من بلاد الفيوم والمهندسون لعمل هذه القطعة الموصوفة وصنعوا بها ما تقدم ذكره يرخونها قليلا قليلا الى ان تصل الى فم الفوهة فتسدها وتمنع الماء من الخروج منها ويعمل الرجال عليها التراب والطين حتى تبقى من جنس البر المتصل بها الى البنيان بحيث يمر البر المتصل يُقسد بسد الفوهة توفر الماء الذي يخرج منها على بلاد الفيوم هذا مع مدد النيل له وقبل ان ينقطع جريانه من فوهة المنهى الذي يجف كل سنة كما وصفت.

The aforementioned block is a long palm log upon which straw and rags are affixed. These are tied up with ropes, so that it becomes very thick. There are strong ropes at its edges, and the ends of these ropes are in the hands of large groups of men on the bank adjacent to the village called al-Lāhūn, and on the opposite bank. They release the ropes little by little, while the water carries the block and pulls it toward the opening located in the dam of al-Lāhūn, in the midst of the structure. The water of the days of the Nile (*ayyām al-Nīl*, i.e., the flood), flows out from this opening, and boats go in and out through it, as they do not want to risk passing over the stone (*ḥajar*) [crest of the dam] for fear of being shipwrecked, due to the power of the water's turbulence. But when the Nile recedes, and the stone is exposed, water continues to escape from this opening. It is then that large groups of men from the villages of the Fayyūm, as well as engineers, gather together to install this block as has been explained. They release it little by little until it comes to the mouth of the opening and blocks it, and thereby prevents the water from escaping. The men pile up soil and clay upon it so that it resembles the bank adjacent to the structure, so much so that a person may cross over the dam from al-Lāhūn to the bank of Qāy, just as he would proceed on the same bank. The purpose of blocking the opening is so that the water, which [otherwise would have] escaped through it, will be available for the villages of the Fayyūm. This occurs at the time when the Nile still reaches it, and before its flow stops at the opening of al-Manhā Canal [i.e., at Dayrūṭ], which becomes dry each year as I described.[62]

Modern accounts of the dam are relatively few, beginning with Evliya Çelebi's description of the *Jisr al-Lāhūn* as "a great bridge of [three?] arches

62. Al-Nābulusī, *VF*, 41.

that stretches like a rainbow in the sky over Joseph's canal."[63] European autopsies supply slightly more information. Paul Lucas writes that the Nile communicated with the Fayyūm through the inlet "in the times of its height (*hauteur*)" by means of a "lock" (*écluse*) on the Baḥr Yūsuf.[64] Savary offers slightly more detail, remarking that the locks (*écluses*) at the entrance to the Fayyūm are opened during the months in which the Nile descends, at which point water enters and is stored in the Fayyūm, a "second inundation" that can later be directed at will for irrigation.[65] P. D. Martin in the *Description de l'Égypte* writes of a three-arched *pont* at the entrance to the Fayyūm. Each archway was 2.8 meters wide and was fitted with a movable weir (*déversoir*) used to modulate the influx of water during the flood.[66] Elsewhere in the *Description*, Jomard writes of a perennially permeable structure that could admit the Baḥr Yūsuf's meager flow during the dry season: "a bridge of stone with three arches, through which waters flow, falling about one meter during the low waters [season]." He adds that another dike runs east-west along the base of the gorge north of al-Lāhūn, "along which the waters of the canal flow, in part, in the time of the flood." Waters not admitted through the dam and inlet during the inundation continued to flow north through an extension of the Baḥr Yūsuf.[67] This notice is reflected in Pierre Jacotin's map of the Fayyūm for the *Description* (see fig. 5 above), which illustrates the features described by Jomard, and in Hanbury Brown's account of the three-arched "old Lāhūn regulator."

This testimony suggests that the dam at al-Lāhūn underwent significant changes over the centuries, transitioning from a structure pierced by sluice gates to a simpler spillway and then back again by the later Ottoman period. As I will argue in chapter 3 below, these changes were in part a response to varying levels of state investment in Fayyūm irrigation, primarily the maintenance of the Baḥr Yūsuf's inlet, which tended to become clogged with silt. Regardless, the balance of the evidence from Strabo to the nineteenth century indicates that the primary purpose of the dam at all times was to capture and store water in the Fayyūm—*tamieuō* in Strabo's Greek and *yakhzan* in Ibn Riḍwān's Arabic. Only near the end of the nineteenth century was the dam redesigned as a regulator in order to control the influx of the newly perennial Baḥr Yūsuf

63. Dankoff, Tezcan, and Sheridan, *Ottoman Explorations*, 352. I have suggested the restoration of the lacuna based on Jomard's and Martin's later accounts. See below with notes 66 and 67.
64. Lucas, *Voyage*, 2:41.
65. Savary, *Lettres*, 2:37.
66. Martin, "Description hydrographique," 21.
67. Jomard, "Mémoire sur le lac de Moeris," 160.

Figure 6. The "old Lāhūn regulator" at the Baḥr Yūsuf's inlet. Brown, *The Fayûm and Lake Moeris* (1892).

precisely. Writing near the end of the century, the French engineer Jacques Chélu remarked that modern restructuring of the Baḥr Yūsuf's flow had indeed effected a dramatic change in the infrastructure at al-Lāhūn, which had previously served merely to retain the waters stored (*emagasinées*) in the depression.[68] The following section explores the phenomenon of water storage in greater depth, arguing that the ability to store and later distribute floodwater was the foundation of the Fayyūm's singular productivity.

Storing the Flood: The Tuṭūn *Basin and Other Reservoirs*

Located in the Tuṭūn Basin are the remains of a massive dike or seawall, sections of which are extant between the modern villages of Iṭsā and Shidmū, the

68. Chélu, *De l'équateur à la Méditerranée*, 384.

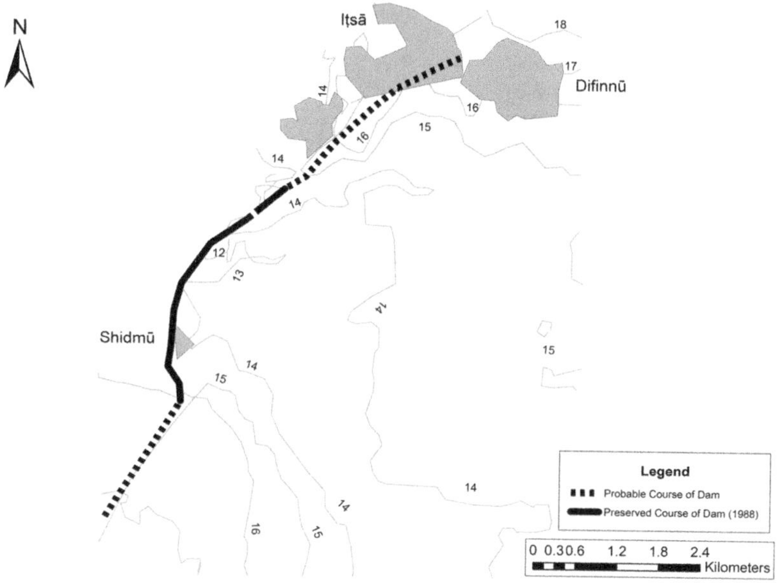

Figure 7. Remains of the Iṭsā-ʿIzbat Abū al-Nūr dike. Redrawn by Julian Thibeau from Garbrecht, "Historical Water Storage," 53, fig. 4.

Figure 8. The Iṭsā-ʿIzbat Abū al-Nūr dike and the Tuṭūn Basin. Detail from Jacotin Pierre, *Carte topographique, Carte 19, Faïoùm* (1818). Image courtesy of David Rumsey Map Collection, David Rumsey Map Center, Stanford Libraries.

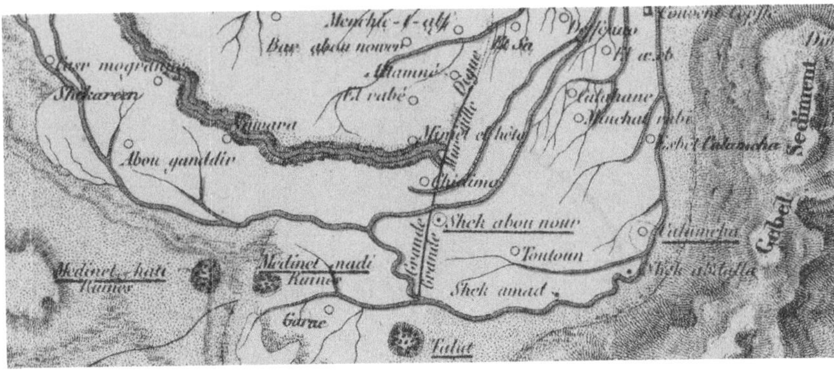

Figure 9. The Iṭsā-ʿIzbat Abū al-Nūr dike and the Tuṭūn Basin. Detail from Linant de Bellefonds, *Carte hydrographique de la Moyenne Égypte* (1854). Image courtesy of gallica.bnf.fr / Bibliothèque nationale de France.

latter an Arabized Egyptian toponym meaning "dam of the water." Originally running more than 7 km between Iṭsā and the village of ʿIzbat ʾAbū al-Nūr, much of the structure now serves as the foundation for the modern road linking the two settlements.[69] The purpose of the dike remains debated. In the most sustained investigation to date, the hydrologists Günther Garbrecht and Horst Jaritz concluded that its original construction was in *opus caementitium*, thus dating its initial construction to the early Roman period, though Dominic Rathbone prefers an earlier but unspecified Ptolemaic date.[70] The mixture of materials and construction styles throughout nevertheless indicates that the dike was continuously repaired well into the nineteenth century, after which point it went out of use.

Sitting largely perpendicular to the mouth of the Wādī al-Nezla, the primary drain in this part of the depression, the dike clearly would have prevented floodwater from escaping from the Tuṭūn Basin and draining into the Birkat Qārūn.[71] Garbrecht and Jaritz accordingly suggest that it enabled the creation of a reservoir of some 114 km^2 every year between October and February/March, which would have been drawn down by local farmers. Overflow would have passed into the small lake in the neighboring Gharaq basin via a canal cut

69. Davoli, *L'archeologia*, 269.
70. Garbrecht and Jaritz, *Untersuchung*; and Garbrecht, "Historical Water Storage." *Contra* Rathbone, "Mapping the South-West Fayyum," 1113.
71. See also the brief description in Bagnall and Rathbone, *Egypt*, 142–43.

Figure 10. Remains of the Iṭsā-ʿIzbat Abū al-Nūr dike near Iṭsā. Photo by author, 2011.

through the rock barrier separating the two basins. The presence of a reservoir in the Tuṭūn Basin, they argue, accounts for one of the region's modern toponyms, "Basin of the Birds" (*Ḥauḍ al-Ṭuyūr*) since a large, semipermanent body of water would have attracted considerable waterfowl. The current colloquial name for the region, *al-Malaʾa*, Garbrecht suggests, refers to "a wide and open area covered with water."[72] Relatively sparse premodern settlement in the region apart from the villages of Tuṭūn and Qalamshāh, they add, lends further support to the idea that this portion of the depression was long given over to other purposes. Building on this theory, papyrologist and archaeologist Cornelia Römer has recently argued that a reservoir impounded behind the dike could have retained water not only for use in the Tuṭūn but also for the canals that served the northwestern stretches of the province, the *Themistou meris* in the Graeco-Roman period. Damage to the wall dated by Garbrecht and Jaritz to the third or fourth centuries CE would therefore seem to explain the

72. Garbrecht, "Historical Water Storage," 59.

Figure 11. Stone-faced section of the Iṭsā-ʿIzbat Abū al-Nūr dike near Iṭsā. Photo by author, 2011.

decline of marginal western settlements like Theadelphia in the later Roman period.[73]

There are several apparent problems with this theory. Archaeologist Paola Davoli has already questioned what economic purpose would have been served by flooding a vast area of prime agricultural land for much of the year. Indeed, Graeco-Roman villages along the southern border canal such as Tebtynis and Kerkeosiris farmed lands to their north, an area that would have been covered by the proposed reservoir.[74] Further, the Tuṭūn Basin was not as depopulated as Garbrecht and Jaritz suppose. Though thinly settled by comparison to the rest of the central Fayyūm, al-Nābulusī identifies a handful of settlements in the area, all of whose fields would have been inundated were the Tuṭūn flooded to the extent proposed by Garbrecht and Jaritz (see map 3).[75] Finally, although the dike was still in use well into the nineteenth century, European travelers and scientists do not describe, in text or in their maps, a reservoir in the Tuṭūn of

73. Römer, "Why Did the Villages?"
74. Davoli, *L'archeologia*, 270.
75. Hayshat Dumūshiyya, al-Qalhāna, Umm al-Sibāʿ, Qambashā, al-Mahīmsī, Tuṭūn, Buljusūq.

the massive size imagined by Garbrecht and Jaritz. Absence of evidence is hardly dispositive, of course, but it does suggest caution.

Drawing on a passage in Robert Hanbury Brown's 1892 study of the Fayyūm, Davoli has suggested that the Itṣā-ʿIzbat Abū al-Nūr dike was indeed designed to retain water in the Tuṭūn but only temporarily for the purposes of traditional flood-recession irrigation. Brown writes that large-scale flood-recession irrigation in the Tuṭūn, a practice that had only recently been abandoned, was responsible for the seasonal variations in the level of the Birkat Qārūn, which received a massive influx of drainage from the Tuṭūn in the early autumn:

> The former manner of conducting the irrigation of parts of the province would have caused a much larger proportionate discharge into the lake, than finds its way to it at present. Considerable areas were enclosed by banks, and inundated under the Basin system, known in the Fayûm as "Malaq," in contradistinction to irrigation by small field channels, a system called "Misqawi." The contents of these small basins, when emptied, flowed into the lake. On the south side of the Fayûm there was, until late years, a large basin known as "Hod-el-Tuyûr" (the Basin of the Birds), which was formed by building an immense wall across a fold contour of R.L. 15.00 [i.e., 15 masl]. The top of this wall is about R.L. 16.00. The bed of the basin is as R.L. 12.00, so we may conclude that, when this wall was built, the lake levels must have been at any rate below R.L. 12.00. This basin was abolished in 1886 and ordinary perennial irrigation introduced over the area formerly included within the basin limits.[76]

Although Garbrecht cites this passage as support for the reservoir theory, Brown is clearly not describing the use of the Tuṭūn basin as a reservoir.[77] Rather, this notice is part of a discussion of traditional flood-recession irrigation in the Fayyūm before the introduction of perennial irrigation via the Ibrāhīmiyya Canal. The description of a "large basin" in the Tuṭūn simply complements his discussion of the many smaller basins previously in use throughout the depression, all of which were drained to the Birkat Qārūn at roughly the same time during the year, thereby causing the surface-level instability and shoreline flooding that British engineers were attempting to remedy. Indeed, other nineteenth-century sources echo Brown's account, thus lending

76. Brown, *The Fayûm*, 96.
77. Garbrecht, "Historical Water Storage," 54, 59, and 65.

support to Davoli's proposition. In the *Description*, Jomard describes the dike, then still in use, as an aid to the irrigation only of the southwest:

> If we go southwest of Medynet el-Fayoum, we first meet, in the village of Begyg [Abgīg], a granite obelisk;[78] farther on, a league and a half, in the same direction, a dike built of stones, of considerable height and thickness. It is regarded as antique, although it was, it seems, rebuilt several times. It is nearly seven thousand meters long and runs by Defennoû and Sedmoueh. Its object is to maintain the waters of the flood at a certain height, and to irrigate the southern part of the province. Excess waters fall into a great ravine called Bahr el Ouaddy [i.e., the al-Wādī drain].[79]

P. D. Martin's brief account of the dike in the *Description* is much of a piece. He writes that because of a sixty-meter-wide rupture somewhere near the village of Shidmū, the dike could no longer retain water for the irrigation of the surrounding territory. In consequence, the structure was altogether "without purpose . . . [since] the waters run through the Ouady [ravine] to flood fruitlessly the lands from [the village of] Nazleh to the Birket-Qeroun."[80] Linant likewise describes the dike as a structure that served only to retain water for large-scale flood-recession irrigation in the area around the villages of Ṭuṭūn and Qalamshā. If the dike and the spillway of this great irrigation basin (*bassin d'inondation*) were damaged, waters would rush into the Maṣraf al-Wādī uncontrolled and be lost, just as Martin had witnessed some decades earlier:

> The other great ravine, the Bahr Neslet or El-Ouadee, rather than originating directly from Bahr Joussef, [originates from] a very large masonry dam at Miniet-el-Heit. Today this dike serves, by retaining the waters that the Bahr Joussef delivers through several channels, to prepare for the flooding of the largest basin of Fayoum cultivation, the one in the area of the villages of Toutoun and Calamcha.[81]

78. The obelisk-like monolith of Senusret I (1920–1875 BCE) at Abgīg, now relocated to Fayyūm City.
79. Jomard, "Description des antiquités du nome Arsinoïte," 449.
80. Martin, "Description hydrographique," 53. In 1801, P. S. Girard describes a major rupture in the dike occasioned by "une inundation extraordinaire" some fifty years before, which had left much of the structure largely useless (Girard, "Mémoire sur les irrigations," 332). This is probably the damage caused by the flood of 1745, which is referred to in Ottoman documentation (Mikhail, "An Irrigated Empire," 581). See also Garbrecht, "Historical Water Storage," 64–65.
81. Linant de Bellefonds, *Mémoires*, 54.

Writing in 1893, former British colonial Inspector General of Egyptian Irrigation Justin Ross discusses the Ḥauḍ al-Ṭuyūr in terms similar to Brown, calling attention to the fluctuation in the level of the Birkat Qārūn caused by the drainage of the Tuṭūn in early autumn as well as the subsequent conversion of this part of the Fayyūm to perennial irrigation and double-cropping:

> On the eastern edge of the Birket el Qurûn there is a wide flat foreshore—the result of many centuries of erosion from the lands by the irrigation canals, etc. Owing to the large amount of waste water from the Bahr Yûsif and to the existence of an old basin, called the Hod et Tuyûr, the level of the lake rose at least 17 feet between 1870 and 1883, and about 10,000 acres of cultivable land were submerged. The measures adopted to remedy this state of affairs were to abolish the basin irrigation of the Hod et Tuyûr, making it *sêfi*—that is, cultivating cotton, etc., in summer.[82]

These issues are also touched upon in several annual Irrigation Reports published by the British colonial Ministry of Public Works. In the 1887 report, the first colonial Inspector General of Irrigation Colin Scott-Moncrieff writes that the recent decision to abolish this basin of 12,000 *faddans* (50.4 km^2) and to replace it with perennial irrigation was an attempt to halt the swamping of lakeshore farmland caused by the yearly draining of the Tuṭūn basin.[83] In an appendix to the report added by Ross, the basin is explicitly referred to as "the 'Malaqah' or the last flood basin in the Fayum." Ross writes that the annual October draining of this massive irrigation basin would cause a 40–60 cm rise in the lake and subsequent land loss along the lakeshore. It was thus decided to make the basin "Misqâwi, or irrigated by water courses both winter, flood time, and summer."[84] By the end of the nineteenth century, British engineers had begun to adopt such traditional Egyptian Arabic agricultural terminology in their own writings, regularly using terms like *ṣayfī* in place of English "summer" to refer to a second annual crop.[85] Their "*malaq*" or "*malaqah*" is likely

82. Ross, "Irrigation and Agriculture in Egypt," 185.
83. Scott-Moncrieff, *Irrigation Report for the Year 1887*, 14.
84. Ross in Scott-Moncrieff, *Irrigation Report for the Year 1887*, appendix C at 63. Justin Ross remarks (p. 88) on his *Report of the Administration of the Department of Irrigation for the Year 1890* (Cairo: National Printing Office, 1891) that the level of the Birkat Qārūn had been falling steadily, opening up more than 10,000 *feddans* of land along its shores since the Ḥūḍ al-Ṭuyūr was converted to perennial irrigation.
85. Derr, *The Lived Nile*, 36–39.

a hypercorrection of Arabic *mala'* (to fill or irrigate by flooding). The local toponym *al-Mala'a* attested by Garbrecht and Jaritz is thus a linguistic remnant from the era of traditional flood-recession irrigation in the Tuṭūn basin.[86] Indeed, the authors cited above indicate that the dike did not create a reservoir in the later nineteenth century but instead enabled traditional Egyptian irrigation practices on a large scale by preventing water from escaping into the Maṣraf al-Wādī. Only after irrigation was complete in the early autumn was the entire region drained by the al-Wādī, thus causing the surface level of the Birkat Qārūn to rise precipitously.

While this much seems clear, al-Nābulusī complicates matters by explicitly describing a reservoir (*al-gharq*, lit. "flooded" or "submerged") in the Tuṭūn and its accompanying dike (*jisr*).[87] In his description of the water supply of the village of Qambashā (mod. Qalamshā) in the southern Tuṭūn, he states that the settlement receives water both from the Baḥr Tanabṭawayh (the southern border canal) but also from *al-gharq* as an allowance or wages (*rizqa*) for watching over the *jisr al-gharq*.[88] Muqrān, a village located somewhere on the western edge of Tuṭūn basin, also received water from *al-gharq*, though only when its water was high (*min a'lā al-mā'*).[89] Yet al-Nābulusī also writes that this *al-gharq* served villages in the far west of the Fayyūm, well outside the confines of the Tuṭūn basin. Diqlāwa, a hamlet of Minyat Aqnā somewhere in the northwest, is said to draw water both from a local canal and from *al-gharq*, sharing this reservoir allotment with the nearby settlement Masjid 'Ā'isha, also of uncertain location.[90] Al-Nābulusī's description of Masjid 'Ā'isha provides the clearest testimony to the existence and functionality of the reservoir. In his account of the settlement's water supply, al-Nābulusī writes that it receives water from "*al-gharq*, known as Qambashā, from water that drains through the sluice gate of the arch (*bāb al-qabw*)."[91] Unfortunately, this notice is insufficient to reconstruct the mechanics of the reservoir in any detail. The "sluice gate of the arch" may be the opening in

86. Since the letter *qaf* is often rendered as a glottal stop in spoken Egyptian Arabic, British engineers probably understood *mala'* as *malaq*. Cf. Flinders Petrie's erroneous restoration of an initial *qaf* in al-Lāhūn, whence the erroneous but enduring "Kahun" in Egyptological scholarship, derived from spurious *al-Qāhūn*. David Warburton, *Architecture, Power, and Religion: Hatshepsut, Amun & Karnak in Context* (Zürich: LIT Verlag, 2010), 71.

87. The first occurrence of *al-gharq* (*VF*, 147) seems only to concern remission of the land-tax (*kharāj*) for the village of Dumūshiyya (Gk. Mouchis, Copt. Tmoushi) because of *al-gharq*, i.e., overinundated and hence uncultivable land.

88. Al-Nābulusī, *VF*, 199.

89. Al-Nābulusī, *VF*, 216.

90. Al-Nābulusī, *VF*, 213.

91. Al-Nābulusī, *VF*, 222.

the dam at ʿIzbat Abū al-Nūr, a site likely identical with the ancient village known both as Arsinoe on the Lock (*epi tou zeugmatos*) and Arsinoe on the Dike (*epi tou chōmatos*). The Baḥr Nazla, the canal that irrigates the contemporary Fayyūm's western margins, still follows this path and may have been the conduit for waters released from the *al-gharq* in al-Nābulusī's day.[92] More than this, however, it is impossible to say without additional evidence.

Eighteenth-century Ottoman sources continue to refer to the so-called dam of *al-Gharaq* as a "huge dam (*sedd-i azīm*) of impressive stature (*bina-i cesīm*) that had been in existence since times of old (*kadim ül-eyyamdan*)." Its stated purpose was to prevent water from "spreading and branching out (*intiṣār ve inṣiʿāb*)" and to retain a "huge lake (*buheyre-i azīme*)" in order to provide water to residents until the next flood, though it is not clear what settlements outside the immediate vicinity of the reservoir, if any, benefited from its waters.[93] Regardless, alongside the dam at al-Lāhūn, the Tuṭūn dike was a critical piece of state-maintained irrigation infrastructure. As a record of repair work in Istanbul dated to 1734 put it, the *al-Gharaq* dike was constantly "exposed to the force of the Nile's crashing waves (*hurūş-i telātum-i emvāc*) and the rush of its deluge (*tezāhüm-i cereyan*)," and therefore required assiduous attention lest it fail.[94] In the first half of the eighteenth century alone almost twenty major repair operations, all largely ineffectual, were undertaken in response to constant petitions from local villagers. By 1746 the dike was almost entirely in ruins and was at last subjected to extensive and costly repairs.[95]

The evidence, in short, is contradictory. On the one hand, nineteenth-century authors write that the dike simply enabled traditional flood-recession irrigation (*malaʾ*) in the Tuṭūn Basin. Earlier testimony, on the other hand, indicates that the dike allowed for the creation of a reservoir that could deliver water to relatively far-flung villages. How to account for this apparent change? It is possible that the initial introduction of perennial irrigation in the 1870s and the subsequent increase in the Fayyūm's aggregate water supply obviated the need for a proper reservoir. The lack of routine maintenance documented in Ottoman evidence may also have contributed to a reduction in the functionality of the structure and its transformation into a simple retaining wall that facilitated traditional flood-recession irrigation.

---

92. Römer, *The Fayoum Survey Project*, 31–33.
93. Mikhail, "An Irrigated Empire," 574–75.
94. Mikhail, "An Irrigated Empire," 576.
95. Mikhail, "An Irrigated Empire," 582 with note 81.

Whatever the changing nature of the Tuṭūn and its dike, it is only the largest instance of water storage in the Fayyūm, since cisterns and reservoirs were in continuous, widespread use in the depression between antiquity and the later nineteenth century. This should come as no surprise, since the storage of water for human and animal consumption was a ubiquitous practice throughout Egypt until the modern period as a result of the yearly decrease in the level of the Nile and the resulting decline in its water quality.[96] The same was true of the water in the Fayyūm's canals. Al-Nābulusī is churlish on the subject, describing the Fayyūm's canal water as "urine dripping from a bladder" and referencing its allegedly deleterious effects on human health.[97] These complaints, however, probably only have concerned the quality of canal water in the waning months of the year. As Wansleben observed in his aforementioned remarks on the flood of 1672, the quantity and quality of the Fayyūm's canal water had visibly (and olfactorily) declined significantly by the closing weeks of the year. While visiting Sinnūris in late July, he was accordingly unwilling to drink from local canals and instead had his water delivered from the capital, likely from the same cisterns previously replenished by the flood.[98] Early in the next century, the English traveler Richard Pococke likewise claimed that already by February "the water of the canals is a little salt and not good, and must be worse till the Nile rises."[99] Condescending though al-Nābulusī's characterization may have been, he may also have been correct about the health hazards posed by the Fayyūm's low and stagnant water late in the year, which would have been a potent vector for various diseases and a breeding ground for malaria-carrying mosquitoes.[100]

While the insufficiency and unsatisfactory quality of canal water seem certain, the nature of Fayyūm reservoirs and cisterns is less so. Village cisterns of Graeco-Roman date are archaeologically attested only at the site of the ancient village of Philoteris on the western edge of the depression. Here the remains of six large cisterns are preserved, embanked by dikes averaging 2 m in height

96. Cisterns at Alexandria: Wansleben, *The Present State of Egypt*, 116; Volney, *Voyage*, 1:6; and Goitein, *A Mediterranean Society*, 68. Cairo: Raymond, *Cairo*, 246–50; Echols and Nassar, "The Canals and Lakes of Cairo," 203–12. Cf. Bagnall, *Egypt in Late Antiquity*, 18. The green water of the final months of the low-water season was at times considered poisonous, though it was more likely simply unpleasant. Derr, *The Lived Nile*, 29.

97. Al-Nābulusī, *VF*, 39: ماء خرج من مثانة

98. Wansleben, *The Present State of Egypt*, 161.

99. Pococke, *A Description of the East*, 58.

100. In the nineteenth century the Fayyūm was second only to the Delta in rates of malaria infection. Scheidel, *Death on the Nile*, 70 and 76–90.

and covering a total area of some 75,000 m^2, representing a total potential volume of at least 150,000 m^3. While the water source of each is not clear, at least two appear to have been filled by one of the village's primary canals during the flood season. Like these neighboring canals, the cisterns were excavated down to a bedrock layer of relatively impermeable limestone. The limestone floors of the cisterns, however, were fitted with drains that channeled their water into subsidiary underground chambers. These chambers were in turn accessed from above ground by wells, which were fitted either with a human-operated *shadūf* or animal-powered *sāqiya*. By this method the cisterns were drawn down for drinking water and small-scale irrigation until the next flood, at which point they would be refilled for another year.[101] This practice of filling local cisterns from a village's principal canal is directly attested papyrologically in *P.Bacch.* 19 (171 CE), a petition from the eastern village of Bakchias. The text describes the *diōryx Patsōntis*—the eastern border canal, which ran northward past the village—as both watering the fields around the village and filling local basins below it (*ta hydrostasia ta hyp' autē katerchetai*).[102] The critical importance of such infrastructure is also suggested in a short papyrus letter published as *P.Fay.* 131 (late third to early fourth century CE). "If the water comes down (*katelthei*)," the writer orders a subordinate, "make every exertion until the cistern (*hydrostasion*) is full."[103]

The capacity of cisterns devoted to the irrigation of specific plots could also be quite significant, since the collapse of one at Karanis in ca. 342 CE, here dubbed a *hydreuma*, left 190 *arourai* in the village dry and uncultivable.[104] Larger reservoirs termed *hypodocheia* are also papyrologically attested on parcels of land under date palms, grape vines, or other garden crops requiring perennial irrigation. Such reservoirs were often paired with adjacent wells that presumably functioned in much the same way as those at Philoteris.[105] The word *hypodocheion* likewise denotes large public reservoirs substantial enough

101. Kopp, "Water Management." See earlier in Römer, *The Fayoum Survey Project*, 234; and Römer, "Philoteris," 297.
102. Cited in Römer, "Philoteris," 297.
103. The term *lakkos* (λάκκος), likewise indicating a cistern or reservoir, is also attested in two papyri definitively assigned by editors to the Fayyūm: *P.Vind.Sijp 10* (fifth–sixth century CE) and *SB* 22.15745 (sixth century CE). Other attestations are uncertain. Bonneau, *Régime administratif*, 56.
104. *P.Col.* 7.174, cited in Bonneau, *Régime administratif*, 61.
105. E.g., *P.Mich.* 5.272 (45–46 CE); *P.Flor.* 2.153 (257 CE). See in general Bonneau, *Régime administratif*, 63–67. A property at Tebtynis described in *P.Mich.* 5.322a (46 CE) comes with a share (*meros*) in a *hypodocheion*, unspecified waters (*hydatos*), fish, mud (likely for use as fertilizer), and well-sweeps (*kēlōneion*, i.e., a *shadūf*). Cf. *PSI* 8.918 (38–39 CE).

to contain fish, the rights to which were leased out by the state.[106] The earliest Arabic evidence for such public reservoirs appears in Abū Isḥāq, who writes that a canal on the eastern edge of the Fayyūm known as the Khalīj al-Awāsī, which took off from the Baḥr Yūsuf and thence flowed to the north, filled lakes or ponds (*birak*, sing. *birka*) at both the village of Bayāḍ on the eastern edge of the depression and at another nearby settlement of uncertain identity called al-Khariba. Water stored in these *birak* was then channeled into several small ditches, which irrigated the surrounding area.[107]

The use of reservoirs and cisterns persisted until the introduction of modern perennial irrigation at the end of the nineteenth century. Colonial-era sources thus provide valuable detail on the purposes of these structures in the last years before their disappearance. Writing in 1893, the British engineer Justin Ross remarked that reservoirs, Arabic *khazzān*, were widespread until the construction of the Ibrāhīmiyya and were used for the irrigation of higher-lying lands, that is, lands ill-served by the canal system during the low-water season:

> Before 1865, when there was no Ibrâhîmîyah canal, the people in the Fayûm used to collect water in small reservoirs called *Khazzân*. These reservoirs held up at their lower end about 20 feet of water. In the winter they were partially emptied by a sluice and the higher parts were cultivated, and at the end of winter they were filled up again. Some of the more important ones were filled in September and retained full during the winter. As the Fayûm has no wells in it, the importance of these reservoirs must have been very great. The last one was abandoned in 1885 and its cultivable area sold.[108]

Jomard provides greater detail in the *Description de l'Égypte* in his account of a large reservoir at Abū Ksā in the northern Fayyūm, also visible on Jacotin's contemporary map. In Jomard's account, the Abū Ksā reservoir served precisely the same functions hypothesized for the dike and reservoir of the Tuṭūn, albeit on a smaller scale: to prevent water from escaping into the lake and to retain it for agricultural irrigation. According to Jomard, the reservoir had openings at different levels, thus allowing it to be exploited continuously

106. *P.Hamb.* 1.6 (129 CE); and *SPP* 8.838 (sixth century CE).
107. Al-Maqrīzī, *Khiṭaṭ* 1:671. This canal is identical with either al-Nābulusī's Baḥr Sharqiyya (mod. Baḥr Sīla) or, perhaps less likely, the ancient eastern desert canal, whose abandoned remains al-Nābulusī dubs the Baḥr Waradān. See "The Margins" in chapter 2.
108. Ross, "Irrigation and Agriculture," 184–85.

as its water level fell. To judge from Jacotin's map, small canals branched from the northern end of the reservoir carrying either (or both) drainage and irrigation water by gravity in the direction of the lake, a functionality that resembles the summary description by Abū Isḥāq of the *birak* at Bayāḍ and al-Khariba:

> Fourteen thousand meters northwest of Medynet el-Fayoum, one encounters the village of Abu Keseh, where there is a very large water reservoir. It is square in shape, fifty meters in length and width. Its construction is of brick with a very hard cement . . . The water of the Nile is introduced into the reservoir during the flood, and it supplies the water necessary for irrigation, by means of openings made at different heights. At the same time, this work functions as a dike to retain the waters of the flood, which arrive at Abu Keseh by one of the nine branches [of the Baḥr Yūsuf] mentioned in the preceding paragraph. Otherwise, since the waters run along too steep a slope, they would not remain on the countryside long enough, and their rapid course could even erode the land. The reservoir permits the distribution of water by degrees and at needs.[109]

Yet even such a large reservoir could be depleted by year's end. Having stopped in Abū Ksā on 9 July 1827, the orientalist Edward William Lane described the structure as a "large reservoir, or tank, lined with brick; at this season nearly dry."[110] A similar structure was also located in the eastern village of Ṭāmiyya (Gk. Tamauis), which supplied water to nearby villages throughout the year.[111] Pococke describes this Ṭāmiyya reservoir in greater detail. At the time of his visit in February of 1737 the eastern border canal that served the village was low and the reservoir therefore served as the primary water source for the surrounding area. He describes it as a brick-built structure, a "sort of pond about half a mile round," which is filled by a feeder canal branching from the main canal. Subsidiary canals take off from the reservoir and carry its water to nearby agricultural land. When full, water overtopped the western edge of the structure and continued its flow through the canal toward the Birkat Qārūn.[112] In his memoirs Linant likewise describes this reservoir as an enormous masonry *massif* bisecting the al-Baṭs (then the *Baḥr Bi-lā-mā'*, Canal without Water), which had at Ṭāmiyya once been joined by the Baḥr Waradān (see

109. Jomard, "Description des antiquités du nome Arsinoïte," 450–51.
110. Lane, *Description of Egypt*, 245.
111. Jomard, "Description des antiquités du nome Arsinoïte," 452.
112. Pococke, *A Description of the East*, 56.

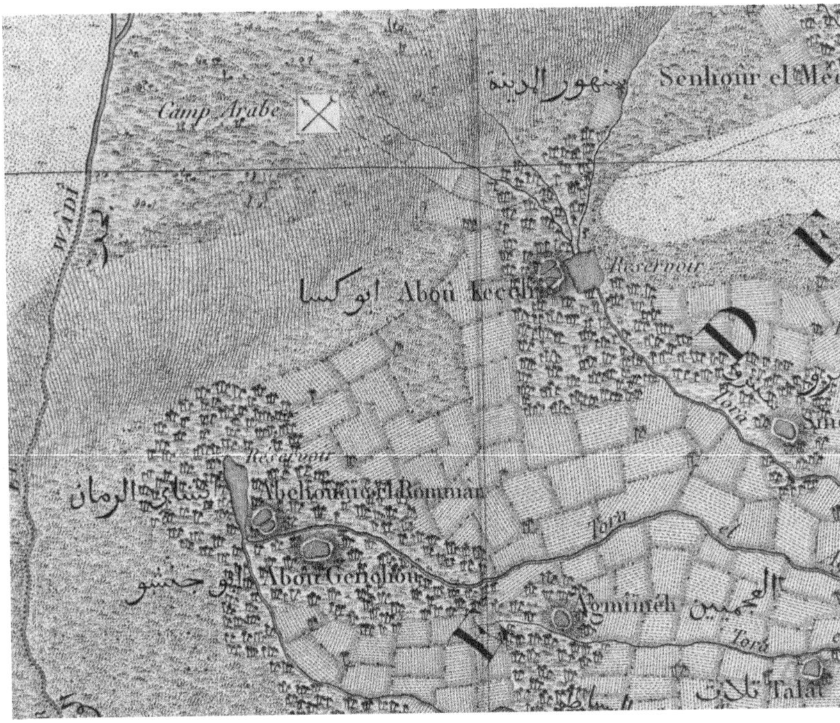

Figure 12. Reservoirs at Abū Ksā (r.) and Ibshawāy (l.). Detail from Jacotin Pierre, *Carte topographique*, Carte 19, *Faïoùm* (1818). Image courtesy of David Rumsey Map Collection, David Rumsey Map Center, Stanford Libraries.

maps 5–6). This structure, he continues, created a reservoir within the body of the canal itself, though it had been destroyed several times in recent decades when damage to the head works at the mouth of the al-Baṭs on the Baḥr Yūsuf had allowed massive amounts of water to rush in uncontrolled.[113] While Jacotin's principal map of the Fayyūm (carte 19 in his Carte topographique de l'Égypte) omits the northeastern corner of the depression, his following carte 20—encompassing the northeastern Fayyūm and the desert to the north as far as Giza—depicts a dike across the al-Baṭs at Ṭāmiyya, which retained a head of water. Linant de Bellefonds' later Carte hydrographique de la moyenne Égypte (1854) depicts much the same, though he labels the feature Réservoir pour les eaux/Krasne-t-tamieh (i.e., Khaznat al-Ṭāmiyya, the reservoir of

113. Linant de Bellefonds, *Mémoires*, 18 and 54.

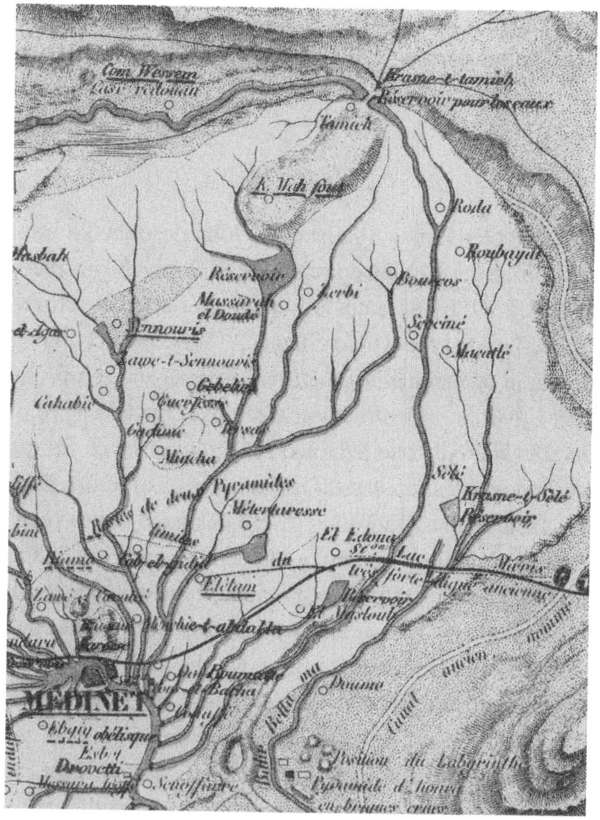

Figure 13. Various reservoirs in the eastern Fayyūm. Detail from Linant de Bellefonds, *Carte hydrographique de la Moyenne Égypte* (1854).

Ṭāmiyya). In this period the reservoir was still able to overtop this dike since Linant depicts the border canal as continuing to flow past Ṭāmiyya toward the lake.[114] He also indicates similar reservoirs in the eastern Fayyūm at the villages of Maʿṣarat Ṣāwī, Sayla, Maslūb, and Maṭar Ṭāris. The maps of both

114. There is an explicit reference to the reservoir in the sixth-century CE papyrus *SPP* 3(2).467, a receipt for taxes paid in wheat on behalf of the so-called ὑποδοχίου (*l.* ὑποδοχείου) Ταμαύεως. The editors translated the otherwise unattested phrase as "granary (*Speicher*) of Tamauis," which makes the nature of the payment obscure. It should instead be understood as a payment of wheat-taxes on lands watered by or at least adjacent to the "reservoir of Tamauis." The reservoir is also probably identical with the area dubbed the "Little Late" (μικρὰ λιμνή: see TM Geo 1256 w. refs.) in early Ptolemaic papyri, as already recognized by Grenfell and Hunt in *P.Tebt.* 2 at: Appendix II, p. 403, w, plate 3. In a 1914 map produced by the Egyptian government, the area comprising the then-defunct reservoir of Ṭāmiyya is still labeled Khazzān Ṭāmiyya: Egypt, *Atlas of Egypt Compiled at the Offices of the Survey Department*. Vol. 2, *Upper Egypt Comprising Maps of the Cultivated Area between Cairo and the Sudan Boundary* (Cairo: Wizarat al-Maliyah, 1914), sheet 99, Mudîrîyet el-Faiyûm, Sinnûris.

Jacotin and Linant clearly represent the canals or ditches that drew from these reservoirs, just as Abū Isḥāq had described centuries earlier.

Conclusion

Writing in 1845, the English traveler James Augustus St. John described a brief respite at Qalyūb, north of Cairo. Sitting under a sycamore, St. John and his party observed "a large pond of water, left by the inundation, which served as a fountain, washing-place, and horse-pond to the whole village."[115] Whether occupying natural hollows in the floodplain or manmade structures, cisterns and reservoirs were vital elements of rural Egyptian infrastructure, since they made a store of fresh water available even after the Nile had declined to its lowest level. So too in the Fayyūm, where the most significant pieces of irrigation infrastructure—the al-Lāhūn dam and the Itṣā-ʿIzbat Abū al-Nūr dike—were designed not to regulate the flow of a truly perennial irrigation system but to capture and contain the flood. The depression's unique canal system then distributed these waters both for immediate use in the irrigation of the annual winter crop and for storage within local cisterns and reservoirs scattered throughout the landscape. The importance of water storage notwithstanding, this chapter has also suggested that the peculiar hydrology of the Baḥr Yūsuf supplied some amount of water to the depression year-round, at least sufficient to keep this principal canal full for the duration of the year. Premodern Fayyūm irrigation was therefore something of a hybrid characterized by both the manipulation of the Baḥr Yūsuf's unusual semi-perennial water flow and the capture, containment, and subsequent distribution of annual floodwaters. This unique infrastructure sustained the garden-like environment so admired by the centuries of travelers quoted in the Introduction. The following chapter more closely investigates the issue of water distribution in order to sketch in outline the patterns of flow throughout the canal system. As we will see, not all parts of the Fayyūm benefited equally from the peculiar hydrology of the depression, especially the far-flung settlements of the margins. These disparities contributed to the further hybridization of a landscape whose topography and soil structure already displayed significant differences between the center and the margins.

115. St. John, *Egypt and Nubia*, 99.

Chapter 2
Hybrid Landscapes

It contains many renowned canals, filled to the brim and overflowing
—Ibn Mammātī, *Qawānīn al-Dawāwīn*

Topography and Soil Structure

Although the Fayyūm's canal system takes advantage of the natural slope of the depression's terrain to deliver water by gravity, the slope of the terrain is shallow in places, particularly in the southern half of the depression, and only becomes steeper in the north once the sea level contour is reached (see map 2). Since water delivery is gravity-dependent, this produces sluggish canal flow in various portions of the system. The problem is most acute along the margins where the terrain is so level and water delivery so slow that it can be difficult to determine at sight which direction water is flowing.[1] Under modern perennial irrigation the increased water supply has combined with this occasional sluggishness to produce considerable waterlogging.[2] The relatively impermeable clay lens atop which the entire depression rests also promotes surface pooling and pond formation, which reduces the volume of soil available to plant roots.[3] Overall, the 2,680 mcm of water that presently enter the Fayyūm each year have created a high water table within 1.5 m of the soil surface. Standing water is abundant and the attendant health hazards—exposed sewage and mosquitoes breeding in pooled water—are a source of state concern.[4] Poor

1. Römer, "The Nile in the Fayum," 174–75 with n. 13.
2. Bassiouny, "Bioenvironmental and meteorological factors."
3. Moustafa, Ibrahim, and Fikry, "Drainage Efficiency and Soil Productivity."
4. Egypt, "Environmental Action Plan." In Sinnūris District the observed water table depth varied

Figure 14. Waterlogging and soil salinization near ʿIzbat Tūnsī. Photo by author, 2011.

drainage also affects soil quality, since excess or standing water swiftly evaporates leaving behind salts and other harmful minerals. Crop yields per *faddān* are therefore generally lower than the national average in consequence of high soil salinity.[5]

All soil in the Fayyūm ultimately derives from Nile alluvium, which was deposited during the annual flood throughout the Pleistocene and Holocene (2,588,000 years BP until present).[6] Yet since the Baḥr Yūsuf carried a smaller silt load than the main channel of the Nile, Fayyūm soils are shallower than those of the Nile Valley and also overlie thick salt deposits, the remains of millennia of lakebed evaporation.[7] Nile alluvium was also not deposited evenly

between 31 and 200 cm, averaging 117 cm, between 2002 and 2009: Shendi, Abdelfattah, and Harbi, "Spatial Monitoring of Soil Salinity," 12.

5. Shendi, Abdelfattah, and Harbi, "Spatial Monitoring of Soil Salinity"; and Moustafa, Ibrahim, and Fikry, "Drainage Efficiency and Soil Productivity."
6. Shendi, Abdelfattah, and Harbi, "Spatial Monitoring of Soil Salinity," 2.
7. Ross, "Irrigation and Agriculture," 185; Willcocks and Craig, *Egyptian Irrigation*, 391: "Since the Nile [silt] deposit is seldom more that 4 or 5 metres in thickness, and generally very much less, while it overlies as a rule bitter salts, [irrigation water] is very liable to be salted."

across the floor of the depression, a phenomenon that resulted in a kaleidoscopic array of soil types and qualities. Al-Nābulusī already recognized these variations, noting that the Fayyūm's soil ranged from pure Nile alluvium to cultivable sandy clay and even to uncultivable rocky sands.[8] Modern analyses likewise divide the depression into three physiographic units—the alluvial, fluvio-lacustrine, and lacustrine plains—whose soil depth, quality, and productive potential decrease at greater distance from the al-Lāhūn inlet.[9] The soils of the central alluvial plain are entirely composed of younger, recently deposited soils (Entisols: Vertic and Typic Torrifluvents) that have not developed from their original parent material due to constant deposition during the flood. This layer covers some 818.3 km^2 of the Fayyūm's inhabited area (40.4 percent of a total 2025.5 km^2) and forms the Fayyūm's least saline and most fertile soils. Concentrated in the central alluvial fan of the depression they represent the average extent of the annual silt deposition during the inundation. The eastern and western edges of the Fayyūm, the oldest marginal terraces of the ancient lake, are covered by a total of some 361.7 km^2 of Typic Calciorthids (17.9 percent), older alluvial deposits which have undergone a considerable amount of calcification, a phenomenon common in regions where evaporation exceeds rainfall, or in the case of the Fayyūm, floodwater influx. Under such circumstances, calcium is not leached from the soil, instead forming hard crusts just below the surface that inhibit root penetration and plant growth. Located along the outer rim of the depression's agricultural zone (the lands of ancient marginal villages), these areas represent a less productive but hardly uncultivable zone. Also of interest are the 75.7 km^2 of Gypsic soils in the northeast (3.7 percent), their chemical makeup explicable only through the earlier presence of a much larger lake that has long since regressed.[10] The soil profile is, of course, significantly more variegated than this bare summary. As a general rule, however, soils grow poorer and the risk of salinization increases at greater distances from the al-Lāhūn inlet, indicating that the majority of alluvial deposition has occurred toward the center of the depression. Indeed, those areas at high risk for salinization are concentrated in the lacustrine and fluvio-lacustrine

8. Al-Nābulusī, *VF*, 34.
9. This paragraph depends on Gad and El Zeiny, "Spatial Analysis." See also Kater et al., "Mineralogical and Chemical Composition of the Main Soil Types."
10. "Anhydrite ($CaSO_4$) is mainly associated with marine evaporites and is rapidly converted to gypsum when exposed to normal soil environment": Ahmet R. Mermut and H. Khademi, "Gypsum Formation in Gypsic Soils," in *Encyclopedia of Soil Science*, ed. Rattan Lal, vol. 1 (London: Routledge), 800.

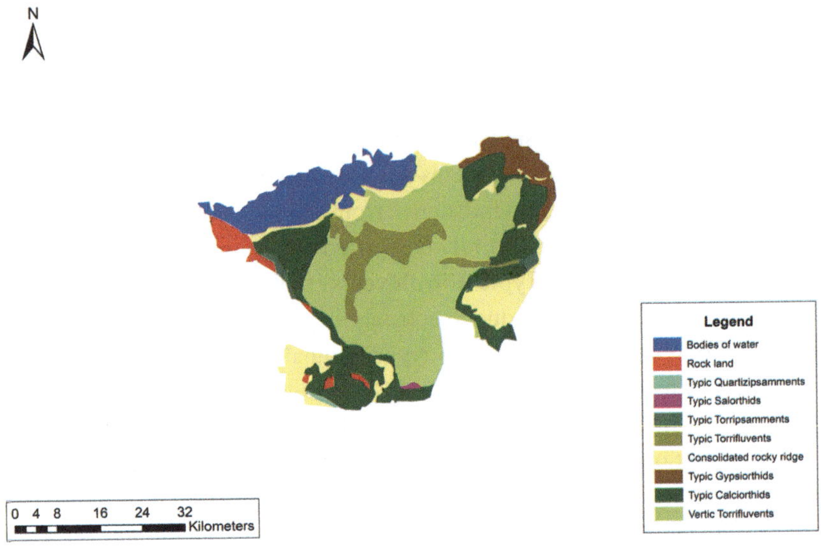

Figure 15. Soil map of the Fayyūm. Redrawn by Julian Thibeau from Gad and El-Zeiny, "Spatial Analysis for Sustainable Development of El Fayoum."

plains along the depression's outer margins.[11] The landscape and topography of the Fayyūm is therefore distinctly hybrid, its marginal and central zones marked by significant disparities in inherent agricultural potential.

Overview of the Canal System

The functionality of the canal system only sharpened the distinctions between center and margins, thereby informing the development of two discrete agroenvironmental zones. Drawing heavily on Abū Isḥāq, Ibn Mammātī, and al-Nābulusī, the following two sections describe the form and function of this hybrid system, highlighting the differences between central and marginal canal flows. It should be noted that this reconstruction is limited to major public canals since it is these that are best documented in the surviving evidence. Unfortunately, the changing toponymy of the region along with occasional

11. See in general Monson, "Salinization."

manuscript corruption make it impossible to plot the route of every attested premodern canal or to identify them conclusively with more contemporary waterways. Indeed, canals are never stable but are instead constantly reshaped by human agency and environmental factors, often effacing evidence of their previous routes.[12] Yet since Fayyūm irrigation has always depended on gravity-driven flow, the paths of principal waterways necessarily follow the slope of the depression's topography. The principal modern public canals are thus at least roughly analogous to their premodern antecedents and help refine our understanding of the system before the transformations of the nineteenth century.

• • •

The Margins

The terminal stretch of the Baḥr Yūsuf within the Fayyūm—the Baḥr or Khalīj al-Aʿẓam in later Arabic authors—was dubbed the Argaitis canal (*Argaitidos diōryx*) in Greek and the Henet of Moeris in earlier Egyptian.[13] Its point of entry into the Fayyūm, the village of al-Lāhūn, retains an Arabized version of its Coptic Egyptian toponym *Lehōne*, signifying the "mouth of the canal."[14]

The first major waterway to split from the Baḥr Yūsuf within the Fayyūm proper is the eastern border canal, which takes off just downstream from al-Lāhūn near Hawwārat al-Maqṭaʿ (al-Nābulusī's Hawwāra al-Baḥriyya and Hawwāra al-Ṣaghīr in the *Description de l'Égypte*). The canal was known by several names during the Graeco-Roman period, among them the Canal of Kleon (*diōryx Kleōnos*) and the Desert Canal of Patson (*oreinē diōryx Patsōntis*).[15] It is identical with al-Nābulusī's Baḥr Waradān (*VF* 46–47), a long-abandoned watercourse whose route took it north along the base of the limestone rim surrounding the Fayyūm, (*taḥt al-jabal*, "under the mountain") until it reached its terminus in the "Fishery Lake" (*Birkat al-Samak*, i.e, the Birkat Qārūn).[16] Linant de Bellefonds' nineteenth-century depictions of the

12. See the case study at Karanis in Cook, "Landscapes of Irrigation," 105–43.
13. Pearl, "ΑΡΓΑΙΤΙΣ and ΜΟΗΡΙΣ"; and Vandorpe, "The Henet of Moeris."
14. Trismegistos (hereafter TM) Geo ID 2024. The Egyptian toponym *Rȝ-tȝ-ḥny.t-n-Mȝ-wr* ("Mouth of the Canal of Moeris") became *Lehōne* (ⲗⲉϩⲱⲛⲉ) in later Coptic Egyptian, whence the Egypto-Arabic *Lāhūn*. See also Peust, *Die Toponyme*, 57.
15. Kleōnos: *P.Petrie* 2.6 (256 BCE), *P.Petrie* 2.36 (241 BCE); Patsōntis: *SB* 6.9437 a (144 CE), *P.Bacch.* 19 (171 CE).
16. For the identification see Shafei, "Fayoum Irrigation," 289 and 300.

abandoned Waradān (maps 5–6) locate its head near Hawwāra and its tail not in the lake proper but rather in the al-Baṭs ravine near the northeastern villages of al-Rawḍa and Ṭāmiyya.[17] The route of this canal was therefore in many respects identical with that of its modern equivalent the Baḥr ʿAbdallah Waḥbī, which takes off roughly 1.5 km upstream from Hawwārat al-Maqṭaʿ.[18] Yet it appears that the head of the ancient eastern canal was excavated some 200–300 m west of Hawwāra, thus bypassing the Labyrinth—the temple and mortuary complex of Amenemhat III (1842–1797 BCE)—while both the Baḥr Sharqiyya as well the modern Baḥr ʿAbdallah Waḥbī cut directly through the site.[19] Regardless, after Hawwāra the canal turns north and east to follow the eastern margins of the depression. It turns westward again near Ṭāmiyya and then flows toward the lake. Recent geoarchaeological work at Karanis/Kaum Aushīm indeed confirms that here, at least, the canal followed the path of the modern Wahbī, a route necessitated by local terrain.[20] The slope of the canal is also quite shallow along much of its length. With an elevation of 24 masl at Hawwārat al-Maqṭaʿ, the Wahbī falls to some 12 masl opposite the site of ancient Bakchias (Umm al-Athl), an average gradient of only 0.024 percent over some 50 km. The terrain nonetheless begins to fall more rapidly after the northern apex of the canal near the small village of ʿIzbat Aḥmad Zakī (11 masl) and its final 15 km stretch reaches 20 mbsl opposite Karanis and 42 mbsl along the eastern shore of the Birkat Qārūn, its average gradient increasing dramatically to 0.35 percent.

The point at which this eastern desert canal/Baḥr Waradān was finally abandoned is unclear. Although al-Nābulusī clearly indicates that it was altogether defunct by the Ayyūbid period, it is possible that it remained functional into the early eleventh century CE. According to Abū Isḥāq, the first canal to separate from the Baḥr Yūsuf was the so-called Khalīj al-Awāsī (plural of *ūsiyya*, from Greek *ousia*, "great estate").[21] This al-Awāsī canal in turn tends to be identified with the Baḥr Sharqiyya ("Eastern Canal"), the easternmost operational waterway in al-Nābulusī's period. Later known as Baḥr Sīla, the canal survives as the modern Baḥr al-Rawḍa.[22] In support of this identification is

17. Cf. Brown, *The Fayûm*, 103–4.
18. Kraemer, "The Meandering Identity of a Fayum Canal."
19. Uytterhoeven and Blom-Böer, "New Light on the Egyptian Labyrinth," 114.
20. Cook, "Landscapes of Irrigation," 147–48.
21. Wickham, "The Power of Property," 72–73.
22. Toussoun, *Mémoire*, 1:260; Salmon, "Répertoire géographique," 32; Gaubert and Mouton, *Hommes et villages*, 172. Toussoun points to a modern canal, El-Aoussia, that branches off from the Baḥr Sīla

Abū Isḥāq's claim that the Khalīj al-Awāsī irrigated a village named Bayāḍ, which is linked to the Baḥr Sharqiyya by al-Nābulusī (*VF* 127) and is described as a dependency of the nearby village of Sīla in an early fourteenth-century cadaster.[23] Yet both al-Nābulusī and Linant de Bellefonds place the tail of the Baḥr Sharqiyya/Baḥr Sīla at al-Rubiyyāt (mod. al-Rūbiyyāt some 6 km southeast of Ṭāmiyya), while Abū Isḥāq clearly states that the al-Awāsī terminated in the al-Baṭs ravine, just as the Waradān does in Linant's maps. Abū Isḥāq also mentions a village (*ḍay'a*) named Khariba ("ruins") among its dependencies, perhaps a reference to al-Khariba al-Kabīr (Greek Philadelphia) or al-Khariba al-Saghīr (ancient identity unknown), the local Arabic names for two abandoned Graeco-Roman villages on the eastern margins that would have been irrigated by the eastern desert canal.[24] Given the tendency for canals to be named after one of their dependencies, it is also noteworthy that al-Nābulusī names a certain al-Lawāsī—an obvious corruption of al-Awāsī—among the abandoned ancient settlements along the Bahr Waradān. Finally, the second canal to branch from the Baḥr Yūsuf according to Abū Isḥāq was the Khalīj Samasṭūs, which deposited considerable alluvial sediment (*abālīz*) on the lands between it and the al-Awāsī. Of its dependencies Abū Isḥāq names only the village Samasṭūs itself, a site named by al-Nābulusī (*VF* 47) among the deserted villages of the eastern Fayyūm. Although al-Nābulusī describes this now-abandoned Samasṭūs as a former dependency of the defunct Baḥr Waradān, this at least confirms that Abū Isḥāq's Samasṭūs canal was well to the east of the Fayyūm.[25] Abū Isḥāq's Khalīj al-Awāsī therefore may be identical with the eastern desert canal/Baḥr Waradān, although this is far from certain. If this identification is accepted, however, the terminus post quem for the final abandonment of the eastern desert canal cannot be earlier than 1031 CE.

After feeding the eastern canal, the terminal stretch of the Baḥr Yūsuf continues to flow west/northwest toward the capital. Just downstream from the

---

at the village of Sīla and thus identifies the village with al-Ūsiyya al-Kubrā, one of the dependencies of the al-Awāsī in 'Abū 'Isḥāq. While a canal named Aousieh is indeed clearly indicated on modern maps, there is no compelling evidence for equating Sīla with 'Abū 'Isḥāq's al-Ūsiyya al-Kubrā. See Audebeau and Commission des domaines de l'État égyptien, *Carte de la Basse-Egypte*.

23. *Al-Rawk al-Nāṣirī* (1313–25 CE), on which see Sato, *State and Rural Society*, 135–52.
24. Lit. "Great Ruined Site" and "Small Ruined Site." See Davoli, *L'archeologia*, 139–43 and 164–65.
25. Samasṭūs, which al-Nābulusī seems to locate south of Umm al-Athl (Bakchias) is probably identical with the poorly attested ancient village of Psimistous, which is located by *SPP* 10.263 (seventh–eighth century CE) in the eastern Fayyūm likewise south of Bakchias. There nonetheless seems to have been a homonymous settlement in the southwest, for which see Berkes and Haug, "Villages, Requisitions, and Tax Districts," 219–20.

mouth of the eastern desert canal it passes by the opening of the al-Baṭs ravine. Jacotin depicts a large dike (*digue*) along the mouth of the ravine (dubbed *Baḥr Bi-lā-mā'* or "Canal without Water"), which seems to serve much the same purpose as the larger dike at the head of the al-Wādī ravine in the Ṭutūn basin—namely, to prevent water from escaping uncontrolled to the lake through the al-Baṭs. According to Edme-François Jomard in the *Description de l'Égypte*, however, this *digue*—"a bridge of ten arches"—did not totally block the mouth of the al-Baṭs. Rather, some water overtopped it during the peak of the flood (*le haut Nil*), falling several meters into the bed of the al-Baṭs/Baḥr Bi-lā-mā' and flowing thence toward the lake. P. D. Martin says much the same elsewhere in the *Description*, writing that when the floodwaters exceeded the height of the structure, they would fall seven meters into the Bahr Bi-lā-mā' ravine and then be carried in the direction of Ṭāmiyya and the Birkat Qārūn. This ravine, he adds, "in receiving only the excess waters of the province, remains dry nearly every year," whence its name.[26] This structure on the al-Baṭs/Baḥr Bi-lā-mā' therefore functioned as a massive release valve that moderated the volume of the Baḥr Yūsuf during the flood in order to prevent the potentially disastrous overflow of the channel.[27] The practice dates at least to the eleventh century, though it likely originated in antiquity. Abū Isḥāq writes simply that the "mouth of the al-Baṭs canal" (*fuwwahat khalīj al-Baṭs*) received the excess waters (*mufāḍil al-miyāh*) flowing through the Baḥr Yūsuf. Gates (*abwāb*) on the mouth remained closed until the water reached a specified height along elevated (*murtafa'a*) lands nearby.[28] Al-Nābulusī elaborates on the process. In his account, during "high Niles" (*al-anyāl al-'āliya*) when there was "fear for the villages of the Fayyūm and their sugarcane," surplus waters were shunted into the al-Baṭs ravine and immediate drained to the lake, thereby raising its surface level and swamping cultivable lands along the shoreline.[29] In al-Nābulusī's period, a drain with an arch barred by two sluice gates (*bābayn*) was located between Ṣunūfar and Qushūsh (mod. Quḥāfa) and it channeled the overflow directly into the al-Baṭs. A drain fitted with a gate is still depicted between

26. Martin, "Description hydrographique," 25.
27. Jomard, "Description des antiquités du nome Arsinoïte," 452–53. So also Martin, "Description hydrographique," 26, where he describes the ravine (*Bahr belâ-mâ*), as receiving only the excess waters (*superflu des eaux*) of the province and thus remaining nearly dry all year. Residual seepage (*suintemens*) in the channel reached as far as al-Rawḍa in the north and fed there the drain leading to the lake (34–35).
28. Al-Maqrīzī, *Khiṭaṭ*, 673.
29. For the phenomenon in antiquity see Hobson, "Agricultural Land and Economic Life."

these two villages on Jacotin's map. Al-Nābulusī also describes a ruined structure at Hawwāra (perhaps an earlier version of Jomard's *digue*) through which vastly more water escaped into the lake, apparently uncontrolled, though seemingly only during years of high flood. All of this water, both the drainage from Hawwāra and the Ṣunūfar/Qushūsh drain as well as other overflow from the Baḥr Yūsuf, combined to increase the volume of the lake over a two-month period during years of high flood.[30]

After passing by the mouth of the al-Baṭs, the Baḥr Yūsuf branches now from its southern bank to feed the Fayyūm's southern border canal, known in antiquity as the Desert Canal of Polemon (*oreinē diōryx Polemōnos*) or the Desert Canal of Tebtynis (*oreinē diōryx Tebtyneōs*). It retained the latter name in the Arab period and is dubbed the Khalīj Tanbaṭāwa by Abū Isḥāq and Baḥr Tanabṭawayh by al-Nābulusī.[31] Both describe the canal as following a path along the base of the Fayyūm's southern desert rim (*taḥt al-jabal*), though Abū Isḥāq also writes that it was equipped with a vaulted drainage outlet or spillway (*mafīḍ bi-qabw*), through which water exited during the flood. This structure has been plausibly connected to the passage of the southern border canal through the small rocky ridge that separates the main portion of the Fayyūm depression from the smaller Gharaq basin in the southwest.[32] A small lake in the Gharaq, no longer extant, once received the overflow and drainage of the canal and it is from this terminus that the contemporary waterway, the Baḥr al-Gharaq, takes its name. The lake is clearly visible on Jacotin's map, there labeled as the "Birkeï Garâh" (Birkat Gharaq) and seems to have occupied most of the basin to the west of Gharaq City.[33] Although the head of the modern Baḥr al-Gharaq lies on the Ḥasan Wāsif intake, a secondary inlet opened in 1905, its premodern head was on the Baḥr Yūsuf according to al-Nābulusī. Jacotin's map shows twin canals branching from the Baḥr Yūsuf near a village named al-Ḥasbah, somewhere in the vicinity of contemporary Dayr al-ʿAzab (Greek Mouchis, al-Nābulusī's Dumūshiyya).[34] The two streams rejoined roughly 3 km south of Dayr al-ʿAzab and continued to run along the Fayyūm's southern rim, eventually terminating in the Birkat Gharaq. In the absence of

30. Al-Nābulusī, *VF*, 212.
31. From the Egyptian toponymy of Tebtynis: Demotic *Tꜣ-nb-Tp-Tn* or *Tꜣ-nb.t-tꜣ-Tn*.
32. Toussoun, *Mémoire* 1:261.
33. Since the 1970's, drainage from the south has been channeled out of the Fayyūm to the southwest where it has filled the Wādī Rayyān depression forming two small drainage lakes. On the Gharaq see Brown, *The Fayûm*, 47–48; and Rathbone, "Mapping the South-West Fayyum."
34. For the latter identification see most recently Winkler, "Mouchis and Its Crocodiles."

the lake, the contemporary Baḥr al-Gharaq terminates along the extreme western edge of the Gharaq basin. From its 23 masl head at Dayr al-ʿAzab, this canal reaches 9 masl over its roughly 44 km modern course, an average gradient of some 0.034 percent.

Confusion as to the premodern route of the southern canal arises from al-Nābulusī's paradoxical claim that, though still operational, this Baḥr Tanabṭawayh was simultaneously silted up and lined with abandoned settlements like the Waradān. The error derives from al-Nābulusī's belief in the parallelism of the Fayyūm's ancient canal system—namely, that it had been encircled by twin canals that emptied into the eastern and western shores of the lake respectively. He therefore regarded the ancient desert canal along the Fayyūm's western margin, by his day long abandoned, as the northerly extension of the Baḥr Tanabṭawayh rather than as a separate waterway. Al-Nābulusī thus correctly traces the route of the southern desert canal from the Baḥr Yūsuf and along the Fayyūm's southern rim, yet he then mistakenly writes that it turned northward, continuing north along the western rim of the depression until it reached the western shore of the Birkat Qārūn. Al-Nābulusī's list of abandoned Tanabṭawayh dependencies consequently includes two well-known ancient sites of the Fayyūm's south—Tanabṭawayh = Tebtynis and Burjtūt = Perkethaut/Philagris[35]—but also Qaṣr Qārūn, the ancient village of Dionysias at the northwestern end of the irrigation system.[36] In reality, the thirteenth-century Baḥr Tanabṭawayh terminated near Ṭalīt (Greek Talei or Talithis), a settlement on the border between the Tuṭūn and Gharaq basins. Ancient villages like Dionysias along the western rim of the Fayyūm had instead been irrigated by a canal similar to the contemporary Baḥr Qaṣr al-Banāt. This canal presently splits from the Baḥr al-Nazla, whose own head is adjacent to that of the Baḥr al-Gharaq at Dayr al-ʿAzab. Unfortunately, the premodern antecedent of the Baḥr Qaṣr al-Banāt is not altogether clear. The most likely ancient candidate is the *diōryx Pseinalitidos*, a public canal that irrigated northwestern border villages like Theadelphia, Euhemeria, and Polydeukia.[37] If this ancient waterway followed a route similar to the modern Baḥr Qaṣr al-Banāt, the roughly 63 km

35. For the toponymy of Philagris/Perkethaut see Clarysse and Van Beek, "Philagris."
36. Al-Nābulusī, *VF*, 46.
37. *BGU* 13.2262 (138–61 CE); *P.Münch.* 3.108 (145 CE); *SB* 16.12597 (145 CE); *P.Fay.* 77–78 (147 CE); *BGU* 4.1076 (148 CE); *P.Hamb.* 1.75 (149 CE); *SB* 16.12320 (153 CE); *P.Mich.* 10.595 (161 CE); *BGU* 4.1077 (163 CE); *BGU* 9.1897 (166 CE); *P.Strasb.* 1.55 (173 CE); *SB* 10.10262 = *P.Brookl.* 11 (206 CE); *SB* 12.10964 (209 CE); *SB* 16.12499 (200–250 CE). Römer, *The Fayoum Survey Project*, 99 and 193.

canal possessed an average gradient of close to 0.046 percent, somewhat steeper than the canals of the eastern and southern margins.[38]

The Center

The obscurity of the western canal notwithstanding, the general outlines of the border waterways are relatively clear. The canals of the Fayyūm's center, however, are less so, becoming visible only during the early Islamic period. Al-Nābulusī hints at the density of this portion of the canal system, writing that five major waterways split from the end of the Baḥr Yūsuf, along with a further fifty-eight additional canals as well as eight small ditches.[39] Evliya Çelebi similarly writes in the seventeenth century that the Baḥr Yūsuf splits into seventy channels that spread out over the entire Fayyūm.[40] By the nineteenth century, according to P. D. Martin in the *Description de l'Égypte*, nine major canals branched from the terminus of the Baḥr Yūsuf on the outskirts of the Fayyūm's capital.[41] According to both Martin's fellow savant, the engineer P. S. Girard, and Linant de Bellefonds three quarters of a century later, the Baḥr Yūsuf terminated in a large, irregular basin (*bassin de distribution*) on the western edge of the capital city, from which branched each of the central Fayyūm's principal public canals. Girard and Linant also write that the mouths of each of these waterways were regulated by brickwork and masonry dikes (Girard's *petites chaussées*) fitted with openings whose widths were calculated to produce a water flow that was proportional to the total number of *feddān*s each canal watered—the canal's command area in contemporary terminology.[42] As discussed in greater detail in chapter 4, this method of proportional water sharing—already attested in our earliest Arabic sources and even then regarded as ancient—ensured that water was distributed equitably by enabling most villages to be assigned a fixed water quota representing their share of the public canal's total flow.

Although the configuration of the central Fayyūm's canal system was surely always in flux, it is possible to sketch out its broad contours between the eleventh and thirteenth centuries CE. Using al-Nābulusī's enumeration of their

38. Archaeological data on the tail of the canal in Römer, *The Fayoum Survey Project*, 107–9.
39. Al-Nābulusī, *VF*, 35.
40. Dankoff, Tezcan, and Sheridan, *Ottoman Explorations*, 352.
41. Martin, "Description hydrgraphique," 23.
42. Linant de Bellefonds, *Mémoires*, 16; Girard, "Mémoire sur les irrigations," 332–33.

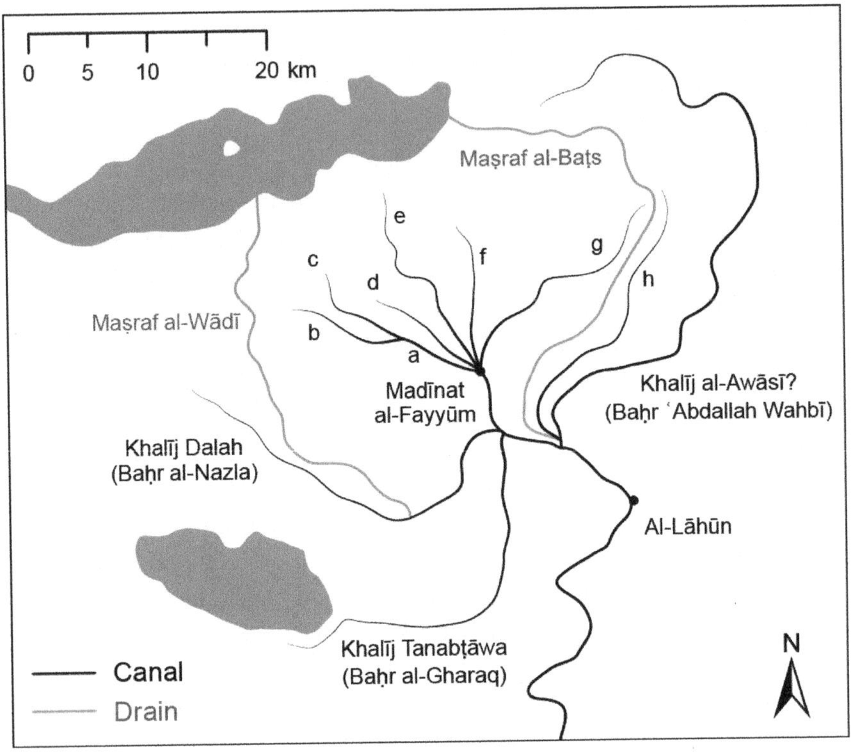

Map 6. Hypothetical outline of the canal system in the eleventh to twelfth centuries CE, based on the descriptions in Abū Isḥāq and Ibn Mammātī. (Cartographer: Eli Weaverdyck).

a: Khalīj Santarīya (Baḥr Gharbiyya)
b: Khalīj Ibshū (Baḥr Ibshawāy)
c: Khalīj Abū Ksā
d: Khalīj Sīnarū
e: Khalīj Majnūna (Baḥr Sanhūr)
f: Khalīj Tabdūd/Tandūd (Baḥr Tirsā/Tandūd)
g: Khalīj Talālah (Baḥr Tanhalā)
h: Khalij Samasṭūs (Baḥr Sayla)

various dependencies, Yossef Rapoport and Ido Shahar have reconstructed the routes of several major public waterways. Evidence from the eleventh and twelfth centuries—the administrative manuals of Abū Isḥāq and Ibn Mammātī—is more cursory but is sufficient for a speculative reconstruction of the system in this period. In addition to the al-Awāsī and Tanabṭāwa, Abū Isḥāq's redacted *dustūr* briefly describes only seven public canals: the Samasṭūs, Dahāla, Dalah, Majnūna, Talālah, Bamwah, and Tabdūd, the last a misreading, as described below.[43] The text was surely far longer before its recension, since Ibn Mammātī writes that Abū Isḥāq did not confine his account to these "royal" (*sulṭānī*) canals—that is, public waterways maintained at state expense. Although Ibn Mammātī's own account is similarly restricted to the best-known (*al-mashhūr*) public canals, it is still twice the length of Abū Isḥāq's.[44] Both documents are nonetheless little more than annotated lists and do not permit us to reconstruct the precise route of each canal described.

Abū Isḥāq's *dustūr* is organized by offtake from the Baḥr Yūsuf. As mentioned above, the first waterway to split from the main canal was the al-Awāsī, which I have tentatively identified as a later incarnation of the ancient eastern desert canal, al-Nābulusī's Baḥr Waradān. Abū Isḥāq's Samasṭūs would therefore be equivalent to the Baḥr Sharqiyya/Sīla (contemporary Baḥr al-Rawḍa), the farthest eastern waterway in al-Nābulusī's period. The third canal, however, the Dahāla, is unidentifiable, since Abū Isḥāq offers no topographical information. Nevertheless, it must have branched from the northern bank of the Baḥr Yūsuf since it directly precedes the Tanabṭāwa, the first offtake from the Yūsuf's southern banks. Linguistic affinity suggests that the following waterway, the Dalah, is identical to al-Nābulusī's Baḥr Dilya, a long canal that took off from the southern banks of the Yūsuf and flowed first southwest then northwest toward the western edge of the depression.[45] This route bears a strong similarity to the early stretch of the modern Baḥr al-Nazla. Since canals were often named from a settlement somewhere along their course, Abū Isḥāq's next canal, the Majnūna, might be linked al-Nābulusī's Banū Majnūn (later Banī Majnūn, now Banī Ṣāliḥ), a village some 6 km northwest of the capital.[46] A possible modern equivalent of the waterway is the Baḥr Banī Majnūn, among whose dependencies was Sīnarū,

43. Al-Maqrīzī, *Khiṭaṭ* 1:672–74.
44. Ibn Mammātī, *Qawānīn al-Dawāwīn*, 229–30, trans. in Cooper, "Ibn Mammātī's Rules," 74–76.
45. Salmon, "Répertoire géographique," 32–33; and al-Nābulusī, *VF*, 47n26.
46. Al-Nābulusī, *VF*, 228 with refs. in n. 301.

some 5 km west of Banī Sāliḥ/Majnūn.[47] Yet since Abū Isḥāq describes the Majnūna as flowing directly to the lake in the north, the Majnūna may have been a predecessor of the modern Baḥr Sanhūr, which irrigates from south to north the villages of Banī Ṣāliḥ/Majnūn, Siliyyīn, Fidaymīn (Fidimīn in al-Nābulusī), and Sanhūr.[48] As for the Talālah, Abū Isḥāq says only that it irrigates quarters of the capital and that the mouth (*fuwwaha*) of the al-Baṭs is on it, which receives its excess waters (*mufāḍil al-miyāh*). There seems to be some confusion in the ordering here, since the position of this canal in Abū Isḥāq's list would place it too far to the west to have any connection to the al-Baṭs. Still, the name of the canal bears strong resemblance to the modern Baḥr Tanhalā, which takes off from the Baḥr Yūsuf at the capital and flows northeast toward the mouth of the al-Baṭs ravine. The Bamwah, however, can almost certainly be linked to a village somewhere near Sanhūr called Bamawayh by al-Nābulusī (cf. the similar shift from Tanabṭāwa to Tanabṭawayh).[49] Indeed, according to al-Nābulusī Bamawayh was irrigated by a "canal known by [the name of] the village" (*baḥr yuʿraf bi-l-nāḥiya*), though he does not name any additional dependencies. Finally, although Abū Isḥāq provides no topographical information for the Tabdūd, the canal is very likely to be identical with the modern Baḥr Tirsā, otherwise known as the Baḥr Tandūd, which follows a northerly course through villages including Ḥarfūsh, Tirsā, and Naqālīfa (al-Nābulusī's Naqalīfa), and eventually terminates in the Birkat Qārūn.[50] The reading Tabdūd, which in the absence of diacritics is identical to Tandūd (ىىدود), is thus surely spurious and should accordingly be amended. The toponym Tandūd later occurs in al-Nābulusī and is closely associated with the region through which the contemporary Baḥr Tirsā/Tandūd flows. Al-Nābulusī records Tandūd as the name of a water-turned stone press at Biyahmū (Egyptian Piamouei, Greek Andriantōn Kōmē), which was also adjacent to Akhṣāṣ Abū ʿUsayya (today Ḥarfūsh).[51] He further refers to freshwater springs (*ʿayun*) in a ravine (*wādī*) at Abū ʿUṣayya "on the bank of Tandūd" (*ʿalā ḥāffat Tandūd*), which are used for

47. Egypt, Ministry of the Interior, *Report on the Epidemic of Cholera*, 76.
48. So Toussoun, *Mémoire*, 1:262.
49. Timm, *Das christlich-koptische Ägypten*, 1:315.
50. Salmon, "Répertoire géographique," 53. The canal is labeled both the Baḥr Tirsā and the Baḥr Tandūd in the 1914 government map in Egypt, Maṣlaḥat al-Misāḥah, *Atlas of Egypt*, Sheet 104 El-Faiyum. It is referred to as the Baḥr Tandūd in a 20 April 2019 Arabic media report on its flooding at the village of Tirsā, accessed 22 December 2020, https://www.albawabhnews.com/3569514.
51. Al-Nābulusī, *VF*, 115 (Biyahmū) and 205 (Minyat Karbīs and Akhṣāṣ Abū ʿUṣayya).

drinking and irrigation when its nearby canal is cut off (*inqaṭa'*). Abū Isḥāq likewise refers to a sweet water spring (*'ayn ḥulwa*) adjacent to this *khalīj* Tabdūd, from which nearby lands are irrigated when the canal is closed. Al-Nābulusī's Tandūd therefore surely refers not only to a stone press but also the otherwise undescribed waterway that powered it, which was essentially equivalent to the modern Baḥr Tirsā/Tandūd.

After replicating the first eight canals in Abū Isḥāq's list (the ninth is corrupt), Ibn Mammātī provides an additional nine entries, though further manuscript corruption makes it difficult to ascertain the proper name of each. While the first is entirely uncertain,[52] the remainder may be resolved as Babīj, Biljāyah (probably spurious),[53] Bīdūd (another corruption of Tandūd since without diacritics it is again سدود), Santariyya, a number of small unnamed waterways (*'adatan khuljān luṭāf*) branching from the Grand Canal/Baḥr Yūsuf, Fidimīn, Abū Ksā, Sanabrū (Sīnarū?), which is connected to the Ibshū canal by aqueduct (*'ibbāra*), and finally the Ihrīt.[54] Derived from Graeco-Coptic *Pepoikion* ("the farmstead"),[55] the initial element Babīj appears in five toponyms in al-Nābulusī's period and it is consequently impossible to assign the waterway to a particular locale.[56] Santariyya was the medieval Arabic name for the Sīwa Oasis and, according to Abū Isḥāq, colloquially described a westerly direction. A speculative identification is therefore the modern Baḥr Gharbiyya ("Westerly Canal"), which branches from the end of the Baḥr Yūsuf at Madīnat al-Fayyūm and briefly flows due west before bifurcating into canals serving Ibshawāy and Abū Ksā. Named for still-extant settlements, the general trajectories of the canals of Fidimīn, Abū Ksā, and Ihrīt are clear. Finally, the manuscript readings *Sntrū* or *Snbrū* are likely corruptions of Sīnarū, as suggested in the most comprehensive modern edition of the text.[57] This is further suggested by the linkage of this waterway to the *khalīj* Ibshū, almost surely referring to the nearby village of Ibshawāy, al-Nābulusī's Ibshāyat al-Rummān (Greek Pisais,

52. سدا or سسدا, which cannot be resolved by comparison with any of the toponyms in al-Nābulusī.
53. No village of this or a similar name in the Fayyūm is attested. Ibn Mammātī elsewhere in his text describes Biljāyah as a village in al-Martāḥiya. Cooper, "Ibn Mammātī's Rules," 98 no. 336.
54. Ibn Mammātī, *Qawānīn al-Dawāwīn*, 230–31. See the translation and notes on MS readings in Cooper, "Ibn Mammati's Rules," 75–76.
55. Salmon, "Le nom de lieu Babîdj"; and Ibn Mammātī, *Qawānīn al-Dawāwīn*, 230–31. Timm, *Das christlich-koptische Ägypten*, 4:1893–94.
56. Babīj Andīr, Babīj Anqāsh, Babīj Faraḥ, Babīj Ghaylān, Babīj Unshū.
57. Ibn Mammātī, *Qawānīn al-Dawāwīn*, 231, trans. in Cooper, "Ibn Mammātī's Rules," 76 with no. 341.

Coptic Pishai).[58] In Jacotin's map the canals serving both Sīnarū and Ibshawāy (as Ibshāy al-Rummān) are indeed immediately proximate though no direct connection between the two is depicted. Finally, a village by the name of Ihrīt lies some 11 km west-southwest of the capital. Al-Nābulusī says that its water is delivered by a canal that branches from the southern bank of the Baḥr Yūsuf but he does not name it.

But whatever the precise complexion of the central system at any point in time, its canals flowed along far steeper terrain than that of the margins. Two canals described by al-Nābulusī clearly exemplify the topography of these waterways. His Sinnūris canal, which watered nine villages north of the capital, opened at 20 masl near Munsha'at al-Ṭawāhīn (mod. Munsha'at ʿAbdallah) and fell to 10 mbsl outside Sinnūris over a length of only 10 km. Its steep 0.3 percent gradient is an increase of 147 to 170 percent over the comparably level slopes of the border canals. So too al-Nābulusī's Dhāt al-Ṣafā' canal, whose head lay at some 30 masl along the Baḥr Yūsuf and reached nearly 13 mbsl near Minyat al-Baṭs (mod. Ṭāmiya) 22 km distant, a gradient of some 0.2 percent. As already seen in the previous chapter, the steep slope of the Fayyūm's central plain was one of the reasons cited by Jomard for the reservoir at Abū Ksā, which was constructed in part to keep floodwaters from rushing along so swiftly that they swept away the soil of the surrounding landscape.[59]

Marginal and Central Canal Flow

Sediments cleared from the beds of modern border canals like the Baḥr ʿAbdallah Wahbī contain numerous bivalves, animals that thrive only in perennial water. In contrast, excavated sediment from the relict premodern border canals shows no such faunal remains, a strong indication that they did not remain full throughout the year.[60] Hellenistic and Roman papyri support this supposition. The apparent emptiness (or insufficiency) of the main canal at Philadelphia in May/June 257 BCE is alluded to in a fragmentary papyrus letter in which the manager of a large estate at Philadelphia discusses the planting of young olives (*elaina phyta*) in the village and mentions waters being brought

58. TM Geo 1836.
59. See chapter 1 above, "Storing the Flood."
60. Cook, "Landscapes of Irrigation," 141 with figure 140 on p. 247.

from Tanis, a village upstream on the eastern border canal.[61] A fragmentary and otherwise obscure official letter from July of ca. 107 CE also refers to the distribution of scarce waters somewhere in the northwest of the Fayyūm. Although there is a lacuna in the papyrus at a critical point, a local official writes to the governor of this portion of the Fayyūm that "the doors of the [sc. reservoirs or canals?], as you know, were raised in your presence as much as the overseers of sowing (*katasporeis*) wished, and they are all nearly out of water, as you know."[62] The papyri also suggest that rather than flowing in force continuously, the border canals received an agriculturally exploitable influx only during the flood. The impending arrival of the waters is mentioned in a letter from the third-century CE Heroninos archive, papers possessed by the manager of the Theadelphian division (*phrontis*) of a large, dispersed Fayyūm estate owned by a wealthy Alexandrian named Aurelius Appianus.[63] On 10 August of 252 CE or 255, Heroninos is ordered to drag in some tree trunks from the fields "before the waters" (*pro tōn hydatōn*)—that is, before the arrival of the flood.[64] The fevered activity surrounding the arrival of the water is hinted at in another letter from the Heroninos archive dated to 19 July 257 CE. Heroninos is here ordered to oversee the preparation of local waterwheels so that "until the canals do not have water there is a *probolē* ("supply" or "reservoir") sufficient for our use."[65] Since the Theadelphian *phrontis* was heavily focused on viticulture, the full exploitation of ephemeral waterways was essential to the perennial irrigation of local vineyards. Concerns over the availability of floodwater are also attested in a petition from Philadelphia dated to 305 CE. The elders of the village of Philadelphia complain that the upstream village of Tanis had somehow been impeding the flow of their shared canal and they request an inspection of irrigation works in Tanis so that they might "be able to enjoy the growth of the flax crop, have drinking water (*potimon hydōr*), sow the plain (*pedion*) of our village, remain in our own village of record (*idia*), and have enjoyment of our

61. *P.Cair.Zen.* 1.59072.
62. *P.Ryl.* 2.81, ll. 5–8: [αἱ τ]ῶ̣ν̣ . . [.]. νων θύραι, ὡς οἶδας, ἐπὶ παρόντος σοῦ [ἐφ'] ὅσον οἱ κατασπορεῖς ἤθελον ἐβαστάχθησαν, [κα]ὶ γὰρ σχεδὸν πᾶσαι ἀφ' ὕδατους εἰσί, ὥσπερ οἶ̣δα̣ς̣. On the office of *katasporeus* see Bonneau, *Le régime administratif*, 168–73.
63. Rathbone, *Economic Rationalism*.
64. *SB* 6.9361 (252 or 255 CE).
65. *P.Flor.* 2.153, ll. 8–13: ἵνα ἕως αἱ διώρυγες ὕδωρ μὴ ἔχουσιν προβολὴ ἡ αὐτάρκ[ης] ἡμῖν τῇ ὑπηρε[σ]ίᾳ γένεται (*l.* γένηται). Cited from Rathbone, *Economic Rationalism*, 224. The document thus describes the filling of reservoirs by waterwheel for later use in the perennial irrigation of vineyards. For the meaning of *probolē* or *probolos* in the context of irrigation see Habermann, *Zur Wasserversorgung*, 153.

property."[66] Notably, the petition is dated to 1 Pachon (26 April), some four months before the beginning of the inundation, and its authors perhaps hoped that any necessary repairs would be accomplished during the remainder of the low-water season. But even during the flood the arrival of water uninterrupted was not a given. In *P.Sakaon* 45 (334 CE), Aurelius Sakaon of Theadelphia complains that during the "time of the waters" (*ton kairon tōn hydatōn*) several upstream cultivators had erected a transverse dike (*emblēma*) in their shared canal, thereby denying Theadelphians access to the flood. Yet even under normal circumstances water delivery to tail end villages may have been uncertain as suggested by the aforementioned *P.Fay.*131 from Euhemeria. The writer's conditional "if the water comes down" hardly indicates a substantial and reliable flow.[67] Unfortunately, it is not clear when during the year this letter was composed.

Canal flow in this part of the Fayyūm could also be exceedingly slow. An undated letter of the Heroninos archive orders its unnamed recipient to have a man named Kopres dam up (*emblēmatisai*) a canal in order to build up water for local holdings. Upon releasing the waters the following day (*apolythēnai*), they would then arrive at olive plantings (*elaiōnas*) somewhere downstream but only after a delay of five days.[68] That it should take a full day for a farmer at Theadelphia to build up a head of water sufficient to irrigate and a further five days for the water to reach the next villages along the canal—Euhemeria and Dionysias, 4.5 km and 15 km distant respectively—suggests both that there was very little water in the canal to begin with by the time it had reached Theadelphia and that its flow was consequently sluggish. The topographical impediments to the water supply of Theadelphia are also referenced in the Sakaon archive, where the village is described as lying "on high ground" (*en hypsēlois topois*) and "at the far end of the district" (*hēmas hysterous einai tou pagou*)—that is, at the end of the canal.[69] As I will discuss in the following section, the risk of water shortages and subsequent irrigation failure was also endemic near the ends of the Fayyūm's canal system, a socioenvironmental phenomenon that informed the agricultural regimes in more marginal settlements.

66. *P.Wisc.* 1.32 = *Pap. Choix* 27.
67. *P.Fay.* 131 (third–fourth century CE).
68. *P.Prag. NS* 52 = *SB* 6.9415 (249–68 CE).
69. *P.Sakaon* 35 (332 CE). On the topography of the village see Römer, *The Fayoum Survey Project*, 107–9.

While there is no papyrological evidence for canal flows in the center of the depression in antiquity, a clear distinction between the center and the margins is indicated by another Heroninos papyrus, a copy of a circular sent to the managers of every Fayyūm division of the Appianus estate containing instructions for the pruning and propagation of vines (266/67 CE). The letter's author Alypios, the general manager of the estate, differentiates between vineyards artificially irrigated by water-lifting (*antlētika ktēmata*), and vineyards "of the plain" (*ta de epipeda* [*ktēmata*])—that is, vineyards watered with or without animal-turned waterwheels.[70] The former are surely the vineyards of marginal villages like Theadelphia described above. The latter were in the central plain (*pedion*) and apparently required no mechanical assistance for perennial irrigation. Topography was key to the perennial irrigation of these vineyards of the plain. As Girard remarks in the *Description*, in the central Fayyūm the location of reservoirs above the level of the lands they served permitted the water to be delivered easily by gravity to the fields below.[71] According to al-Nābulusī, the flow of central canals was also strong enough to power stone sugarcane presses at Abū Ksā, Biyahmū, Dhāt al-Ṣafā᾽, Sinnūris, Fānū (broken during al-Nābulusī's inspection), and Sanhūr as well as water mills in Akhṣāṣ al-Ḥallāq, Bamawayh, Dhāt al-Ṣafā᾽, Sinnūris, Fānū, and Akhṣāṣ Abū῾Uṣayya. In the present day, water-powered waterwheels continue to be used in the central core of the alluvial plain where the slope of the terrain is most dramatic. And while water-delivery powered entirely by gravity is possible throughout much of the depression, marginal fields cannot be irrigated by gravity. Here, fields lie well above the level of the canals that serve them and water must be raised by gas or electric pumps (see figs. 15).[72]

Their more reliable and vigorous flow notwithstanding, even the canals of the center were not fully perennial. Our earliest source, Ibn ῾Abd al-Ḥakam, writes that the patriarch Joseph established Fayyūm irrigation in such a way that water was diverted back and forth between "alternating" (*muṭāṭiyya*) canals and so-called "elevated" or "upper-level" canals (*murtafi῾*) according to

70. *P.Flor.* 2.148 (266–67 CE).
71. Girard, "Mémoire sur l'agriculture," 16.
72. Price, "The Evolution of Irrigation." See also the short documentary by Raymond Collet, *La dernière sakieh du Fayoum*, which describes the last wooden animal-powered *sāqiya* still in its original place, though no longer in service, which was used for the irrigation of land located two meters above the level of its adjacent canal, accessed 19 May 2021, https://www.cealex.org/ressources-documentaires/videotheque/sakieh-fayoum.

an hourly schedule, day and night.[73] Al-Masʿūdī clarifies that the Fayyūm's canals were divided into three categories: "elevated" (*murtafiʿ*), "alternating" (*muṭāṭīʾ*), and so-called *muṭāṭīʾ al-muṭāṭīʾ*, the latter being "a term used by the Egyptians, meaning the lower (*al-munkhafiḍ*)."[74] The system therefore comprised canals that irrigated high-lying land, canals for lower-lying land, and canals subject to a schedule of opening and closing. Following Ibn ʿAbd al-Ḥakam, it would seem that water was diverted back and forth between the alternating and upper-level canals in the attempt to ensure that canals irrigating higher lands received a supply during the low-water months. That such a system was already regarded as ancient in the ninth century CE suggests that Arabic authors are describing a practice with roots in antiquity.[75]

In Abū Isḥāq's account, a number of the Fayyūm's principal waterways were regulated on a fixed schedule between Hatūr to Barmūda on the Coptic calendar (late November–late April), months during which the water level was progressively falling. The heads of these "alternating" canals (*muṭāṭiyya*) were regulated by one or more gates (*abwāb*) dating to the time of Joseph (*Yūsufiyya*), which were opened and closed in rotation. The gates of the Baḥr Tanabṭawa, for instance, were closed from 10 Hatūr to the end of the month (20 November–9 December). They were then opened on 1 Kihak for twenty days (10–30 December) then closed until Epiphany (*Laylat al-Ghiṭās*, 6 January), after which they were again opened until the end of Ṭūba (7 February). They were then closed from 1–20 Amshīr (8–28 February) and reopened until 10 Baramhat (20 March). At this point Abū Isḥāq says that the gates were opened until 10 Baramūda and left in place thereafter.[76] The schedule was broadly similar along the so-called *khalīj* Tabdūd (*l.* Tandūd). This canal also possessed a freshwater spring (*ʿayn ḥulwa*) from which nearby lands were irrigated even when the canal was closed. The spring had appeared at a time when canal water had been scarce (*ʿudima al-māʾ*) at which point it had been dug out to create a well. This canal followed the same schedule as the Tanabṭāwa until late in the year, when it was reopened from 20 Amshīr (28 February) until 20 Baramhāt (29 March). It was then reopened thirty days later on 20 Baramūda (28 April) and apparently left open for the rest of the year.[77] These practices survived into

73. Ibn ʿAbd al-Ḥakam, *Futūḥ Miṣr*, 16.
74. Al-Masʿūdī, *Murūj*, 2:369–70. Citations and translation in Rapoport and Shahar, "Irrigation," 17–18.
75. Rapoport and Shahar, "Irrigation," 17; Ibn ʿAbd al-Ḥakam, *Futūḥ*, 16.
76. Al-Maqrīzī, *Khiṭaṭ*, 1:672.
77. Al-Maqrīzī, *Khiṭaṭ*, 1:674.

the modern period, since Evliya Çelebi impressionistically refers to the alternating breaching and sealing of the levees on each of the canals that branched from the Baḥr Yūsuf. When the water level in the channel had reached a specified height, he writes, levees were breached and the Fayyūm was flooded. After the water had dried up, the levees were once again opened to release water into the system.[78] Jomard later records that each canal issuing from the end of the Baḥr Yūsuf was governed at its mouth by a gate (*porte*) that could be raised and lowered in accordance with the needs of their dependent villages.[79] Both Jomard and Çelebi insist that water was equitably shared, though disputes arose when customary rules (Çelebi's *kanun*, Jomard's *usages*) governing water sharing were violated.

The system to which Ibn ʿAbd al-Ḥakam alludes and which Abū Isḥāq describes seems to have been an attempt to ensure that at least some water was reserved for the "elevated" canals during the time in which the Fayyūm was disconnected from the flow of the Baḥr Yūsuf and the aggregate water supply was consequently reduced. Still, that the gates of alternating canals were simply left open from the end of April/Baramūda suggests that the waters ran low late in the year, a situation that would have primarily affected villages on higher ground or toward the ends of canals. In consequence, the water regime of a canal and a village's position along it will have exerted a strong influence on the agricultural potential of Fayyūm settlements. Comparison between the papyri and al-Nābulusī indeed confirms that the hydrology and hydraulics of Fayyūm canals as well as a settlement's position along its waterway—whether closer to its head or to its tail—deeply influenced agricultural practices throughout the depression.

Canal Flow and Cropping Patterns

It has long been acknowledged that the papyri preserve no evidence for regular double-cropping along the Fayyūm's margins. As the previous section argued, the border canals delivered water less plentifully and effectively in the dry season, making it impossible to irrigate fields here perennially, except those equipped with reservoirs that had been filled during the flood. While the sow-

78. Dankoff, Tezcan, and Sheridan, *Ottoman Explorations*, 352–53.
79. Jomard, "Description des antiquités du nome Arsinoïte," 448.

ing of a second, summer wheat (a "three-month wheat," *trimēnon pyron*) is attested in Philadelphia during the reign of Ptolemy II, it is described as requiring irrigation "by hand" (*apo cheiros*) or, failing that, by as many *shādūfs* (*kēlōneia*) as needed.[80] There are also several references in Roman papyri from Tebtynis to land given over to double-cropping in wheat (*eis sporan kai episporan*, "for sowing and resowing") but these are rare exceptions.[81] Even on the division of the large estate of the Alexandrian Aurelius Appianus at Theadelphia in the third century CE there is only one clear attestation of double-cropping.[82] Indeed, lessees of agricultural land along the Greek and Roman margins generally paid rent to landowners and state taxes in kind (wheat and barley) during the harvest month of Payni, indicative of a single yearly crop.[83]

The design of fields along the margins was thus adapted to traditional flood-recession irrigation and plots were embanked by ring dikes (*perichōmata*), which transformed each field into a small inundation basin.[84] Ring dikes are frequently mentioned in Fayyūm land leases in the clauses that enumerate the duties of the lessee.[85] In these texts, the act of flood-recession irrigation (*potismos*) is regularly paired with the responsibility to maintain the dikes surrounding the fields (*chōmatismos* or *perichōmatismos*) and occasionally to build and maintain *emblēmata*, which diverted water from a feeder canal onto a nearby field.[86] The texts of such agreements accordingly structure the agricultural year into four distinct periods of field- and canal-preparation; irrigation; growth; and the harvest of a single field crop. The steps involved are clearly if tersely enumerated in an early Roman petition in which a lessee reports that after taking possession of twenty-four *arourai* of public land, he subsequently embanked it (*chōmatisas*) and then began to flood it (*apo merous limnasas*) for the sowing (*katasporas*) of the current year.[87] That the margins of the Fayyūm

80. *P.Cair.Zen.* 2.59155 = *SB* 3.6733 (256 BCE). Cited from Monson, "Salinization," 134.
81. Monson, "Salinization," 133–34.
82. Rathbone, *Economic Rationalism*, 235, citing *P.Flor.* 2.194 (259 CE).
83. E.g., *SB* 6.9110 (26 CE); *SB* 16.12539 (26 CE); *P.Mich.* 12.633 (ca. 30 CE). Bagnall, *Egypt in Late Antiquity*, 22.
84. See Rathbone, *Economic Rationalism*, 119–228, for basin irrigation on the Appianos estate.
85. Customary duties other than those specified in the texts are simply referred to as τ[ὰ ἄλλα π]άντα ὅσα καθήκι (*l.* καθήκει) in *P.Cairo Isid.* 101 (300 CE), *P.Gen.* 1.78 (third century CE?); τὰ ἄλλα ὅσα καθήκει: *P.Tebt.* 2.378 (265 CE), and *SB* 6.9269 (297 CE).
86. The pairing of περιχωματισμούς and ποτισμούς appears in numerous other texts throughout antiquity. See e.g., *SB* 16.13017 (24 BCE); *P.Mich.* 12.633 (30 CE?); *P.Tebt.* 2.378 (265 CE). In *P.Tebt.* 1.105 (103 BCE) the lessee is to return the land to the lessor free of rushes and weeds as well as fully embanked (κεχωματισμένην).
87. *P.Mert.* 1.3 (3 CE), ll. 11–15: ὧν (sc. ἀρουρῶν) καὶ ἀντιλαβόμενος καὶ χωματίσας καὶ ἀπὸ μέρους λιμνάσας ταύτας τῆς εἰς τὸ αὐτὸ ἔτος κατασπορᾶς.

were originally designed for this traditional form of Egyptian irrigation rather than perennial irrigation is also evident in early Ptolemaic papyri, most importantly *P.Lille* 1.1 (258 BCE). The document preserves an illustration of a 10,000-*aroura* gift-estate (*dorea*) at Philadelphia granted by Ptolemy II to his finance minister (*dioikētēs*) Apollonios, the whole of which was subdivided into forty smaller plots of 250 *ar.* each, dubbed *perichōmata.*[88]

The water supply was nonetheless insecure toward the tail ends of the canal system. The risk of lands being unflooded (*abrochos*) and correspondingly uncultivable was therefore endemic. Frequently accounted for in land leases of both Ptolemaic and Roman date, such failures of the water supply entitled the lessee to a release from the rent.[89] The problem is visible in the land surveys conducted by Menchēs, scribe of the southern village of Kerkeosiris in the late second century BCE. Here, derelict land (*hypologos*) was categorized as either unflooded over a long period of time (*chērsos*), underwater—that is, flooded but not drained (*embrochos*), or salted (*halmyris*).[90] Beginning in the middle of the second century CE, flood-failure is evidenced at the field level in a new document type, the so-called *abrochia*-declaration: official attestations that one's land was unflooded and thus eligible for a remission of annual imposts.[91] It nonetheless remains all but impossible to assess the prevalence of water shortages in downstream villages, since we lack year on year records of the water supply from any Fayyūm village. A fourth-century CE partial land survey from Karanis nonetheless suggests that a marginal village's cultivable territory might be a veritable patchwork of both flooded and unflooded fields. The fragmentary register published as *P.Cair.Isid.* 6 (300–305 CE) contains a list of landholders in Karanis and the plots of private or state-owned land that they farm, most of them quite small. Grain-bearing *arourai* reported in the survey are classified either as sown (*sporimē*), irrigated and productive, or as unflooded (*abrochos*), normally productive but unflooded at the time of the survey. While numerous lacunae make precise figures irrecoverable, of the roughly 762 *arourai* of grain-land clearly attested just under 195 *ar.* (25.5 percent) are classified

88. Thompson, "Irrigation and Drainage," 111 and 118–20. See also *P.Col.* 3.54 (256–255 BCE) for reference to 100 *arourai* of "seed land" (γῆς σπορίμου) in a περίχωμα at Philadelphia. On gift-estates (*doreai*) see Andrew Monson, "Dorea," in *The Encyclopedia of Ancient History*, ed. Roger S. Bagnall (London: Wiley-Blackwell, 2012).

89. E.g., *P.Yale* 1.51 (184 BCE); *P.Tebt.* 1.106 (101 BCE); *P.Ryl.* 4.601 (26 BCE); SB 16.13017 (24 BCE); *P.Tebt.* 2.374 (131 CE).

90. Verhoogt, *Menches*, 112–13.

91. Habermann, "Aspekte des Bewässerungswesens."

as *abrochos*. While it would be imprudent to place too much interpretive weight on such a small sample, especially one whose representativeness cannot be determined, this landscape scarcely resembles that of the central Fayyūm, as we will see further below.

Yet while flood-irrigation remained the primary means by which fields were irrigated, additional water could be raised from canals, reservoirs, and wells by waterwheels (Greek *mēchanai*, Arabic *sawāqī*). Yet such expensive machinery was restricted to the wealthy, such as the Appianus estate mentioned above.[92] Demanding substantial investment in materials (rare or even imported woods,[93] metal, ceramic pots) in addition to construction and maintenance, waterwheels also required one or two animals to provide power, food and water for those animals, one or more attendants to monitor the work, a constant supply of replacement parts, and eventually new animals. The expense and effort involved in procuring or maintaining a waterwheel are documented in another papyrus of the Appianus estate, which records forty-eight days of labor for the manufacture of one 5.5 cubit (2.9 m) component, probably the *tympanon*, the wheel that held and turned the long garland of pots constantly lowered into and raised from the well or reservoir.[94] Other texts from the archive detail the types of work done by carpenters and their wages, all tasks ranging from two to five days of labor, the latter earning the workman a sizable cash wage.[95] Single owners of one or more waterwheels are thus certain to have been the proprietors of large estates or at least relatively wealthy farmers who could finance the purchase, operation, and maintenance of a wheel out of their own capital.[96] The majority of the Fayyūm's farmers, smallholders and lessees alike, likely irrigated only at the time of the flood, though this does not rule out the possibility of shared use of communally owned waterwheels, a phenomenon well documented in later periods.[97]

Still more marginal, however, were the *drymoi* (sing. *drymos*, "marshes"), which could be put under wetland crops such as rushes, reeds, or papyrus. Located at various points along the edges of the depression, *drymoi* were ini-

---

92. Eyre, "The Water Regime."
93. On the use of wood, at times imported, in irrigation machinery in the Ottoman period see Mikhail, *Nature and Empire*, 124–69.
94. *P.Prag. Gr.* 1.44 recto, ii 1–12.
95. Rathbone, "*Mēchanai* (Waterwheels) in the Roman Fayyum," 256–59.
96. *Pace* Malouta and Wilson, "Mechanical Irrigation," we see more intensive use of these machines.
97. The *sāqiya* described in the documentary referenced at n. 72 above was communally owned and farmers took turns using it each with their own draught animal. See also chapter 4 below for the fictional account from al-Sharqāwī, *Al-Arḍ*.

tially the remains of natural marshlands along the shores of the ancient lake but by the Graeco-Roman period they no longer received water naturally and required artificial irrigation to be cultivated.[98] Indeed, the lack of irrigation might render them entirely uncultivable. In a petition of the year 144 CE one Ptolemais, lessee of a state-owned marsh at Theadelphia, complained that the customary amount of water had not been provided by local officials, thereby threatening his *drymos* with desiccation and unproductivity.[99] In 145/46 CE, Ptolemais drafted a similar petition, which supplies further insight into the water supply of such marshes. In this later text, Ptolemais claimed that a local official had failed to supply water to the *drymos* when the Nile was still at its height (*akmazontos tou Neilou*) and before the water had been directed to local fields (*edaphē*). The marsh eventually received some water later in the year, yet Ptolemais complains that it had come "from drainage" (*apo apochymatōn*) and was thus of little quantity and also contained no *gonos* ("seed") or *mētra* ("womb").[100] The perplexing terminology here suggests folk-wisdom concerning the fertility and productivity of silt-laden Nile floodwater. Indeed, if Ptolemais in this year received only drainage from local inundation basins such water would have been largely devoid of fertilizing silt and otherwise contaminated with salt and other minerals leeched from the soil of other fields.[101]

It goes without saying that this single-cropped landscape scarcely resembles the perennial, paradisiacal garden so memorably described by later Arabic authors. This was a function of greater water availability in the center, the portion of the depression depicted by the observers enumerated in the Introduction. Here, the double-cropping referred to by authors like al-Bakrī is well represented. Al-Nābulusī's survey contains 26 entries in which villages are explicitly said to have practiced double-cropping—that is, the cultivation of both a flood-irrigated winter field crop (*shatawī*) and a summer field crop (*ṣayfī*). While these settlements amount to just under 21 percent of his 124 separate village entries, it should be noted that "cultivation" (*zaraʿ*) in al-Nābulusī refers to field crops, principally wheat (*qamḥ*) and barley (*shaʿīr*), but also broad beans (*fūl*) and sometimes chickling vetch/grass pea (*jullubān*). A village may thus have produced only a winter field crop or even little field produce altogether yet still have been under heavy, perennial cultivation. The descrip-

98. Bonneau, "Le *drymos* (δρυμός), marais du Fayoum."
99. *P.Wisc.* 1.34.
100. *P.Mich.* 11.617 (145–46 CE).
101. On irrigating with drainage in the contemporary Fayyūm see Barnes, *Cultivating the Nile*, 162–67.

Table 2.1. Explicit attestations of double-cropping (*shatawī* and *ṣayfī*) in al-Nābulusī

Al-ʿIdwa (*VF* 69)	Dimūh al-Lāhūn (*VF* 153)
Ibrīziyā and al-Zarbī (*VF 74*)	Dhāt al-Ṣafā'(*VF* 155)
Abhīt (*VF* 76)	Sinnūris (*VF* 159)
Al-Qubarā' (*VF* 80)	Sirsinā (*VF* 165)
Al-Rūbiyyūn (*VF* 105)	Sīnarū (*VF* 170)
Biyahmū (*VF* 113)	Shisfa (*VF* 174)
Tuṭūn (*VF* 134)	Fanū and Naqalīfa (*VF* 190–91)
Jarfis (*VF* 137)	Qambashā (*VF* 199)
Khawr al-Rammād (*VF* 141)	Munsha'at Ibn Kurdī (*VF* 208)
Kharāb Jundī and al-Maṣlūb (*VF* 142)	Munsha'at al-Ṭawāḥīn (*VF* 210)
Dumūshiyya (*VF* 145)	Minyat al-Baṭs (*VF* 226)
Difidnū (*VF* 148)	Minyat al-Dīk, Banū Majnūn, Shalmaṣ (*VF* 228)
The Dinfāras of Jardū and Ihrīt (*VF* 150)	Hayshat Dumūshiyya (*VF* 236)

tion of Akhsāṣ al-Ḥallāq near Sinnūris is telling. Heavily invested in perennially irrigated, water-intensive pears, apples, grapes, and roses, the village sold considerable produce in the Fayyūm's capital as well as Bahnasā (ancient Oxyrhynchus) and other major urban centers of Fusṭāṭ, Cairo, Damietta, and Alexandria. Al-Nābulusī even compares its landscape to the orchards of Damascus thanks to the shade of the trees and the canals and streams that flowed ceaselessly night and day. Such abundance notwithstanding, no field crop is recorded.[102] It should also be noted that al-Nābulusī often fails to mention whether or not a village's field crop was produced in a single or double harvest. The village of Miṭr Ṭāris, for instance—a "bride among the brides of Fayyūm"—contained considerable orchards and vineyards, streams flowing day and night, and produced a substantial cereal harvest. Al-Nābulusī nevertheless does not mention whether or not this field cultivation was perennial or merely *shatawī*.[103] The number of villages under both winter and summer field cultivation may therefore have been higher than the survey suggests.

Of the twenty-three villages in which al-Nābulusī states that only a *shatawī* crop was grown, eight cultivated considerable additional water-intensive produce, for example, orchard crops. A further fifteen settlements, however, had only scattered ancillary cultivation.

Tellingly, the entries for Dimūh al-Dāthir, Ṭimā, and the joint settlements

102. Al-Nābulusī, *VF*, 78.
103. Al-Nābulusī, *VF*, 217–18.

Table 2.2. Winter field-crops and extensive additional cultivation from al-Nābulusī

Village Name	Cultivation patterns
Jardū (*VF* 139)	Winter cereals (*al-ḥubūb al-shatawiyya*), dates, vine, grapes, acacia, sycamore.
Khawr al-Rammād (*VF* 141)	Winter cereals, acacia, sugarcane, vegetables, and waterwheels.
Shallāla (*VF* 176)	Winter wheat, barley, broad beans. Dates, figs, and other trees.
Shidmūh (*VF* 180)	Dates, vine, plantations, acacia, winter crops. Summer crops were cultivated until sugarcane increased.
Ṣunūfar (*VF* 182)	Winter crops, dates, trees, sycamores, many orchards.
Ṭubhār (*VF* 185)	Winter crops and nothing else (*al-shatawī lā ghayr*), orchards, vineyards, dates, figs.
Fidimīn (*VF* 197)	Winter crops, dates, figs, olives, sugarcane.
Minyat Karbīs, and Akhṣāṣ Abū ʿUṣayya (*VF* 205–6)	Minyat Karbīs: Dates, carob, sycamore, figs, orchards. Akhṣāṣ Abū ʿUṣayya: Watered in the winter from the nearby Fanū canal. One orchard irrigated by waterwheel.

of Babīj Ghaylān and Kawm al-Raml are described as being irrigated like "the countryside" (*al-rīf*, referring to the Nile Valley), meaning that they were watered only during the flood. Clearly, then, aggregate water availability had a significant effect on cropping patterns in these villages. Al-Nābulusī further specifies that a number of settlements received water only during "the days of the Nile" (*ayyām al-Nīl*—that is, the flood) and thus irrigated their fields with the "irrigation of the [Nile Valley] countryside" (*rayy al-rīf*) rather than with the combination of canals and waterwheels peculiar to the Fayyūm (*saqī*). Although he does not always state that these flood-irrigated settlements cultivated only a winter crop, lack of perennial water would have made it impossible to irrigate a summer crop on a significant scale.

An additional proxy for water-stress (and perhaps also soil salinity, on which see further below) is the predominance of barley, which requires less water than wheat and tolerates saltier soils. As Yossef Rapoport notes, most of the villages described by al-Nābulusī seem to have hewed as closely as possible to a 2:1 ratio of wheat to barley. Yet a number of settlements produced a preponderance of barley and little to no wheat or other field crops, these figures expressed as a percentage of their assessed land tax (*kharāj*). Although al-Nābulusī does not record the water supply of all of these settlements in detail, cultivation in several was clearly influenced by localized water pressure, particularly in the villages along the Sharqiyya canal—al-Rubiyyāt/Maqtūl, Sīla,

Table 2.3. Winter field-crops and minimal/no additional cultivation from al-Nābulusī

Village Name	Cultivation patterns
Al-Ṭārima (*VF* 93)	Young dates and water specifically for winter crops (*al-shatawī khāṣa*).
Al-Qalhanā (*VF* 101)	Winter crops only, sugarcane sown in nearby Dumūshiyya.
Bilāla (*VF* 110)	Dates and sugarcane.
Babīj Ghaylān and Kawm al-Raml (*VF* 130)	Winter crops and flax cultivated like the Egyptian countryside (*rīf*).
Tuṭūn (*VF* 134)	Winter cultivation and some cotton.
Disya (*VF* 143)	Dates, *sidr* (Christ's thorn jujube), and acacia but winter crops only (*al-shatawī wa-lu ghayr*).
Dimūh al-Dāthir ("the extinct", *VF* 152)	Some land irrigated by the flood like the countryside, some by *saqī* like the Fayyūm. Winter cultivation only, no trees, dates, vine, gardens, or orchards. Recently reoccupied after abandonment.
Sīla (*VF* 168)	Dates and acacia, water specifically for winter crops (*al-shatawī khāṣa*).
Sīnarū (*VF* 170)	Winter crops, some sugarcane. Vineyards failed for lack of water.
Shāna (*VF* 178)	Winter cereals, primarily barley but no summer crops (*laysa li-hā ṣayfī*).
Shushhā (*VF* 179)	Winter cereals, a few date palms, no trees or vineyards.
Ṭimā (*VF* 183)	Winter cereals cultivated with the "irrigation of the countryside from the Nile" (*rayy al-rīf min al-Nīl*), not canal-irrigated (*saqī*) like the Fayyūm.
Ṭubhār (*VF* 185)	Only winter crops. Some orchards and vine irrigated from sources elsewhere during the low-water season. See further in Table 2.5.
ʿAnz (*VF* 188)	Winter cultivation only (*al-shatawī lā ghayr*).
Qushūsh (*VF* 202)	Winter cultivation only (*al-shatawī khāṣa*), dates and *sidr*.

Bayaḍ, and Shāna—of which only the first maintained a 50/50 split between wheat and barley and thus does not appear in the table below. A few villages are also said to cultivate only field crops (*zaraʿ*) and nothing else: Būr Sīnarū, Bandīq, Babīj Andīr, Tirsā, and Ḥaddāda. All but Būr Sīnarū already appear somewhere in Tables 2.1–2.5.

Finally, al-Nābulusī notes a small handful of settlements that suffered from some form of water stress in the low-water season. Although it is not always explicit, the common feature among this final set of villages was their elevation or location toward the end of their canal(s). Both situations placed such settlements at a comparative disadvantage during the low-water season.

Although our ancient and medieval evidence differs in character and emphasis, both corpora demonstrate that the hydraulic regime of the Fayyūm's

Table 2.4. Villages watered only during the inundation in al-Nābulusī

Village Name	Irrigation and Cultivation Patterns
Al-Lāhūn and Umm al-Nakhārīr (*VF* 95)[1]	Al-Lāhūn: Watered during the days of the Nile "like the villages of the countryside without [perennial] irrigation after the fashion of the Fayyūm" (*uswatan bilād al-kūra min ghayr saqī ʿalā ʿādat al-Fayyūm*). Umm al-Nakhārīr: "Its lands are watered like the villages of the countryside" (*turwā arāḍīhā uswatan bilād al-rīf*).
Al-Ḥammām (*VF* 97)[2]	Watered by a canal fed during the "days of the Nile [flood]" (*ayyām al-Nīl*) from a canal running east from the dam (*bunyān*) at al-Lāhūn, like the irrigation of the countryside (*rayy rīf*). Flax and field crops.
Al-Haysha, attached to al-Lāhūn (*VF* 99)[3]	Cultivates with the water of the Nile (*māʾ al-Nīl*).
Bāja (*VF* 109)	Watered during the days of the Nile by a canal shared with Minyat al-Usquf (see below). No significant field crops, some sugarcane and vegetables watered by wheel.
Bandīq (*VF* 129)	Watered from the Waradān canal during the "days of the Nile as the countryside is watered, not like the Fayyūm" (*turwā ayyām al-Nīl ka-mā turwā al-rīf ghayr al-Fayyūm*). Field crops only.
Dumūshiyya (*VF* 145)	Cultivates both winter and summer crops but possesses one meadow (*marj*) that cultivates flax, cucumber, wheat, and barley "with the water of the Nile just as the countryside is cultivated" (*bi-māʾ al-Nīl ka-mā tuzraʿ al-rīf*).
Dimūh al-Dāthir (*VF* 152)	Some of its lands (*baʿḍ arāḍīhā*) are cultivated like the countryside (*ka-l-rīf*) and partially by perennial irrigation (*saqī*) like the lands of the Fayyūm (*k-arāḍī al-Fayyūm*).
Sidmant (*VF* 173)	"Watered by the Nile [flood] and cultivated as the countryside is cultivated" (*turwā min al-Nīl wa tuzraʿ ka-mā yuzraʿ al-rīf*).
Ṭimā (*VF* 183)	Cultivates winter cereals (*al-ḥubūb al-shatawiyya*) and irrigates with the "[flood] irrigation of the countryside from the Nile, not by [canal] irrigation like the lands of Fayyūm" (*tuzraʿ rayy al-rīf min al-Nīl lā min saqī k-arāḍī al-Fayyūm*).
Minyat al-Usquf (*VF* 205)	Watered during the flood (*ayyām al-Nīl*) by a canal shared with Bāja.
Nāmūsatayn (*VF* 234)	Watered with "the irrigation of the countryside" (*rayy al-rīf*).
Hawwārat Dumūshiyya (*VF* 235)	Watered by the Nile (*turwā bi-māʾ al-Nīl*). Field crops.
Hawwāra al-Baḥriyya (*VF* 238)	Watered by the Nile (*turwā bi-māʾ al-Nīl*). Field crops, date palm shoots, *sidr*, figs, and sycamore.

[1] Technically exterior to the Fayyūm depression.
[2] See previous note.
[3] See previous note.

Table 2.5. Villages with a predominance of barley in al-Nābulusī

Village Name	Irrigation and Cultivation Practices
Al-Lāhūn and Umm al-Nakhārīr (*VF* 95)[1]	Al-Lāhūn: Watered during the days of the Nile "like the villages of the countryside without [perennial] irrigation after the fashion of the Fayyūm" (*uswatan bilād al-kūra min ghair saqī ʿalā ʿadatan al-Fayyūm*). Umm al-Nakhārīr: "Its lands are watered like the villages of the countryside" (*turwā arāḍīhā uswatan bilād al-rīf*).
Al-Ḥammām (*VF* 97)[2]	Watered by a canal fed during the "days of the Nile [flood]" (*ayyām al-Nīl*) from a canal running east from the dam (*bunyān*) at al-Lahūn, like the irrigation of the countryside (*rayy rīf*). Flax and field crops.
Al-Haysha, attached to al-Lāhūn (*VF* 99)[3]	Cultivates with the water of the Nile (*māʾ al-Nīl*).
Bāja (*VF* 109)	Watered during the days of the Nile by a canal shared with Minyat al-Usquf (see below). No significant field crops, some sugarcane and vegetables watered by wheel.
Bandīq (*VF* 129)	Watered from the Waradān canal during the "days of the Nile as the countryside is watered, not like the Fayyūm" (*turwā ayyām al-Nīl ka-mā turwā al-rīf ghayr al-Fayyūm*). Field crops only.
Dumūshiyya (*VF* 145)	Cultivates both winter and summer crops but possesses one meadow (*marj*) that cultivates flax, cucumber, wheat, and barley "with the water of the Nile just as the countryside is cultivated" (*bi-māʾ al-Nīl ka-mā tuzraʿ al-rīf*).
Dimūh al-Dāthir (*VF* 152)	Some of its lands (*baʿḍ arāḍīhā*) are cultivated like the countryside (*ka-l-rīf*) and partially by perennial irrigation (*saqī*) like the lands of the Fayyūm (*k-arāḍī al-Fayyūm*).
Sidmant (*VF* 173)	"Watered by the Nile [flood] and cultivated as the countryside is cultivated" (*turwā min al-Nīl wa tuzraʿ ka-mā yuzraʿ al-rīf*).
Ṭimā (*VF* 183)	Cultivates winter cereals (*al-ḥubūb al-shatawiyya*) and irrigates with the "[flood] irrigation of the countryside from the Nile, not by [canal] irrigation like the lands of Fayyūm" (*tuzraʿ rayy al-rīf min al-Nīl lā min saqī k-arāḍī al-Fayyūm*).
Minyat al-Usquf (*VF* 205)	Watered during the flood (*ayyām al-Nīl*) by a canal shared with Bāja.
Nāmūsatayn (*VF* 234)	Watered with "the irrigation of the countryside" (*rayy al-rīf*).
Hawwārat Dumūshiyya (*VF* 235)	Watered by the Nile (*turwā bi-māʾ al-Nīl*). Field crops.
Hawwāra al-Baḥriyya (*VF* 238)	Watered by the Nile (*turwā bi-māʾ al-Nīl*). Field crops, date palm shoots, *sidr*, figs, and sycamore.

[1] Technically exterior to the Fayyūm depression.

[2] See previous note.

[3] See previous note.

Table 2.6. Villages with periodic water stress in al-Nābulusī

Village	Irrigation and Cultivation Patterns
Ibshāyat al-Rummān (*VF* 91)	On the extreme western edge of the Fayyūm; people drink from a spring-water well (*b'ir ma'īn*) in the summer (*al-ṣayf*) when the canal water (*mā' al-baḥr*) recedes from them because of its scantiness (*li-qillatihi*).
Bāja (*VF* 109)	Receives water raised by animals ("on cattle's necks," *al-a'nāq al-abqār*) after the flood. No significant field crops; orchards, trees, sugarcane, and vegetables all watered by wheels.
Dimashqīn al-Baṣl (*VF* 151)	In "the days of summer" (*ayyām al-ṣayf*) water delivered "on cattle's necks." Field crops and flax irrigated by the Nile.
Ṭubhār (*VF* 185)	See Table 2.3 above. Lands are elevated (*'ālīya*) and water reaches it only with effort (*bi-kulfa*). Becomes scanty during the recession of the flood (*iḥtirāq al-mā'*, "the burning up of the water") and may dry up altogether, at which time people drink from wells.
Abū 'Uṣayya (*VF* 205)	People drink from wells when water is cut off (*inqiṭa' al-mā'*). Some field crops.

canal system had a profound effect on irrigation and agricultural practices. It is indeed remarkable how closely the cultivation patterns of medieval central villages irrigated only by the Nile flood resemble those of the ancient margins, whose canals likewise seem to have provided abundant water only during the inundation and which were therefore able to produce only a single annual field crop. The comparably greater water availability throughout the center, however, helped to support perennial cultivation of both field and fruit crops on a wide scale, enough to prove that the effusive praise of Arabic authors and early modern visitors was firmly grounded in agroenvironmental reality. It remains, however, to discuss briefly an issue that has only recently attracted scholarly attention: the connection between artificial irrigation and soil salinity.

Salinity and Marginality

While artificial irrigation increases agricultural productivity, it often simultaneously increases soil salinity, particularly in arid regions with high evaporation rates. Traditional Egyptian flood-recession irrigation practices retarded salinization by first soaking the soil to dissolve salts and minerals and then by returning the mineral-impregnated water to the river upon drainage. Yet if a basin was insufficiently filled and its water allowed to evaporate before drainage, a visible salt crust would be left on the surface along with a higher but usually invisible salt content in the upper horizons of the soil, both of which would diminish the productive potential of the field. European scientists frequently addressed this problem during the British colonial period. In 1887 the engineer Julien Barois noted that poorly drained fields showed "a whitish efflorescence and true salt deposits, which render any cultivation at such places impossible." Brackish infiltration from nearby canals and drains was another source of increased soil salinity. Thus, Barois continued, farmers throughout the country often preferred that their feeder canals lie at a lower level than their fields. This, of course, required such farmers to raise water mechanically but that was a small price to pay to avoid the "fatal action of infiltration water," which steadily reduced the productivity of a field by increasing its overall salinity.[104]

The problem of salinity became more salient following the introduction of modern perennial irrigation and the subsequent exponential increase of the water supply and the concomitant demand for artificial fertilizer.[105] Still, the Nile's water remains very fresh—containing <1000 mg/liter of total dissolved solids (TDS). At Aswan current TDS levels reach only ca. 150 mg/liter and a mere 250 mg/liter at Cairo 950 km downstream.[106] At the beginning of the twentieth century, chemist Alfred Lucas investigated the soil and water of the Fayyūm and remarked that the water quality was quite high, with only one sample—obtained at the tail end of a major drain—containing soluble materials in quantities sufficient to render it unsuitable for agricultural use.[107] The floodwaters that entered the Fayyūm in antiquity were thus surely of similar purity. Nevertheless, the imposition of an irrigation regime in this enclosed arid basin assured a steady if protracted build-up of salts. The degree to which this phenomenon affected the Fayyūm in antiquity, particularly along the more susceptible margins, is therefore of significant interest. Papyrologist and historian Andrew Monson has recently argued that the outer rim of the Fayyūm was characterized by relatively high marginality and low-to-average productivity

104. Barois, *Irrigation in Egypt*, 19–20.
105. Fertilizer: Derr, *The Lived Nile*, 6.
106. Kotb et al., "Soil Salinity in the Nile Delta," 247.
107. Lucas defines "unsuitable" water as having 300 parts soluble materials (of which ~50 percent are salts) per 100,000 parts water: Lucas, *A Preliminary Investigation*, 9.

in antiquity.[108] High marginality may also have contributed to the comparably higher proportion of state land in the Fayyūm, a reflection of the relative undesirability of much of its agricultural area. These agroenvironmental conditions may in turn have informed the occasional reshuffling of public land (*diamisthōsis*) among public tenants, a social practice that helped to share the risks of cultivating marginal lands more equally among the community of public tenant farmers.[109] Among the reasons for this lower productivity was higher than average soil salinity, a problem that he suggests became worse over time, resulting in progressively lower yields and thus contributing to the eventual abandonment of these settlements at the end of antiquity.[110] Unfortunately the attestations of salinity in the papyri are scattered and difficult to quantify and qualify. That the phenomenon was familiar to ancient farmers is beyond question. There are four attestations in the papyri of a small settlement near Herakleia somewhere in the vicinity of Pisais/Ibshawāy bearing the name Halmyra ("salted") or Halmyras *epoikion*.[111] While there is no further information about the settlement, the name suggests that it may have specialized in salt production. Jomard does indeed report considerable salt-production in this part of the Fayyūm, activity that continues in the present day on an industrial scale.[112] Jacotin's map also shows salt production (*saline*) in the south of the Fayyūm west of the Birkat Gharaq, the image likely representing a saltern, a set of basins in which salt was produced by evaporation. Al-Nābulusī likewise reports the existence of a disused saltern (*mallāḥa*) in Dumūshiyya (ancient Mouchis), which had been repurposed as a fishery. Water had formerly been raised for the saltern from a nearby spring-well, but the facility went out of use when the price of salt fell too low to cover expenses.[113] More frequent in the papyri, however, are attestations of land rendered under- or unproductive by salinity. The term *halmyris* ("salted land") appears periodically in the land surveys from the second-century BCE archive of Menches, village scribe of the southern settlement Kerkeosiris.[114] As Andrew Monson has shown, the archive docu-

108. Monson, *From the Ptolemies to the Romans*, chap. 3.
109. Monson, "Communal Agriculture." Cf. Rowlandson, "The Organization of Public Land."
110. Monson, "Salinization."
111. *P.Lond.* 3.901, 5[a] (75–125 CE); *BGU* 1.227 (second century CE); *BGU* 13.2242 (second century CE); and *BGU* 3.790 (198–99 CE). Possibly also in *SB* 1.5338 (fourth–seventh century CE) reading Ἀ[λμυ]ρᾶς.
112. Römer, *The Fayoum Survey Project*, 327–28; and Thompson, "The Fayum in the *Description de l'Égypte*," 167–70.
113. Al-Nābulusī, *VF*, 145.
114. E.g., *P.Tebt.* 1.60–64, 66, 74, 75, 83–85.

ments the dramatic increase in salted royal (*basilikē*, i.e., state-owned) land in the village during the latter half of the century. By 132 BCE, some 326 *arourai* were salted and a further 173 waterlogged. The total had risen to 595 *ar.* of salted land (25 percent of the royal land in the village) and 275 *ar.* of waterlogged land by 118 BCE.[115] Paradoxically, basin irrigation was a major culprit in some of the cases of increased salinity. Water overflow from various basins and drainage from nearby villages had inundated fields at Kerkeosiris causing much of the attested damage. Menches also reports the bursting (*ekptōma*) of a massive basin, the "great *perichōma*" in the nearby villages of Theogonis and Talithis (now Ṭalīt), which had swamped local fields and made them useless. In all, 20 percent of the village's land was reported as derelict and unproductive in this period due to various causes, salinity included.[116]

While there are no Roman-era sources of such comprehensive and granular detail as Menches' village land surveys, papyri of Roman date contain scattered attestations of salt land. A representative text is *P.Col.* 4.95 (mid-third century CE), possibly from Philadelphia. The text is a brief account of fifteen *arourai* of land, some of which produced fodder and of which five *ar.* were dry and salt crusted (*halismoi xērou*). The much longer land survey *P.Lond.* 2.267 (114 CE), records considerable amounts of salted and untaxed land (*halmē aphoros*) in Soknopaiou Nesos. Lying north of the lake in the desert, this village was in many ways unique and should not be regarded as representative.[117] "Dry salt land" (*chersalmē*) is also recorded near the village of Ibion Argaiou (*P.Oxy.* 6.918, second century CE) and in an unknown location in a text broadly dated to the first to fourth centuries CE (*SB* 14.11913). So also the short survey fragment *P.Strasb.* 8.788 from Theadelphia (157–58 CE), which notes the presence of "dry salt land" (*halmē kai chersos*) in the village. Yet the most complete description of both saline conditions and the methods by which they were remedied is *P.Hamb.* 1.12 of 209–10 CE, which preserves a survey of land described as 213 3/32 arable *ar.*, with 2 1/4 *ar.* devoted to a brick yard and 2 *ar.* to a threshing floor. The text epitomizes some sixteen years of irregularly performed surveys during which these 213 3/32 *ar.* were reclaimed from some 259 *ar.* of unproductive land. During the first year of the survey 263 13/16 *ar.* were assessed at a reduced rent: the 4 1/4 *ar.* devoted to the brickyard and threshing floor, as well as 44 21/64 *ar.* uncovered late by the flood, 186 1/4 *ar.* salted

115. Monson, "Salinization," 129.
116. Verhoogt, *Menches*, 120.
117. Provenance: Monson, "Salinization," 130n34.

(*chersalmē*) and 28 55/64 *ar.* dry over the long term (*chersos*). Roughly 79 percent, that is, was thoroughly uncultivable, 96 percent if the nearly 45 *ar.* uncovered late are included in the total. The results of the second year of the survey are lost in a lacuna and there is no mention of a survey in the third or fourth year. After the fifth–seventh years during which no surveys were performed, the surveys of the eighth to seventeenth years list the land as under water (*hyph' hydōr*)—that is, inundated to wash the soil of salt, before it was finally restored to productivity.

Since soil salinity tends to afflict lands that are either underwatered or poorly drained, it is at least possible that the problem was less acute in the central Fayyūm where water was more abundant and the steep terrain better facilitated drainage. Indeed, al-Nābulusī's only reference to salt is his description of the abandoned saltern at Dumūshiyya. Two rare papyri describing central villages in antiquity likewise hint at relatively low levels of salinity and marginality. *P.Bagnall* 9 (early second century BCE) is a register of unused land in five more centrally located villages: Kerkesoucha, Psenaryo (Sīnarū), Tanchoris, Ptolemais Hormou (al-Lāhūn), and Hauēris (Hawwāra). While the location of Tanchoris is unknown, Kerkesoucha seems to have been somewhere in the north/northeast and thus the most marginal.[118] The categories of unused land in these villages include not only salt land (*halmyris*) but also canals, roads, hills, waterlogged land (*embrochos*), dikes, rocky land, unwatered high lands (*abrochos hypsēlos*), long-term dry land (*chersos*) and several other categories. For four of the five villages the total amount of unused land is preserved almost in full: Kerkesoucha: 264.66 *ar.*; Psenaryo: 198.22; Tanchoiris: 108.16; and Ptolemais Hormou 184 +. Salt land is attested only at Psenaryo (26.16 *ar.*), Ptolemais Hormou (22 +), and Haueris (figure lost). Unfortunately, the total amount of arable land in these villages cannot be known so the ratio of productive to derelict land cannot be calculated. Still, villages elsewhere in the Fayyūm whose cultivable areas are either attested or are estimable range from 4,000 to 11,000 *ar.*[119] If these five settlements were anywhere within that range, their total unproductive land was tiny by comparison to near-contemporary Kerkeosiris, some 3.4–6.6 percent in the case of Kerkesoucha, a figure that also includes uncultivable features like canals, roads, and rock in addition to derelict fields.

118. Tanchoiris: TM Geo 2249; Kerkesoucha: TM Geo 1067.
119. See the list in Bowman, "Agricultural Production in Egypt," 237.

We may compare these figures with those attested at the village of Psenhyris in a survey of 80 CE (*P.Congr.* 15.15), a rare glimpse of the land of one of the Fayyūm's central villages in antiquity.[120] The site is today often identified with Sinnūris, a large village some 12 km north of the capital, though the name bears greater linguistic affinity with nearby Sanhūr, 9 km due west of Sinnūris.[121] Either identification places the village squarely in the Fayyūm's wet and fertile central "green belt." Perhaps unsurprisingly, the survey accordingly reports only 45 1/4 *ar.* of long-term dry land (*chersos*) split over nine parcels of between 2 and 11 3/4 *ar.* Of these derelict lands only an unspecified portion of thc 6.5 *ar.* of *chersos* in parcel five and the 3 *ar.* of *chersos* in parcel 8 are described as "salted out" (*exēlmēkuia/-kota*). If the agricultural territory of the village fell between the 4,000 and 11,000 *ar.* range, its 45 1/4 derelict *ar.* would amount to between 0.4 percent and 1.13 percent of its cultivable land, of which salted land was an insignificant fraction.

Conclusion

Resting on the slender foundations of two fragmentary papyri, the suggestion that villages in the central plain of the Graeco-Roman Fayyūm contained less marginal land than the better-attested villages of the depression's outer rim is little more than impressionistic. Yet when we compare the landscapes glimpsed in *P.Bagnall* 9 and *P.Congr.* 15.15 with that of Karanis in *P.Cair.Isid.* 6, which describes fully one-fourth of the village's land as *abrochos*, the differences are stark. While any conclusions must remain tentative, the evidence strongly implies significant environmental disparities between the center and the margins in antiquity. Therefore, rather than regarding as mutually exclusive the annually irrigated, single-cropped landscapes reliably attested in the papyri and the perennially watered, double-cropped lands described by later Arabic authors, this chapter suggests that the Fayyūm was fundamentally a hybrid agroenvironment, even in the Graeco-Roman period. These landscapes were shaped by hydrological and hydraulic distinctions between the Fayyūm's central canals and those of its outer margins, whose disparate flow regimes established two different sets of agricultural parameters in the depression. The

120. *P.Congr.* 15.15.
121. Banaji, *Agrarian Change*, 247.

remainder of this book now turns away from these more technical matters and toward human entanglements with the Fayyūm's water, beginning with the role(s) of the state. While the following chapter's diachronic comparison between the early Ptolemaic, Roman, and late Ayyūbid states has programmatic aims that extend beyond the confines of the present study, its account of Roman intervention in Fayyūm irrigation is central to the arguments to follow. Indeed, it is here in the entanglement between the Roman state, rural society, and flowing water that a clearer understanding of the predicament faced by late fourth-century Theadelphia, the subject of the final chapter, is to be sought.

Chapter 3
Governing Flow

If you want managed water, you've got to have people on the ground with the motivation to manage it
—Robert Benedetti[1]

Reintegrating the State

The canal system described in the previous two chapters was large, complex, and fragile. Its linchpin was the dam at al-Lāhūn, which first admitted and then contained the waters of the flood. Of secondary importance, it seems, was the dike between Itṣā and ʿIzbat Abū-l Nūr, which retained water for the benefit of settlements in the Tuṭūn Basin and, to some extent, beyond. The delivery and disposal of these stored waters further demanded a dense network of clear, unobstructed, and reinforced canals and drains. Stretching up to 70 km in length, the border canals were of particular concern, since blockages or other damage at any point along their routes threatened every downstream community with water shortages. Rigorous maintenance of all these features, both proactive and reactive, was therefore essential, since the failure of any of this infrastructure would have been highly disruptive, at least in the short term.

The fragility of the Fayyūm's water infrastructure, coupled with the system's origins as a political project of the early Ptolemies, suggests that the Egyptian state may have had a sustained role in its care and maintenance. Earlier scholars were categorical on this point, arguing that Egyptian irrigation had

1. Quoted in Patricia McBroom, "California Delta as National Heritage Could Help Save Its People," *California Spigot*, Tuesday 31 January 2012, accessed 22 March 2021, https://californiaspigot.blogspot.com/2012/01/california-delta-as-national-heritage.html.

since time immemorial depended upon strong centralized governance.[2] Citing both the opinions of Napoleon Bonaparte[3] and the environmental-determinist maxim of modern Egypt's first prime minister Nubar Pasha—"The Egyptian Question is the Irrigation Question!"—Ptolemaic historian Dorothy Thompson epitomized the *communis opinio* in 1971, writing that "control of the irrigation system, of the dikes and channels, of the height of the flood and the extent of cultivation has always been a characteristic operation of any successful Egyptian government."[4] The claim was scarcely controversial at the time. Beginning with the Napoleonic *Description de l'Égypte*, Western scholars and popular writers routinely characterized the administration of irrigation as a primordial function of the central government, an argument that justified European colonial intervention in Egypt as a necessary corrective to centuries of allegedly lax and ineffectual Ottoman governance.[5] This narrative later informed Sinologist Karl Wittfogel's famous model of the hydraulic state (aka "hydraulic despotism"), which posits that the need for farmers to come together regularly for collective maintenance on shared irrigation infrastructure was generative of social and bureaucratic complexity in early societies.[6]

While such arguments retain some hold on the popular imagination,[7] contemporary Egyptologists have decisively rejected them, thereby sidelining the central government in accounts of ancient Egyptian irrigation. Such revisionism was long in the making. Already in 1976, the geographer and archaeologist Karl Butzer argued decisively against the claim that irrigation had required

2. Westermann, "The Development of the Irrigation System of Egypt," is a classic example.
3. "In no country [sc. other than Egypt] does the administration have such an influence on public prosperity. If the administration is good, the canals are well dug, well maintained, the rules of irrigation are justly carried out, and the flood more extensive. If the administration is bad, vicious, or weak, the canals are blocked with mud, the dikes poorly maintained, the rules of irrigation violated, the principles of the system of inundation impeded by sedition and the personal interests of individuals or localities. The government has no influence on the rain or snow that falls on Beauce or on Brie; but in Egypt the government has an immediate influence on the extent of the flood which takes their place. This is what differentiates the Egypt administered by the Ptolemies from the Egypt already in decline under the Romans and ruined under the Turks." Napoleon Bonaparte, *Correspondence de Napoléon Ier*, tome 29, *Oeuvres de Napoléon Ier a Sainte-Hélène* (Paris: Imprimerie Impériale, 1870), 463. Cited by Crawford, *Kerkeosiris*, 34, from the excerpt in Alexandre Moret, *The Nile and Egyptian Civilization*, trans. M. R. Dobie (London: Routledge and Kegan Paul, 1927).
4. Crawford, *Kerkeosiris*, 106. Revisited in Thompson, "Irrigation and Drainage," 107.
5. Haug, "Civilizing the Past."
6. Karl Wittfogel, *Oriental Despotism: A Comparative Study of Total Power* (New Haven: Yale University Press, 1957).
7. E.g., Steven Solomon, *Water: e Epic Struggle for Wealth, Power, and Civilization* (New York: Harper Collins, 2010); and Brian Herbert, *Dreamer of Dune: The Biography of Frank Herbert* (New York: Tor, 2003), 172.

large, complex, and centralized Egyptian governments.[8] The papyrologist Danielle Bonneau likewise remarked in 1980 that there was never any independent branch of the central administration during the Graeco-Roman period whose remit was to manage and maintain rural waterworks.[9] Recent scholarship has accelerated this turn away from central authority and now emphasizes local agency in the management of irrigation works. In a summary of current thinking, Ptolemaic historian Joseph Manning has written that although Egypt possessed a centralizing principal in the figure of the pharaoh, it simultaneously lacked the administrative machinery necessary to supervise irrigation directly, let alone despotically. Reducing the pharaohs and later Ptolemies to the role of director, Manning identifies local elites and the growing bureaucracy as the primary actors in the work of water management, concluding that "there never was any connection between irrigation and centralized state power outside the concern for revenue."[10] In Egyptologist Juan Carlos Moreno García's similar assessment, "irrigation management was usually an internal community affair." The role of the central state was accordingly "limited to calculating the expected amount of taxes based on the level of the seasonal flood."[11] This consensus is nowhere more clearly articulated than in Moreno García's edited collection *Ancient Egyptian Administration* (2013), which traces the evolution of Egyptian governance from the pharaonic to the late period. Across twenty-two chapters and more than one thousand pages of text, detailed analysis of water management appears only in a study of local administration during the pharaonic Middle Kingdom.[12]

This radical interpretive turn has been a valuable corrective. It has shattered exaggerated earlier assumptions about the managerial capabilities of ancient states and highlighted the underappreciated role of Egypt's rural population in shaping the country's agricultural landscape, a phenomenon I will address in the following chapter. These gains notwithstanding, current scholarship has been too quick to deny that state actors had any sustained and interventionist role in Egyptian irrigation. To be clear, I do not mean to valorize the unqualified *étatisme* of earlier writers; rather, I simply suggest that the wholesale rejection of statism has inadvertently fostered a false dichotomy between

8. Butzer, *Early Hydraulic Civilization.*
9. Bonneau, "La haute administration."
10. Manning, *Last Pharaohs*, 44, repeated in Manning, *The Open Sea*, 103. See also Manning, "Irrigation et État," esp. 617–23.
11. Moreno García, "Ancient States and Pharaonic Egypt," 215.
12. Willems, "Nomarchs and Local Potentates."

state power and local agency, which has in turn forestalled a more nuanced appreciation of the variety of nondespotic entanglements between central governments, water, and rural society throughout premodernity. A new approach is needed, one that acknowledges the administrative limitations of premodern states while also respecting their mobilizatory capabilities as well as their significant financial interest in sustaining, perhaps even expanding, the productivity of Egypt's irrigated countryside.

The arguments in this chapter are deeply informed by Alan Mikhail's studies of eighteenth-century Ottoman Egyptian water management, which reveal a deep and complex entanglement between nature, rural society, and the state.[13] Since the Ottoman state extracted surplus grain from Egypt to support less-productive regions of the empire, its fiscal policy was necessarily deeply invested in successful Egyptian irrigation. This critical link between the irrigated Egyptian countryside and hungry populations abroad thus gave the Ottoman state powerful incentive to intervene directly in the day-to-day business of water management. Yet no such direct intervention ever materialized. Instead, the Ottomans practiced a form of "coordinated localism" that capitalized upon spontaneous local rituals of communal dike- and canal-maintenance. The state supported these efforts by encouraging farmers to participate in the annual work, by coercing free riders, and by coordinating laborers and resources for projects of larger scale or immediate importance. Officials were also responsive to the petitions through which villagers begged assistance with projects that could not be accomplished with local resources alone. In their petitions villagers might caution (or threaten) that annual cereal revenues would inevitably suffer if state assistance was not forthcoming.[14] Such rhetoric reminds us that Ottoman involvement in Egyptian irrigation was not motivated by paternalist concerns for the Egyptian peasantry per se but by a desire to ensure the uninterrupted production of taxable agricultural surplus.

Students of ancient Egypt must acknowledge Mikhail's work, for it convincingly demonstrates that a simple concern for revenue could in fact draw state governments deeply into the lives of everyday rural irrigators, an intimate relationship that nonetheless never amounted to (or even aspired to) totalizing control over rural water flows. But while Mikhail's coordinated localism suggests new avenues of approach to water management in Egyptian antiquity, we

13. Mikhail, "An Irrigated Empire"; Mikhail, *Nature and Empire*; and Mikhail, *Under Osman's Tree*, 19–110.

14. Kelly, *Petitions*, 37.

must still proceed with caution. In the first place, Mikhail's conclusions were based upon tens of thousands of administrative documents from the Ottoman archives in Cairo and Istanbul. Even when the relatively numerous papyri are accounted for, premodern evidence pales in comparison, and it is consequently impossible to illuminate any earlier period with comparable clarity. Second, and still more critically, Mikhail's arguments illuminate the agrofiscal policies of a single state over a relatively brief period—the "long eighteenth century" of 1675–1820 CE. By contrast, historians of ancient and early Islamic Egypt are confronted with a long succession of states whose scope, ambitions, and institutions differed, often considerably.[15] Nor did rural policy remain unchanged even during single historical periods such as the three centuries of Ptolemaic rule or the seven centuries of Romano-Byzantine imperium.[16] It is therefore highly unlikely that any single model of state-society-nature entanglement will possess universal explanatory power for the whole of premodernity. We must instead be prepared to craft flexible models that acknowledge both the major *lacunae* in our evidence and the reality of change over time. While we will never be able to match the depth and detail of Mikhail's work, we can nonetheless offer occasional glimpses of a complex and evolving array of entanglements between premodern states, rural society, and the irrigated landscape.

In consideration of these limitations, this chapter addresses Fayyūm water governance through a diachronic and comparative survey of the early Ptolemaic, Roman, and late Ayyūbid periods. It is of course already well established in the scholarship that state power was the driving impetus behind the Hellenistic reclamation of the Fayyūm and the construction of its innovative canal system.[17] This early Ptolemaic intervention was essentially political in nature, driven by the nascent kingdom's need to consolidate its control over the countryside and to carve out new agricultural lands on which to settle large numbers Graeco-Macedonian veterans. In contrast, Rome's interest in rural Egypt was primarily extractive. Like the later Ottomans, imperial Rome sought to maximize agricultural production so that surplus grain could regularly be directed to the imperial capital (first Rome, later Constantinople) and, at times, other

15. Cf. Manning, "Irrigation et État," 619, on the *histoire discontinue* of the Egyptian state. For analyses of political and institutional change between states see, e.g., Monson, *From the Ptolemies to the Romans*; and Conermann and Şen, *The Mamluk-Ottoman Transition*.

16. E.g., the administrative changes between the early and late Roman periods: Banaji, *Agrarian Change*.

17. See, e.g., Thompson, "Irrigation and Drainage."

non-Egyptian urban centers. State penetration consequently manifested itself in markedly different ways between the Ptolemaic and Roman periods. Chiefly concerned with the consolidation of the new canal system, early Ptolemaic administration was centralized under the direction of a single centrally appointed official—the *architektōn* or chief engineer—who was responsible for hiring, equipping, instructing, and dispatching the manpower necessary to maintain, repair, and even expand the system. Under Roman rule some three centuries later, however, the existence of long-established village communities in the Fayyūm's countryside made a form of coordinated localism possible. Like the Ottomans, Rome devolved primary responsibility for the annual maintenance of the Fayyūm's water works onto the system's principal beneficiaries, country farmers. Their yearly labors were nonetheless administered as a corvée, which was closely supervised and minutely documented by a complex administration that emanated from the provincial capital in Alexandria and reached all the way down to an array of local village liturgists—individuals performing compulsory public services or *leitourgiai* for the state—who ultimately oversaw and enforced this annual labor. As we will see further below, it was through these compulsory annual labors that rural subjectivity in Roman Egypt was constituted.

Notwithstanding the distinctions between Ptolemaic and Roman water governance, the principal aim of both administrative structures was to ensure the soundness of the canal system at all points, thereby enabling water to flow freely from its head at al-Lāhūn to its multiple far-flung tails. In contrast to such interventionism, the late Ayyūbid state had a minimal fiscal stake in Fayyūm agriculture and thus little incentive for sustained involvement in rural irrigation. The canal system in this period was instead maintained largely by spontaneous local self-organization, enabling the continued irrigation of a flourishing, if much reduced, agricultural countryside. This durability of now-ancient Fayyūm irrigation practices even in the absence of state coordination accordingly suggests that state power had always been an instrumental rather than existential component of Fayyūm irrigation—essential to the preservation of the region's original Ptolemaic plan but not to the survival of irrigated agriculture itself. Indeed, the most significant problems attested by al-Nābulusī are infrastructural rather than socioenvironmental, indicating that the decay of certain elements of the Fayyūm's physical irrigation infrastructure was the most consequential result of the retreat of the state after antiquity.

Central Coordination: The Early Ptolemaic Architektōn

Our evidence for the office of *architektōn* and its functions derives from the archive of Kleon and Theodoros, a collection of public and private papers assembled by these two successive *architektones* between 260 and 237 BCE.[18] Despite their significant responsibilities, Kleon and Theodoros possessed neither a permanent labor force nor did they impose on farmers the obligations of corvée. Rather, they made extensive use of private contractors drawn from the local population. Whenever maintenance or repairs were needed anywhere within the canal system, the *architektōn* and the Fayyūm's chief financial officer (the *oikonomos*) jointly offered a contract for the work at public auction in the nome capital. The winning bidder was then responsible for assembling laborers, who were paid cash wages, and accomplishing the work. A contract concluded by Theodoros in the year 246/245 BCE is a representative example of this type of contract (van Beek, *P.Petrie Kleon* 91, no. 6). After an introductory dating formula, the contract briefly describes the task at hand, in this case the removal of sand from a clogged canal that passed through the villages of Berenike Nea and Persea, two settlements located somewhere just to the north of the capital city.

> ἐξεδόθη ἐκ [τοῦ] βασιλικοῦ ὑπὸ κήρυκα διὰ Ἑρμαφίλου οἰκονόμου [[καὶ . . .]] παρ[όν]τος Πετοσίριος τοῦ βα(σιλικοῦ) γρ(αμματέως) καὶ Θεοδώρου ἀρχιτέκτονος· ἐν τῶι [Β]ερενίκης τῆς νεᾶς ποταμῶι τῆς Ἀριστάρχου νο(μαρχίας) τῶι [ἄγο]ντι ἀπὸ τοῦ κατὰ Πόαν θησαυροῦ παρὰ κ[ώμ]ην τὴν καλο[υμένη]ν Περσέαν [. . .] τῆς ἄμμου ἀνακαθᾶρα[ι ἀπὸ τοῦ] ἀποδιχθέντ[ος] διὰ τὸ [ὕ]φαμμον εἶναι.

> A contract was given out from the Treasury by auction through the *oikonomos* Hermaphilos, in the presence of Petosiris, royal scribe,[19] and of Theodoros, *architektōn*, for the following work: to clear the canal near Berenike Nea in the nomarchy of Aristarchos,[20] which comes from the granary near Poan and runs

18. Van Beek, *Archive (P.Petrie Kleon)*. In papyrological usage, *archive* usually refers to documents intentionally collected in antiquity. A *dossier*, by contrast, refers to papers relating to a single subject or individual(s) assembled by modern scholars. See in general Katelijn Vandorpe, "Archives and Dossiers," in *The Oxford Handbook of Papyrology*, ed. Roger Bagnall (Oxford: Oxford University Press, 2009), 216–55.
19. The *basilikos grammateus* or "royal scribe" was the nome's chief recordkeeper.
20. The nomarchy (*nomarchia*) was an early Ptolemaic administrative subdivision of the Arsinoite

past the village called Persea [. . .], from sand starting from the indicated place because it is silted up.[21]

Detailed instructions follow specifying the spatial remit of the project in cubits (*pēcheis*). The contractor is further cautioned to maintain a safe distance of three cubits between his own activity and the dikes (*chōmtata*) of the canal—the earthwork embankments that prevented canals from overflowing and destructively inundating adjacent farmland during the flood season. He was nonetheless simultaneously responsible for reinforcing these dikes to ensure their soundness. All the tools required for the work would be provided from public stores and were to be returned after completion. Should the contract be unfulfilled, or its terms violated, the work could then be offered at auction again or wage-laborers hired on a day-by-day basis. In either event, the contractor would be penalized 150 percent of his original bid and compelled to reimburse the state for any additional expenses.[22]

This and the other surviving contracts in the archive indicate that Kleon and Theodoros offered every significant aspect of irrigation work at auction, from the construction, repair, and cleaning of canals, the repair and reinforcement of dikes, to the construction, maintenance, and occasional demolition of the bridges that spanned the Fayyūm's many waterways.[23] It is these documents that underlie contemporary impressions of water governance in the early Ptolemaic Fayyūm. Writing forcefully against older descriptions of Ptolemaic *dirigisme*, Joseph Manning has argued that the contracts of Kleon and Theodoros reveal nothing of "oriental despotism" or hydraulic *étatisme*. Labor was instead freely solicited at public auction, fairly compensated, and overseen with a "light touch, incentives to perform, and a desire for efficiency."[24] Moreover, even when faced with labor shortages, Kleon and Theodoros did not press residents into service but instead offered tax incentives to encourage par-

nome. Willy Clarysse, "Nomarchs and Toparchs in the Third Century Fayum," in *Archeologia e papiri nel Fayyum: Storia della ricerca, problemi e prospettive. Atti del Convegno internazionale, Siracusa, 24–25 maggio 1996.* Quaderni del Museo del Papiro, Siracusa, 8 (Syracuse: Instituto internazionale del papiro, 1997), 29–54.

21. Van Beek, *Archive (P.Petrie Kleon)*, 91, no. 6, ll. 209–15, trans. Bart Van Beek.
22. On Ptolemaic labor contracts in general see S. Van Reden in James G. Keenan, Joseph G. Manning, and Uri Yitach-Firanko, *Law and Legal Practice in Egypt from Alexander to the Arab Conquest* (Cambridge: Cambridge University Press, 2014), 408–10.
23. Van Beek, *Archive (P.Petrie Kleon)*, 90 and 91.
24. Manning, *Land and Power*, 107.

ticipation.[25] On this reading, although the Ptolemaic reclamation of the Fayyūm was an impressive demonstration of the state's power over the land of Egypt, it did not produce a uniquely "hydraulic" model of state governance.

While this argument is correct on its merits, its interpretive utility is constrained by its focus on refuting the classic image of a *dirigiste* and despotic Ptolemaic state, most famously articulated in the work of Michael Rostovtzeff.[26] It therefore does not do justice to the degree to which the state, through the person of the *architektōn*, was entangled with Fayyūm irrigation even at the local level. Though his powers were hardly totalizing, the *architektōn* could ill afford to ignore any problem, however small, since failures at any point in the highly interconnected canal system would invariably produce a cascade of troubles downstream. The engineer's gaze was therefore both micro- and macroscopic, encompassing the minutiae of canal- and dike-work in single villages as well as the regulation of water flow throughout the system.

The local aspect of this remit is indicated in a letter addressed to Kleon in 257 BCE (*P.Petrie Kleon* 17). The sender was a man named Panakestor, the manager of the 10,000-*aroura* gift-estate at Philadelphia belonging to the early-Ptolemaic finance minister (*dioikētēs*) Apollonios (see chapter 2 under "Canal flow and Cropping Patterns."). Panakestor here chides Kleon for ignoring the problems at Philadelphia and paying too much attention to a nearby area called the "Little Lake" (i.e., the reservoir at Tamauis, on which see chapter 1 under "Storing the Flood" at n. 114). "You should not have continued on your way," Panakestor writes, "but should instead have come by for a moment and, having observed that the land is not irrigated, you should have asked why we do not irrigate. For you have not been appointed to direct only the works of the Little Lake, but those at this (land) as well." "Now then," he continues, "come meet us tomorrow at the sluice and give instructions on how the water is to be diverted (*angkōnizein*, i.e., from a feeder canal into local channels) for we are inexperienced. We will provide you with workmen and other supplies, however much you order." Panakestor concludes by threatening to write to the *dioikētēs* Apollonios himself, informing him of Kleon's supposed negligence.[27]

25. Van Beek, *Archive (P.Petrie Kleon)*, 17 (254 BCE).
26. M. Rostovtzeff, *A Large Estate in Egypt in the Third Century B.C.: A Study in Economic History* (Madison: University of Wisconsin Press, 1922).
27. Trans. modified from Bart Van Beek. ll. 3–8: Οὐκ ἔδει μὲν οὖν σε π[α]ραπορεύεσθαι, ἀλλὰ καὶ

While the letter can be read as simply an instance of a well-connected bully browbeating a lesser official,[28] Panakestor was still correct that Kleon's responsibilities encompassed the entire nome, including the problems of a single estate in a single village. Indeed, a flurry of outgoing correspondence from the year 242–241 BCE sees the *architektōn* direct his attentions to problems in multiple locales. In one letter, Kleon begs another official to send reeds to reinforce flood-damaged sluice gates at Ptolemais Hormou. In another he discusses opening additional gates at Ptolemais Hormou to increase the overall supply of water to the canal system during the flood. A further letter sees him sharply order a subordinate to ensure that an earthen dike has been erected in a village called Psenaryo (Ar. Sīnarū) so that floodwaters do not escape into a nearby ravine. In still other letters he turns to unfinished dikes and canals in villages such as Pseonnophris (Ar. Sanūfar) and Sebennytos.[29]

It is the *architektōn*'s preserved accounts, however, that most clearly reveal both the breadth and depth of his entanglement with Fayyūm irrigation. Microscopically detailed, these texts record the accomplishment of various irrigation works throughout the nome, precisely enumerating the volumes of earth excavated, the numbers of tools disbursed,[30] and the wages paid to laborers.[31] One account even tallies the number of rushes (probably indicating bales or bundles) used to reinforce dikes and embankments in various parts of the Fayyūm both before and during the flood, a number well in excess of one million.[32] Simultaneously hyperlocal and regional in scope, the accounts reflect a deep and ongoing engagement with the functionality of the canal system at every point along its course.

We glimpse, albeit obliquely, the rationale underlying this hyperlocalism in a badly preserved letter of 26 August 256 BCE (*P.Petrie Kleon* 18). While the

πρὸς ἡμᾶς παραβαλεῖν ὥρ[α]ς μόρ̣ιο̣ν καὶ τεθεα̣μένον σε μὴ βρεχομένην τὴν̣ γῆν ἐπερωτῆσαι [δι]ὰ̣ τίν' αἰτίαν οὐ βρέχ[ομ]εν. Οὐ γὰ[ρ] μόνον τέταξαι τὴν μικρὰν λίμνην ἀρχιτεκτο̣νεῖν [ἀλλὰ] καὶ ταύτην. Ἔ̣τι ο̣ὖν κα̣[ὶ] ν̣ῦ[ν] σ̣υ̣νάντησον ἡμῖν αὔριον ἐπὶ τ̣ὴν ἄφ[ε]σιν καὶ ἀρχιτεκ<τ>ό[νη] σ̣ο̣[ν] ὡς δεῖ τὸ ὕδωρ ἀ̣γκ̣ω[νίζειν· ἡμεῖς γὰρ] ἄπειροί ἐσμεν. Σώματα̣ δ̣ὲ̣ κ̣α̣[ὶ] τὴν λοιπὴν [χ] ορηγίαν ἡμεῖς παρέξομέν σοι, ὅ[ση]ν ἂν συντάσσῃς.

28. So Naphtali Lewis, *Greeks in Ptolemaic Egypt* (Oxford: Oxford University Press, 1986), 42–43.
29. Van Beek, *Archive (P.Petrie Kleon)*, 88 (242–241 BCE). See the summary of these and other letters in Thompson, "Irrigation and Drainage," 114–16.
30. Van Beek, *Archive (P.Petrie Kleon)*, 94 (250–248 BCE).
31. Van Beek, *Archive (P.Petrie Kleon)*, 92 (258–257 BCE).
32. Van Beek, *Archive (P.Petrie Kleon)*, 95, col. 2 (260–236 BCE). Possibly for use in παραφρυγανισμός (*paraphryganismos*), the revetment of canal embankments with bundles of reeds. Van Beek, *Archive (P.Petrie Kleon)*, pp. 23–25.

names of the sender and the recipient are lost, its date places it within Kleon's term as *architektōn*, which terminated only in 250/249 BCE.

> [————τὰς θύρα]ς τὰς ἐν Μοντίλαι κεκλιμένας καὶ τὴν διώρυγα [——
> —]. . [————]. κατὰ Βουκόλων Κώμην κέκλεικα [————]ιν. Ἐπι[. . .].
> . ọῦ̣ṿ ὅ̣τι τὰ χώματα καὶ αἱ διαβάθραι οὐ μὴ [————] το βουλεύσα̣σθαι, εἰ
> δεῖ μίαν θύραν κλεῖσαι ἐν Πτολεμαίδι. Ἔρρωσο. (ἔτους) κθ Ἐπεὶφ ε̣.

> [-——the sluice gate]s in Montila are closed and the canal [-— -] I have shut [the gates?——-] near Boukolōn Kōmē [-— Know that] the dikes and the bridges cannot [-— -] to discuss whether it is necessary to close one sluice gate (*thyra*) in Ptolemais. Farewell. Year 29, Epeiph 5.[33]

Although the context and original intent of the letter are irrecoverable and the surviving text seems unpromising, its toponymy is revealing. As described in the previous chapter Ptolemais Hormou and its sluice gates—here referring to either the control works at the al-Lāhūn inlet or some other important piece of nearby infrastructure—lie at the head of the canal system. The so-called Montila canal was also one of the larger waterways in the Fayyūm and it is continuously attested in the papyri until the third century CE.[34] Its route began somewhere near the center of the nome just south of the capital and passed by several villages in the area: Hiera Nesos, Kerkeēsis, Ptolemais Melissourgōn, and Tebetny, the last of which survives as Difinnū, al-Nābulusī's Difidnū, roughly 8.5 kilometers south-southwest of the capital. The tail of the Montila was somewhere in the Fayyūm's northwest, perhaps near the village of Boukolōn Kōmē, a northwestern settlement whose precise location is unknown. Later Roman papyri also locate a stretch of the Montila near Theadelphia in the far northwest. The route of the canal thus bears at least a superficial similarity to the Baḥr al-Nazla, one of the modern Fayyūm's major border waterways. Consequently, this fragmentary letter documents an attempt by the *architektōn* to regulate the flow of water through an entire subsection of the canal system, from the headworks at Ptolemais Hormou, though a major public canal, as far as a village near the northwestern terminus of the system. It is for this reason that Kleon and Theodoros paid such close attention to seemingly minor local

33. Van Beek, *Archive (P.Petrie Kleon)*, 18, trans. Bart Van Beek.
34. *P.Tebt.* 3.828 (second century BCE); *P.Berl.Leihg.* 13 (113–38 CE); *P.Berl.Frisk.* 1 (155 CE); *P.Col.* 5.1 (161/180 CE).

concerns: only by ensuring the soundness of infrastructure throughout the depression could they guarantee that water would flow freely and without obstruction from the head of the canal system to its various tails.

Environmental Subjectivity: The Roman Penthēmeros

Despite its centrality to early Fayyūm irrigation, the figure of the *architektōn* disappears from the papyri following the conclusion of the archive of Kleon and Theodoros in 237 BCE. When and how the office was abandoned is unknown. The classic narrative of the historical trajectory of the Ptolemaic state treats the period after the reign of Ptolemy III (246–222 BCE) as one of internal disintegration and decline, of which the disappearance of the *architektones* might be regarded as symptomatic.[35] It is more likely, however, that the position was necessary only during the canal system's initial consolidation. As the Fayyūm's villages grew in size and became well established, they will have developed their own local rituals of annual dike- and canal-maintenance, thereby obviating the need for an official whose primary responsibility was to procure and dispatch the manpower and materials necessary to maintain the water infrastructure throughout the depression. Indeed, by the last decade of the third century BCE, one of the stated duties of the *oikonomos*, a nome-level financial official, was to inspect the entirety of the canal system, from major public canals (*diōryges*) to their offtakes (*epirrhyseis*) to local village channels (*hydragōgoi*), and to ensure that the network was clean, free of obstruction, and in a good state of repair.[36] The role of the *oikonomos* is thus here envisioned as purely supervisory, suggesting that the labor necessary to keep the system in working order came from below.

This sort of independent and spontaneous communal activity, well-documented in small-scale irrigation systems throughout the globe, was a recurring event in the Egyptian agricultural calendar so routine that it escaped widespread documentation, at least until the Roman period.[37] That rituals of communal maintenance were already in existence by the later Ptolemaic period is nonetheless suggested by a second-century BCE petition addressed to a cer-

35. On Ptolemaic decline narratives, Manning, *Last Pharaohs*, 76–77.
36. *P.Tebt.* 3.703 (ca. 210 BCE), ll. 29–40.
37. See, e.g., the essays in Jonathan B. Mabry, ed., *Canals and Communities: Small-Scale Irrigation Systems* (Phoenix: University of Arizona Press, 1996), particularly part 1, pp. 33–116.

tain Teōs, the village scribe (*kōmogrammateus*) of a settlement located somewhere in the Fayyūm's northeast called Attinou Isieion (*SB* 18.13735). The sender of the petition is one Protomachos, son of Protomachos, a "*katoikic* cavalryman" or military settler, who possessed a plot of land (*klēros*) in the area, ostensibly on long-term loan from the crown. According to Protomachos, it was customary for all the holders of *klēroi* and other landholdings along a canal extending from Attinou Isieion to collectively dig out accumulated silt, which was then used to embank the canal to strengthen it against a breach during the flood:

> Τεῶτι κωμογραμματεῖ Ἀττίνου Εἰσι(ήου) παρὰ Πρωτομάχου τοῦ Πρωτομάχου τῶν κατοίκων ἱππέων. ὄντος ἐθιζμου ἔτι ἄνωθεν τοὺς ὑποκειμένους κλήρους καὶ τὰς ἄλλας γᾶς τῇ φερούσῃ ἐκ τοῦ Ἀττίνου Εἰσιήου διώρυγι ἥ ἐστιν ποτίστρα, τούτους δὲ ἀνασκάπτιν τὸν ἐν τῇ διώρυγι χοῦν ἐπὶ τὰ χώματα πρὸς τὸ μὴ κατακλυσθῆναι τὰς γᾶς, διὸ ἐπιδίδωμι ὅπως προσκαλεσάμενος τούς τε γεωργοῦντας καὶ τοὺς κυρίους τῶν τọ́[πων] διὰ τοῦ παρ' ἐμοῦ γεωργοῦ συṇṭάξῃς ποιήσασθαι τὰ σκάμματα καὶ τὴν στέγ[ν]ωσιν τῶν χωμάτων τῆ[ς] διώ[ρυ]γος \πρὸ τῆς ὕδατος ἐμβọḷῆς/ ἢ ὅτι ποτιζομένης τῆς παρ' ἐμοῦ γῆς συνβοηθ̣[οῦσι ἡμ]ῖν· βρạχῆναι ἐγώ τε καὶ ὁ παρ' ἐμοῦ γεωργὸς ἂν αἴτι[ο]ι ὦμεν ὑποτάξαι δὲ καὶ το[ῦ] [ὑπ]ομνήματος τὸ ἀντίγραφον ἵν' ὑπάρχῃ ἐν χρ[ημα]τισμῶι· τυνχάνω δ' ἐπὶ τῶν αὐτῶν ἐπιδεδωκὼς τὰ ὑπομνήματα τῷ στρατηγ[ῷ καὶ τῷ] βασιλικῷ γραμματεῖ. εὐτύχει.

> To Teōs, *kōmogrammateus* of Attinou Isieion, from Prōtomachos son of Prōtomachos, one of the *katoikic* cavalrymen. It has long been customary for the *kleroi* and other lands lying adjacent to the canal that extends from Attinou Isieion, which is an irrigation conduit,[38] to dig out the mud for the dikes, so that the lands are not inundated. For this reason I request that, after I have summoned the tenant farmers and the holders of land allotments through my own tenant farmer, you order that the digging out and the firming up of the dikes of the canal be completed before the influx of the water, or that they should come to my aid once my land has been flooded. Should I and my tenant-farmer be held responsible for being flooded, (I ask you) to forward a copy of this petition so that it be put on record. I have already submitted petitions on these matters to the nome governor (*stratēgos*) and the royal scribe (*basilikos grammateus*).[39]

38. *Potistra* refers to a conduit that delivers irrigation water. The term is frequently paired in documents with *ekxysis* (ἔκχυσις), a drainage conduit. See Bonneau, *Le régime administratif*, 26–27.
39. The chief of records for the Arsinoite Nome/Fayyūm.

We must treat cautiously the reference to long-established custom (*ethismou eti anōthen*), since Prōtomachos is here both soliciting state intervention on his own behalf and attempting to absolve himself of future liability should local irrigation works fail. The language of the request nonetheless suggests that annual maintenance of this canal was not typically subject to state coordination. This is not to say that state coordination of dike- and canal-maintenance was unknown in the Hellenistic period, for the practice of corvée labor in irrigation long predated the Ptolemies and is duly attested in early Ptolemaic papyri.[40] Rome, however, would establish a more systematic model of water governance, whose broad outlines resemble later Ottoman coordinated localism. In brief, several months before the beginning of the flood season, the provincial government in Alexandria began to exhort nome-level officials to see to it that the public irrigation infrastructure of their nomes was prepared for the arrival of the waters. Governors then forwarded these exhortations to local officials, often liturgists, who coordinated and oversaw maintenance work at the local level. Local officials might also at times provide material assistance with tasks too large, complex, or resource-intensive for rural communities to carry out on their own.

The beginnings of Roman coordinated localism are obscure. Several well-rehearsed literary notices report that the first emperor Augustus put his soldiery to work cleaning rural canals several years after the Roman annexation of Egypt in 31 BCE. Such was the success of the endeavor, the geographer Strabo enthusiastically asserts, that floods low enough to have caused famine in previous years were hereafter sufficient.[41] While it would be rash to overinterpret such propaganda as a wholesale "reorganization of the irrigation system,"[42] there are hints that the Roman administration was indeed concerned with the integrity of rural water infrastructure even in these early years. In a papyrus of 25 BCE, several Egyptian-named individuals swear by the emperor Caesar that they will strengthen and line with brushwood the dikes of three public embankments (*dēmosia chōmata*) near a village by the name of Korphotoi in the Herakleopolite nome. For this work they acknowledge the receipt of 120 silver drachmas from public funds.[43] A papyrus from the nearby Oxyrhynchite nome

40. Willems, "Nomarchs and Local Potentates," 343. A third-century Theban papyrus, *UPZ* 2.157 (241 BCE), lists three classes of individuals exempt from corvée: the elderly who guard dikes, the too-young and the too-old, and the incapacitated (*adynatoi*). Clarysse and Thompson, *Counting the People*, 2:42. For Ptolemaic compulsory labor in general see further below at n. 52.

41. Strabo 17.1.3. Cf. Suetonius, *Vita Augusti* 18.2; and Dio Cassius 51.18.1.

42. So Westermann, "Aelius Gallus," restated in Capponi, *Augustan Egypt*, 18.

43. *BGU* 16.2590 = *SB* 16.12312 (1 Aug. 25 BCE). Trans. Brashear, "Before the Penthemeros," 29.

some two decades later also references a rural official pressing (*ochlēsai*) men into service on nearby embankments.[44] It is nonetheless worth noting that in both cases the laborers were locals, dependents of the very infrastructure on which they were deployed. This suggests that Rome had already in this early period come to rely on local agency rather than the centralization implied in the literary notices of canal-clearance under Augustus.

As we will see below, systematic reliance on local actors was indeed already well-established in the Fayyūm by the middle of the first century CE if not earlier. Unfortunately, no contemporaneous papyri of Alexandrian provenance survive to shed light on the development and consolidation of this practice at the provincial level. The earliest such surviving text is an as-yet unpublished circular of 115 CE, which was sent by the reigning prefect of Egypt Marcus Rutilius Lupus to all nome governors (*stratēgoi*) of Egypt. The surviving copy (*antigraphon*) of this letter, badly fragmented, is addressed to Claudius Apollonios, *stratēgos* of the Hermonthite nome, and Archias, former *stratēgos* of the Oxyrhynchite. The prefect writes that the evident disrepair of Egypt's irrigation works, which had been revealed to him on a recent inspection of the countryside, was leading to the neglect of irrigation in many areas. He accordingly directs nome governors to attend to the restoration of local infrastructure, reminding them of earlier, albeit unspecified, commands (*entolōn*) of the emperor Trajan on the subject.[45]

While the circumstances prompting this second-century circular may have been unique, such hortatory correspondence was a regular product of the central administration by the following century. The most complete example is a circular of 278 CE addressed by the *dioikētēs* (financial minister) Ulpius Aurelius to the governors of the Fayyūm and the Heptanomia, the latter comprising the Nile Valley between the apex of the Delta and Thebes (*P.Oxy.* 12.1409). Composed in late March or early April, the middle of Egypt's dry season when the Nile was at its lowest ebb, the letter orders *stratēgoi* to begin coordinating yearly efforts within their districts in preparation for the coming of the summer flood:

> Οὔλπιος Αὐρήλιος στρατηγοῖς κ[αὶ] δεκαπρώτοις Ἑπτανομίας καὶ Ἀρσινοΐτου [χαίρειν. τοῦ καιροῦ τῆς τῶν] χωμάτων ἀπεργασίας καὶ τῆ[ς] τῶν διωρύχων ἀνακαθάρσεως ἐνεστη[κότος παραγγέλλειν ὑμῖν ἀναγ]καῖον ἡγησάμην διὰ τῶνδε τῶν γραμμάτων ὡς χρὴ σύμπαντας τοὺς γẹ[ωργοὺς—ca.18

44. *P.Ryl.* 4.603 (ca. 7 BCE).
45. Egerton Papyrus 13 with Micucci, "The Egerton Collection."

> -] ταῦτα ἀπεργάζεσθαι ἤδη μετὰ πάσης προθυμίας ἐπὶ τὰ διαφέροντα αὐτοῖς π̣.[. . . . . .] π̣ρὸ[ς τὸ δ]η[μοσίᾳ τε] πᾶσιν καὶ ἰδίᾳ ἑκάστῳ συμφέρον· τὴν γὰρ ἀπὸ τῶν ἔργων τούτων γεινομένην ὠφ[ελί]α̣ν πάντας ε̣[ἰδέναι πέ]πεισμαι.
>
> Ulpius Aurelius to the *stratēgoi* and administrators of the Heptanomia and Arsinoite nomes, greetings. Since the time for the building up of the dikes and the cleansing of the canals is at hand, I thought it necessary to announce to you by this letter that all of the cultivators and [. . .] ought now to build these up with all zeal on the [sc. lands/fields?] belonging to them, convinced that everyone is aware of the benefit that comes from these tasks.[46]

The *dioikētēs* next commands the *stratēgoi* to encourage local farmers and to select overseers (*epimelētai*) from either local magistrates or private citizens. These overseers would, in turn, compel everyone to perform their work personally and were not permitted to accept cash in exchange for physical labor. This would, he continues, ensure that:

> [ὥστε ἐπε]νεχθῆναι εἰς τὸ τεταγμένον ὕψος τε καὶ πλάτος τὰ χώματα καὶ τοὺς διακόπους ἀποφραγῆναι πρὸ[ς τὸ δύνα]σθαι ἀντέχειν τῇ ἐσομένῃ εὐτυχῶς πλημύρᾳ τοῦ ἱερωτάτου Νείλου, τάς τε διώρυχας ἀνα[καθαρῆ]ναι μέχρι τῶν καλουμένων γνωμόνων καὶ τοῦ συνήθους διαστήματος, ἵν[α ε]ὐμαρῶς [τὴν] ἐσομέν[ην τῶν] ὑδάτων εἴσροιαν ὑποδέχοιντο πρὸς ἀρδείαν τῶν ἐδαφῶν, τούτου κοινωφ[ελ]οῦς τυγχ[άνοντος.]
>
> The dikes are raised to the established height and width and the breaches are blocked up, in order that they may be able to withstand the blessedly impending flood of the most sacred Nile, and so that the canals are cleansed up to the so-called standards and the accustomed dimensions, in order that they may easily bear the coming onrush of water for the irrigation of the fields, this being for the common good.[47]

He concludes with a warning to any persons who attempt to exact bribes or shirk their responsibilities: both their lives and property are at stake for "endangering measures intended for the safety of all of Egypt."[48] Similar remarks are

46. *P.Oxy.* 12.1409 (278 CE), ll. 7–12.
47. *P.Oxy.* 12.1409, ll. 15–19.
48. *P.Oxy.* 12.1409, ll. 21–22: ὡς λυμαινόμενος τοῖς ἐπὶ τῇ σωτηρίᾳ συνπά[ση]ς τῆς Αἰγύπτου προηρ[ημέ]νοις.

recorded in another surviving, albeit far more fragmentary, circular of the same century (*SB* 14.11349). Although both the author and recipients of this text are lost, the letter is likewise addressed to nome *stratēgoi*, who are informed that they "must in this affair display clearly your special zeal, so that I may know how and in what way the *stratēgos* who wishes to preserve his own safety assists the crop of this prosperous season." The author then orders each *stratēgos* to ensure "that the dikes and canals in his nome be entirely completed according to the relevant regulations," at which point the text becomes fragmentary.[49]

We owe the survival of such directives to the administrative machinery that circulated them throughout the countryside. *P.Oxy.* 12.1409, a copy of the Alexandrian original, was originally produced in the offices of the governor of the Oxyrhynchite nome some eighty kilometers south of the Fayyūm. After appending his own introduction, the governor then forwarded the copy to the local officials responsible for the actual coordination and oversight of the annual works. A copy of a similar circular from the southern Egyptian city of Panopolis (Akhmīm) dated 16 February 300 CE reveals that such *communiqués* were then to be posted "not only in the city [*polis*, i.e., the nome capital] but also in each of the chief villages (*mētrokōmiais*) of the nome so that all may know the commands."[50] Addressed by a late-Roman overseer (*procurator*) of the Lower Thebaid to nome governors in this region of southern Egypt, this later Panopolite circular contains exhortations similar to those *P.Oxy.* 12.1409: governors are to ensure that local dike inspectors (*chōmatepeiktai*) oversee the customary maintenance of dikes and embankments and attend to any other works that might be of benefit or that had been long neglected. Unlike the earlier exempla, however, the Panopolite circular indicates that communication and exhortation might flow from the bottom up as well as from the top down. Indeed, the *procurator* encourages landowners, tenant farmers, and local tax officials (*dekaprōtoi*) to call attention to problems that required prompt state intervention:

---

49. Ll. 5–11: χρὴς τοὺς ̣στρατηγοὺς ὑμᾶς ἐνταῦθα τὴν ἑαυτῶν σπουδὴν διατικνοί[ε]σθαι, ὅπως εἰδείην πῶς καὶ τίνα τρόπον ὁ τὴν ἰδίαν σωτηρίαν [δ]ι̣α̣φυλάξαι βουλόμενος στατηγὸς τῇ τῆς εὐ[ετ]η̣ρίας φορᾷ συνα[γ]ωνίζεσθαι. κελεύω οὖν ἕκαστον στρατηγὸν τὰ ἐν τῷ ὑπ' αὐτὸν νόμον χώματα κ̣α̣ὶ δ̣ιῴ[ρυγ]α̣ς̣ ἐ̣ντελῶς ἀπεργάζεσθαι κατὰ τὰ πε[ρὶ το]ύτου δ̣[ιατετ]υ̣π̣ωμένα. Trans. J. David Thomas, who reads εὐ[ετ]η̣ρίας in l. 8.: "Notes on Two Yale Papyri," *Chronique d'Égypte* 51 (1976), 315–19 at 319.

50. *P.Panop.Beatty* 2, l. 228: μὴ μόνον ἐν τῇ πόλει ἀλλὰ κ̣αὶ ἐν ταῖς τοῦ νομοῦ μητροκωμίαις ἀντίγρα(φα) καθ' ἑκάστην π̣ρ̣ο̣[τιθ]ε̣ναι, ὡσὰν ἅπαντες εἰδῖεν τὰ προστεταγμένα.

ἀκόλουθον δε ἐνόμισα καὶ προγράμματι προκαλέσασθαι τοὺς ἑκ̣ασ̣τ[α]χοῦ κτήτοράς τε καὶ γεωργοὺς ὁμοῦ δὲ καὶ δεκαπρώτους . . . ὥστε εἴ τι τοιοῦτο συμφέρειν αὐτοῖς νομίζοιεν, προσειέναι τοῖς τε στρατ̣ηγοῖς καὶ χωματεπείκταις ἔτι μὴν καὶ συνοψισταῖ̣[ς] ὑποδικνύ[ντες] καὶ κ̣ακεινα χρησίμως δεῖν ἀπεργασθῆναι ἅτινα ἀμεληθῆναι εἰς δεῦρο συμβέβηκεν. οἶμαι γὰρ καὶ κακείνους μεμνημένους τ̣ῶν παρ' ἐμοῦ ἐντολῶν μὴ ἐν δευτέρῳ θήσεσθαι τ̣[ὴν τοι]αύτην ἐπιμέλειαν.

I have now thought it appropriate in addition by public notice to appeal to landowners and farmers everywhere, together with the *dekaprōtoi* . . . asking that if they should consider any such work profitable to them, they should apply to the governors and dike inspectors and surveyors, indicating those works that could usefully be undertaken but which have until now been neglected.[51]

As these circulars indicate, neither the upper levels of provincial bureaucracy nor even the nome *stratēgoi* had any direct personal role in dike- and canal-maintenance. While their exhortations may have inaugurated the annual efforts, it was local officials who actually oversaw the labor and local farmers who undertook it. Unfortunately, the urban provenance of many of our papyri has obscured the day-to-day work of Roman coordinated localism in the Egyptian countryside outside the Fayyūm. In general, corvée labor was apportioned among rural villages according to the amount of land each possessed. Each laborer was in turn responsible for moving five *naubia* (sing. *naubion*) of earth for local embankments (19.5 m^3).[52] The abundant village papyri of the Fayyūm,

51. *P.Panop.Beatty* 2, ll. 224–27. Trans. modified from T. C. Skeat.

52. Sijpesteijn, *Penthemeros-Certificates*, 18–21. The administration of the *naubion* corvée is described in the third-century CE administrative glossary *P.Oxy.* 38.2847r, ll. 21–26: ἕκαστος τῶν ἐπιχωρίων ἀνασκάπτι (*l.* ἀνασκάπτει) πέντε ναύ̣βια εισ̣. . φαλ̣. . . κατὰ τὴν ἐπιβολὴν τῶν χωμάτων, ἐλάττ[ονα δὲ] ὡς ᾇν ἡ χρεία ἀπαιτῇ̣. ναύβιον δέ ἐστιν μέτρον̣ [γ]ῆς ὀρυσσόμενον τρεῖς πήχεις ἔχ[ο]ν ̣τ̣οῦ π̣λά[του]ς καὶ μήκους καὶ βάθους ("Each country resident digs up five *naubia* for . . . with regard to the building up of the dikes, or fewer whenever need demands it. A *naubion* is a measure of excavated earth having three cubits in breadth, length, and depth"). Cf. the petition from *naubion* laborers *P.Oxy.* 12.1469 (298 CE), which describes the imposition of *naubia* on villages. Following the completion of the work local officials drew up a report for submission to the nome *stratēgos*, e.g., the Oxyrhynchite reports *P.Oxy.* 49.3475 (220 CE); and *P.Col.* 10.289 (331 CE). Laborers could also receive receipts similar to those issued for the *penthēmeros*, e.g., *O.Bodl.* 2.1699 (first/second century CE); *O.Mich.* 1.273 (188 CE); *P.Col.* 7.168 (373 CE). Since the word *naubion* is Egyptian in origin (Demotic *nby*) and the *naubion* corvée appears already well-developed in early Demotic and Greek Ptolemaic receipts issued to laborers, the origins of the system are likely pre-Ptolemaic. Muhs, *Tax Receipts*, 57–60, e.g., *O.Bodl.* 1.242 (256/5 BCE).

however, preserve significant evidence for the ground-level administration of its annual dike- and canal-corvée. Here Roman coordinated localism took the form of the *penthēmeros* ("five-day"), so called from the five days of work on public dikes and canals demanded each year from able-bodied adult males. This unique approach to the maintenance of the Fayyūm's equally unique canal system was consolidated early in the Roman period, appearing in developed form already in the 40s CE.[53] This labor-tax on most able-bodied adult males was linked to each individual's village of record (Greek *idia*, Latin *origo*) and was thus usually discharged on the public dikes and canals that served his own village. At the level of village administration, local officials maintained a list of male residents of the village obliged to perform dike- and canal-work (*chōmatika kai diōrychika erga*, often abbreviated to *chōmatika erga*) in a given year.[54] This work largely consisted of clearing accumulated silt and brushwood from the beds of canals (*diōryches*), material then used to firm up the canals' earthwork embankments (*chōmata*), as alluded to in the Ptolemaic petition *SB* 18.13735 discussed above.[55] After discharging their obligatory five days (or so) of labor on public infrastructure, workers were issued a receipt, of which many specimens survive. The standard formula lists the name and filiation of the worker, his official village of record (*idia*), and the canal on which he had worked. The same information was simultaneously recorded and preserved in government ledgers.[56] This second-century certificate from the northeastern village of Karanis is a representative example of the corpus:

> ἔτους ὀγδόου Αὐτοκράτορος Κ[α]ίσ[α]ρ[ο]ς Τίτου Αἰλίου Ἁδριανοῦ Ἀ[ντ]ωνε[ίν]ου Σεβαστοῦ Εὐσεβοῦς. εἴργ(ασται) ὑπ(ὲρ) χω(ματικῶν) ἔργ(ων) τοῦ διελ(ηλυθότος) ζ (ἔτους) Ἁδριανοῦ κ ἕως κδ ἐν ὀρινεῖ (*l.* ὀρεινῇ) διώ(ρυγι)

53. Sijpesteijn, *Penthemeros-Certificates*, 10–12. For recently published certificates and relevant commentary see Claytor, "*Penthemeros* Certificates from the Granary C123, Karanis"; Claytor, "More *Penthemeros* Certificates from Karanis"; and Abd-Elhady, Gad, and Hartenstein, "Five *Penthemeros* Certificates from the Cairo Museum."

54. *BGU* 2.618 (213/4 CE)′ and *BGU* 7.1634 (229/30 CE), from the villages of Mendēs and Dinnis, respectively.

55. For an early comparative study see Boak, "Notes on Canal and Dike Work." The word ἀφυλισμός (*aphylismos*), meaning to remove brushwood, is sometimes used of the work of cleaning out canals. Bonneau, *Le régime administratif*, 129–30. Later Arabic sources describe oxen dragging a shovel-like box called a جرافة (*jarāfa*) along the beds of canals to dredge them. Balls of mud and brush from the cleaning were then used to reinforce and repair embankments. Silt from canal beds was also used as a fertilizer in nearby fields. Borsch, *The Black Death*, 34 with notes 35–36. Cf. the operation of *paraphryganismos* described at n. 32 above.

56. *P.Mich.* 6.380 (=*P.Cair.Mich.* 2.12e), and 6.381 (170 CE?); *P.Münch.* 3.136 (early second century CE).

Πατσών[τ]εως Καρα(νίδος). Πνεφερῶ(ς) Πεθέω(ς) το(ῦ) Πετεσούχ(ου) (μητρὸς) Θαισᾶτο(ς). (hand 3) Κέλερ σεση(μείωμαι).

The eighth year of the Emperor Caesar Titus Aelius Hadrianus Antoninus Augustus Pius. Has worked in fulfillment of his obligation of the past seventh year with respect to the embankments, Hadrianos 20 to 24, in the desert canal of Patsōntis, (on behalf of) Karanis: (*2nd hand*) Pnepheros, son of Petheus and Thaisas, and grandson of Petesouchos. (*3rd hand*) I, Celer, have signed the certificate.[57]

That the labor Pnepheros performed was a form of coordinated localism rather than random, unadulterated coercion is clear from the name of the canal upon which he worked, the desert canal of Patsōntis (i.e., the eastern desert canal), which was the main public waterway serving Karanis. Residents of Karanis are therefore often attested at work on this canal in *penthēmeros* receipts,[58] as were residents of settlements along the canal upstream from Karanis including Bakchias[59] and Philadelphia.[60] Residents of Karanis[61] and Soknopaiou Nēsos[62] also appear at work on the *Epagathianē diōryx*, a canal whose route is not known but must have been somewhere in the northeast. Likewise in the southern Fayyūm, residents of Tebtynis are attested at work on the southern desert canal, which watered their village and other settlements along the southern margins.[63] Residents of the western villages of Theadelphia,

57. *SB* 6.9437a (144 CE).

58. *SB* 8.9924a (114/5 CE); *P.Sijp.* 42a (130 CE); *SB* 6.9437a (144 CE) and c (146 CE); *P.Cair.Goodsp.* 25 (161 CE); *BGU* 3.722 (161/2 CE); *P.Mich.* 6.419 (162 CE); *BGU* 15.2519 (164 CE); and *P.Mich.* 6.381 (second half of second century CE). Citations from Claytor, "*Penthemeros* Certificates from the Granary C123, Karanis," 67, no. 34.

59. *P.Strasb.* 3.165 (120 CE); *P.Strasb.* 4.249b (122 CE); *SPP* 22.162 (124 CE); *P.Ryl.* 2.210 (131 CE); *P.Strasb.* 3.158 (126 CE); *P.Strasb.* 3.159 (127 CE); *P.Strasb.* 3.162 (130 CE); *SB* 18.13178 (139 CE).

60. *SB* 16.12860 (87/88 CE); *P.Phil.* 20 (118 CE); *BGU* 1.264 (191 CE).

61. Claytor, "*Penthemeros* Certificates from the Granary C123, Karanis," no. 10 (131 CE); *P.Mich.* 6.418 (157 CE); *P.Mich.* 6.380 (=*P.Cair.Mich.* 2.12e) and 6.381 (170 CE?); *SB* 6.9496 (late second century CE).

62. *SB* 16.12300 (111–12 CE); *BGU* 15.2518 (119 CE?); *SPP* 22.161 (128 CE); *SPP* 22.8 (132 CE); *P.Lond.* 3.841a (139 CE); *P.Dub.* 9 (140 CE); *BGU* 3.876 (152 CE); *SPP* 22.162 (154 CE); *P.Coll. Youtie* 1.43 (157–58 CE): *PSI* 9.1045 (154 CE); *BGU* 2.593 (158 CE); *P.Coll.Youtie* 1.44 (158–59 CE); *SPP* 22.160 (159 CE); *BGU* 3.877 (159 CE); *BGU* 3.723 (159–60 CE); *P.Lond.* 2.321c (159–60 CE); *SB* 12.11068 (160 CE); *P.Lond.* 3.841b (160 CE); *P.Lond.* 2.325b (161 CE); *P.Grenf.* 2.53c (162 CE); *P.Coll.Youtie* 1.45 (163 CE); *P.Lond.* 2.321b (165 CE); *P.Grenf.* 2.53e (178 CE); *P.Grenf.* 2.53f (178 CE).

63. Tebtynis residents at work on the *oreinē diōryx Tebtyneōs*: *SB* 18.13983 (112 CE); *P.Kron.* 66 (117 CE); *P.Kron.* 56 (120–51 CE); *P.Kron.* 57 (120 CE); *P.Kron.* 58 (123 CE); *P.Kron.* 60 (131 CE);

Polydeukia, and Euhēmeria likewise worked along the *Pseinalitidos diōryx*, the western desert canal.[64] Albeit larger in scale, such work was in substance identical to the practices described in the Ptolemaic petition *SB* 18.13735: the communal maintenance of a canal by its dependents.[65] The *penthēmeros* simply institutionalized these annual rituals, thereby transforming informal, collective obligations to one's irrigation community into a formal, individual obligation both to the state and to one's *idia*. It was thus a distinctly rigid form of coordinated localism that sought to maintain extant patterns of settlement and of water flow by encumbering farmers with an annual quantum of labor to be discharged principally, though not exclusively, in their village of record.[66] By striving to fix populations of obligatory laborers in place at every inhabited point within the Fayyūm's canal network, the *penthēmeros* worked to impose perpetual stability upon a fluid and thus inherently unstable waterscape, thereby ensuring maximal agricultural productivity and more predictable returns to the Roman fisc.

Yet the coercive character of the *penthēmeros*, along with its emphasis on maintaining water flow throughout the whole of the ancient canal system, is clearest in instances in which laborers were dispatched to work on critical elements of public infrastructure elsewhere in the Fayyūm, distances of up to 50 km from their home villages. Although such laborers can at times be found at work on the desert canals of villages other than their own,[67] they are most frequently attested at work on two central features of the canal system: the Argaitis canal (the terminal stretches of the Baḥr Yūsuf inside the Fayyūm) and the control works at Ptolemais Hormou/al-Lāhūn.[68] While it is plausible that residents of the villages nearest these critical features were responsible for much of their maintenance, we possess no such evidence.[69] Surviving *penthēmeros*

P.Kron. 61 (106–31 CE); *SB* 18.13985 (132 CE); *SB* 18.13986 (132 CE); *P.Kron.* 69 (153 CE); *SB* 18.13978 (138–61 CE); *SB* 18.13980 (140–41 CE).

64. Theadelphia: *SB* 16.12597 (145 CE); *P.Münch.* 3.108 (145 CE); *P.Fay.* 77 (148 CE) and 78 (147 CE); *BGU* 4.1076 (148 CE); *P.Mich.* 10.595 (161 CE); *SB* 12.10964 (209 CE). Euhēmeria: *P.Hamb.* 1.75 (149 CE). Polydeukia: *BGU* 13.2262 (138–61 CE); *SB* 16.12320 (153 CE); *BGU* 4.1077 (163 CE).
65. Cf. Mikhail, *Nature and Empire*, 175, for the similar Ottoman system.
66. This interpretation inspired by Grey, *Constructing Communities*, 191–92.
67. *P.Mich.* 12.655 (57/58 CE) records a Theadelphian at work on the southern *oreinē diōryx Polemōnos*. *SB* 10.10262 = *P.Brookl.* 11 (206 CE) also seems to show a laborer from Philadelphia on the *Pseinalitidos diōryx*, although the reading is a restoration.
68. On the Argaitis canal see Pearl, "ΑΡΓΑΙΤΙΣ and ΜΟΗΡΙΣ," 27–34.
69. Although several texts from the second-century CE archive of Petaus, *kōmogrammateus* of Ptolemais Hormou (*P.Petaus* 49–51), concern the administration of dike- and canal-work, none references the maintenance of the dam or other infrastructure near the village.

receipts from marginal villages nonetheless attest residents of Bakchias, Karanis, and Tebtynis at work along the Argaitis canal.[70] Indeed, the first papyrus ever published, the so-called *Charta Borgiana* or Schow Papyrus, preserves a list of 181 residents of Tebtynis at work in Ptolemais Hormou from 10–14 Mecheir of 193 CE.[71] Villagers from Tebtynis, Narmouthis, Theadelphia, Soknopaiou Nesos, and Karanis are also variously attested at work in Ptolemais Hormou on a sluice dubbed "six-gated" (*hexathyros*). The structure cannot be identified with certainty but it was presumably a critical component of the system's headworks.[72] A receipt from the Lund papyrus collection also shows a resident of Tebtynis at work on a "gate" (*pylē*) at Ptolemais Hormou, though the identity of the structure is again uncertain.[73] Laborers from Bakchias, Karanis, and Sokopaiou Nesos are also attested at work on a so-called "desert (canal) of Ptolemais Hormou" (*oreinē Ptolemaidos Hormou*), whose identity is also uncertain but might refer to the end of the Baḥr Yūsuf alongside the Lāhūn gap.[74]

That such mobilization could be viewed as a hardship best avoided is illustrated by a petition of 171 CE from the eastern village of Bakchias (*P.Bacch.* 19)

70. Bakchias: *P.Strasb. Gr.* 1.16 (119 CE); *P.Strasb. Gr.* 1.18 (120 CE); *P.Strasb. Gr.* 3.156 (122 CE); *P.Strasb. Gr.* 3.163 (128 CE); *P.Strasb. Gr.* 3.160 (130 CE); *P.Strasb. Gr.* 3.161+164 (130 CE); *P.Strasb. Gr.* 3.167 (143 CE); *P.Strasb. Gr.* 3.168 (144 CE); *P.Mert.* 2.69 (147 CE). Karanis: *P.Got.* 1 (140 CE). Tebtynis: *P.Stras.* 4.249c (129 CE); *SB* 1.5124 (193 CE), ll. 444-55 and 487-97.; *SB* 10.10550 (209 CE).

71. *SB* 1.5124. See also *P.Mich.* 6, pp. 55–56. There is confusion in the scholarship as to whether the text records the names of villagers from Tebtynis sent to Ptolemais Hormou or vice versa. For the latter interpretation see e.g., Thompson, "Irrigation and Drainage," 107–8; and Litinas, "Habent sua fata fragmenta," 399. I hold with the former view, exemplified by Youtie and Pearl in their introduction to *P.Mich.* 6.30; Pearl, ΕΞΑΘΥΡΟΣ 225–27; and Bonneau, "Ptolémaïs Hormou," 322, and regard τὰ χωματικὰ ἔργα Τεπτύνεως in l. 2 as a reference to the labors owed by residents of Tebtynis. Accordingly, the locales that follow—Ptolemais Hormou, the *diōryx Pholēmeōs*, and *diōryx Argaitidos*—are the places to which these individuals were dispatched. For the sake of consistency, if the abbreviated canal names are indeed to be resolved as (ἐν τῇ) + dative as suggested by Sijpestejn (*Penthemeros-Certificates* 66, cf. *BL* 5.95) rather than in the genitive as the *ed. pr.*, then Πτολεμαΐδ( ) Ὅρμ( ) in l. 3 must likewise be resolved as (ἐν τῇ) Πτολεμαΐδ(ι) Ὅρμ(ῳ).

72. Pearl, "ΕΞΑΘΥΡΟΣ." Tebtynis: *P.Kron.* 65 (136 CE); *SB* 18.13979 (143 CE); *P.Kron.* 68 (150 CE); *P.Kron.* 69 (153 CE); *SB* 18.13987 (153 CE); *PSI* 16.1528 (163 CE); *SB* 18.13989 (163 CE); *SB* 16.12674 (169–72 CE). Narmouthis: *BGU* 13.2258 (138 CE). Theadelphia: *PSI* 15.1519 (46 CE); *SB* 16.12316 (123 CE); *SB* 16.12317 (134 CE); *SB* 16.12598 (146 CE); *SB* 16.12599 (146 CE); *P.Sorb.* 1.59 (148 CE). Soknopaiou Nesos: *P.Lond.* 2.139B (51 CE). Karanis: *SB* 16.12299 (101–2 CE); *P.Wisc.* 2.79 (108 CE); *SB* 6.9231 (145 CE).

73. Todd M. Hickey, "A *Penthēmeros* Certificate from the Reign of Caracalla (*P.Lund inv.* 12)," *Zeitschrift für Papyrologie und Epigraphik* 178 (2011): 240–42.

74. See Claytor, "*Penthemeros* Certificates from the Granary," 71 with n. 37. Bakchias: *P.Grenf.* 2.53d (167 CE); *P.Fay.* 79 (197 CE). Karanis: *P.Mich.* 6.381 (second half second century CE); *P.Col.* 7.168 (373 CE). Soknopaiou Nesos: *P.IFAO* 1.32 (69 CE). On the identity of the desert canal of Ptolemais Hormou so already Youtie and Pearl in *P.Mich.* 6.380.

in which several local priests complain to the governor of their division (*meris*) of the Fayyūm that they were being compelled, contrary to custom, to perform the requisite annual dike works far from the village. The priests accordingly beg to be allowed to perform the annual work on the nearby desert canal of Patsōntis, so that they might remain in the village and perform their daily rites in the local temple:

> Ποτάμωνι στρα(τηγῷ) Ἀρσι(νοίτου) Ἡρακ(λείδου) μερίδος παρὰ [Πε]τεύριος Πετεύριος καὶ Σισόιτος [Ὀρ]σενούφεως καὶ τῶν λοιπῶν ἱε[ρ]έων ἱεροῦ τοῦ ὄντος ἐν κώμῃ Βακχιάδι. ἐπεὶ ἔθος ἡμεῖν ἐστιν ἄγεσθαι εἰς χωματικὰ ἔργα οὐκ ἐν ἄλλοις τόποις, εἰ μόνον ἐν διώ[ρυγι] Πατσώντε[ω]ς λεγομένῃ ἀ[φ' ἧς] τά τε περὶ τὴ[ν] κώμην ἐδάφ[η ὑδρ]εύεται καὶ ἰς [τ]ὰ ὑδροστάσια τὰ ὑ[π' αὐ]τῇ κατέρχεται, νῦν δὲ ὁ ὑπὸ το[ῦ] αἰγιαλοφύλακος κατασταθεὶς ἐκβολεὺς βιάζεται ἡμᾶς παρὰ τὸ ἔθος ἐν ἄλλοις τόποις μακρόθεν τῆς κώμης ἐργάζεσθαι, ἀξιοῦμεν ἐὰν σοὶ δόξῃ κελεῦσαι αὐτὸν ἀποστῆναι τῆς καθ' ἡμῶν ἐπηρ[ί]ας ἰς τὸ δύνασθαι ἡμᾶς ἐν τοῖς σ[υ]νήθεσι τόποις ἐργαζομένους π[λη]σιάζουσι τῇ κώμῃ καθ' ἑκάστην ἡμέραν τὰς τῶν θεῶν θρησκείας ποιεῖσθαι γεινομένας ὑπέρ τε δ[ι]αμονῆς τοῦ κυρί[ο]υ ἡμῶν αὐτοκ[ρ]άτορος Αὐρηλίου Ἀντωνείνου Καί[σαρος κ]αὶ τοῦ ἱερωτάτου Νείλου τε[λείας] ἀναβάσ[ε]ως ἵν' ὦμεν βεβ[οηθημένοι.] Π[ε]τεῦρις (ἐτῶν) μ ἄση(μος) Σισόις (ἐτῶν) λε ἄση(μος) (ἔτους) ια Αὐρηλί[ο]υ Ἀντωνείνου Καίσ[αρος τοῦ] κυρίου Παῦνι κ.

To Potamōn, *stratēgos* of the Herakleides *meris* of the Arsinoite nome, from Peteuris son of Peteuris and Sisois son of Orsenouphis and the rest of the priests of the temple that is in the village of Bakchias. Since it is customary for us to be taken to dike works not in other places except on the canal called of Patsōntis, by which the fields around the village are watered and which runs down to the *hydrostasia* beneath it, but the *ekboleus* appointed by the *aigialophylax* is now forcing us, contrary to custom, to work in other places far from the village, we request, if it seems fitting to you, that you order him to desist from this abusive treatment of us, so that it is possible for us, while working in the accustomed places, to remain near the village in order to perform the daily rites for the gods on behalf of the *genius* of our lord emperor Aurelius Antoninus Caesar and the fulfillment of the rise of the most holy Nile, in order that we might be relieved. Peteuris, aged 40 years, no identifying mark. Sisois, aged 35, no identifying mark. The eleventh year of Aurelius Antoninus Caesar the lord, Payni 20.

Although priestly status sometimes guaranteed exemptions from the corvée, whether or not this particular petition was successful cannot be known.[75] The average villager, however, had few options for avoiding the *penthēmeros*. Violent refusal of corvée labor may have been uncommon, since it is attested only once and in a papyrus of the Oxyrhynchite nome rather than from the Fayyūm.[76] Later comparative evidence suggests that passive avoidance was probably a more common strategy, a phenomenon generally referred to in Graeco-Roman Egypt as *anachōrēsis* or "flight from the land."[77] Although *anachōrēsis* is usually regarded as the flight from taxation or liturgical burdens, farmers might seek to avoid rural labor as well. This, at least, was the opinion of the emperor Caracalla, who declared to the Alexandrians in 215 CE that "those who flee from their own rural districts (*tas chōras tas idias*) in order to avoid performing rural labor (*ergon agroikon*)" are unquestionably to be "expelled" (*ekblēsimoi*) from the city and compelled to return to their villages.[78]

Caracalla's rhetoric notwithstanding, there is no evidence for systemic avoidance of or resistance to the *penthēmeros*. Communal solidarity and even simple self-interest must account, at least in part, for the general willingness of villagers to maintain the canals upon which they and their fellow villagers depended. But beyond simply representing a successful form of coordinated localism, I have already suggested that the *penthēmeros* was central to the production of rural subjectivity in the Roman Fayyūm. Indeed, by reducing individual human bodies as units of rural labor inextricably bound to their villages of record yet simultaneously transferable at will throughout the nome, the *penthēmeros* transformed the internal local rituals of independent irrigation

---

75. On exemptions see Sijpesteijn, *Penthemeros-Certificates*, 9. See also the discussion of priestly exemptions in *P.Mich.* 6.381 (second half second century CE). For the exemption allegedly granted a Bakchias priest on grounds of poor eyesight see *P.Mich.* 11.618 (166–69 CE). Exemptions granted by the *stratēgos* to priests of Soknobraisis in Bakchias: *P.Bacch.* 21 (178 CE).

76. In *P.Oxy.* 38.2853 (245/6 CE) local irrigation supervisors from either the Thinite or Herakleopolite nome claim that they were attacked by farmers whom they attempted to compel to work on a local canal.

77. On *anachōrēsis* in general see Naphtali Lewis, "A Reversal of a Tax Policy in Roman Egypt," *GRBS* 34 (1993): 101–18.

78. *P.Giss.* 40 ii (215 CE), ll. 24–25: οἵτινες φεύγουσι τὰς χώρας τὰς ἰδίας ἵνα μὴ ἔ̣ρ[γον] ἄ̣γροικον ποιῶσι. Cf. Mikhail, "Unleashing the Beast," 343 with n. 126 on corvée labor in the eighteenth century: "Peasants were forced to clear mud from canals or to bring in crops . . . Many peasants, however, not surprisingly, found even this local form of forced labor to be objectionable and attempted to escape from it." Mikhail cites a seventeenth-century comic verse reported by the contemporary scholar Yūsuf al-Shirbīnī that satirizes peasant avoidance of forced labor: "And on the day when the corvée descends on the people in the village / Umm Waṭīf hides me in the oven."

communities into external obligations not only to the state and to one's *idia* but also to the Fayyūm as a whole. The difference is critical: informal social obligations to one's community were vacated upon that community's dissolution; obligations owed to the state were not. Yet as we will see later in this book, only when the local labor obligations Fayyūm irrigators discharged on behalf of the state were in harmony with the needs of local communities could this administrative system retain its coherence. Still, in becoming the obliging subjects of an imperial environmental regime that placed significant demands on their bodies and severe constraints on their mobility, Fayyūm villagers entered into an intimate relationship with Roman authority, thereby earning the right to make claims on the state. This sense of rights emerging from obliging subjectivity has been documented elsewhere in the Roman east. As historian Ari Bryen has recently written, "inhabitants of the empire had a sense of rights, and were capable of persuading the Roman imperial government that those rights ought to be respected." These rights were nonetheless not grounded in any legal status such as citizenship but were instead quasi-contractual and "attached to subjects to the extent to which they were willing to be obedient and to pay what was required of them."[79] This sense of mutual give and take—of local acquiescence and attendant state solicitude—is hinted at in the early fourth-century Panopolite circular cited above, which encourages rural residents to request state assistance with matters of local—and consequently fiscal—benefit. It is clearer still in several Fayyūm petitions that solicit state intervention on precisely the terms that Bryen describes. In two second-century papyri from Theadelphia (see chapter 2 under "Canal Flow and Cropping Patterns"), one Ptolemaios complains about the water supply to the patrimonial brushwood-marsh he leases in the village. He addresses the first to Aelius Herakleitos, assistant to the Roman *procurator* in charge of the emperor's patrimonial property in Egypt,[80] and the second, the following year, to the *stratēgos* of the Themistos and Polemon divisions of the Fayyūm.[81] The complaints in each are substantively identical: Ptolemais alleges that the marsh has not received its customary amount of water and consequently risks drying out and becoming unproductive (*aphoros*) for the coming three years. Protesting his blame-

79. Bryen, "Citizenship and Its Alternatives."

80. *P.Wisc.* 1.34 (144 CE). On the office of *procurator usiacus* see Franziska Beutler, "Wer war ein *Procurator usiacus*? Die Verwaltung des Patrimoniums in Ägypten in der ersten Hälfte des 2. Jahrhunderts," *Cahiers du Centre G. Glotz* 18.1 (2007): 67–82.

81. *P.Mich.* 11.617 (145–46 CE).

lessness more forcefully in the second petition, Ptolemias recalls his "many confrontations" (*pollas antikatastaseis*) with the Fayyūm's chief irrigation official (the *aigialophylax* or "shore guard"[82]), all of them fruitless. He accordingly concludes each petition with the request that his addressees inform local officials that they alone will be held to account for any shortfalls in revenue.[83]

A fourth-century petition from Philadelphia, *P.Wisc.* 1.32 (305 CE), more concretely embodies the mutual obligations between state and subject. Addressing themselves to the *stratēgos* of the Fayyūm, the twin heads of village administration (*komarchoi*) assert their eagerness to work for the benefit of the Roman fisc yet claim that their longstanding debt to the treasury is the fault of the neighboring village of Tanis, which lay upstream from Philadelphia on the eastern border canal. They accordingly request an inspection of irrigation works in Tanis, in order that they may receive the water they need, pay an array of taxes, and—crucially—remain in their village of record (*idia*). This barely veiled threat of *anachōrēsis* is here double-edged, for it would represent not only the loss of revenue-generating rural land but also, by the removal of vital laborers from their appointed position, damage to the integrity of the irrigation system at Philadelphia and, by extension, all points downstream:

> [ὑπατείας τῶν κυρίων ἡμῶν Κωνσταντίου καὶ Μαξιμιανοῦ τῶ]ν ἐπιφαν[εστά]των Καισάρων τὸ [ε]. Αὐρηλίωι Ἀπίωνι στρατηγῶι Ἀρσινο[είτο]υ παρὰ Αὐρηλίων Παμουτίου Ἀθιόει καὶ Ἀρριανοῦ Ἀ[π]ύγχεως ἀμφοτέρων κωμαρχ[ῶν] κώμης Φιλαδελφίας. ὑπὲρ λυσ[ιτε]λίας τοῦ ἱερωτάτου ταμίου καὶ τοῦ συστῆναι τὴν ἡμέτεραν [όρ]μὴν ποιούμεθα. στρατηγῶν ἄριστε, μακρόθεν τυγχά[νο]μεν κατοικοῦντες τὴν ἡμέτεραν κώμην, πάνπολλα δὲ χρέως [ἔχο]ντες τῷ ἱερωτάτῳ ταμίῳ ὑπὲρ τῆς ἡμετέρας κώμης οὐ μὴν [ἀλλ] ἀ καὶ ὑπὲρ μέρους Τάνεως. ὑ(πὲρ) τοῦ οὖν εὐθέως δύνασθαι τὰ ὕδ[ατα] καὶ ἡμᾶς καταλαμβάνιν καὶ ὄνησιν ἔχειν καὶ τοῦ πᾶσαν τὴν γῆν ἡμῶν γεωργῖν καὶ δύνασθαι τελῖν τὰ ὑπὲρ αὐτῶν δημόσια τελέσματα καὶ ἀννωνιακη καὶ παντοίας ἐπιβολὰς, ἐπιδίδομέν σοι τάδε τὰ ἔνγραφα ἀξιοῦντες ἐπιστῖλέ σαι δι' ἑνὸς τῶν πε[ρ]ὶ σὲ ὑπηρετῶν τῇ κρατίστη βουλῇ διὰ τοῦ ἐνάρχου πρυτάνεως Αὐ[ρ]ηλί[ο]υ Καστορίωνος, ὅπως ἐρή[σ]ηται συνοψιστὴν τὸν τὴν σύνοψιν τῶν ῥίθου καὶ λιθικῶν ποιούμεν (*l.* ποιούμεν<ον>) τῶν ἐν κώμη

82. Boneau, *Le régime administatif*, 240–44.

83. *P.Wisc.* 35 (144 CE) preserves a letter to the *aigialophylax* in response to Ptolemaios' initial petition. Official correspondence from several years later documents continued water shortages in the *drymos*, now leased to one Pamphilios. *P.Wisc.* 31 (149 CE).

Τάνει τοῦ [ὑμ]ετ[έρ(ου)] νομοῦ, ἵνα δυνηθῶμεν καὶ ἡμῖς τῆς τοῦ λίνου παροχῆς ἀπολαύειν καὶ πότιμον ὕδωρ ἔχειν καὶ τὸ πεδίον τῆς ἡμε[τ]έρας κώμης κατασπίριν καὶ ἐν τῇ ἰδίᾳ στῆναι καὶ ὄνησιν ἔχ[ει]ν τῶν ἰδίων. διευτύχ(ει).

In the consulship of our lords Constantius and Maximianus, the most illustrious Caesares, the fifth time. To Aurelius Apion, *stratēgos* of the Arsinoite from Aurelius Pamutis, son of Athiois and Aurelius Arrianos, son of Apynchis, both *kōmarchoi* of the village of Philadelphia. We act for the profit of the most sacred treasury and for the organizing of our village. Best of the *stratēgoi*, for a long time we inhabit our village having a great many debts to the sacred treasury, truly not because of our village, but in fact because of the role of Tanis. In order now that we, too, can get the water forthwith and benefit and cultivate all our land and can pay the state-taxes for this and the *annona* and all kinds of extra charges we submit this document to you, requesting that you, by means of your servants, inform the most exalted council through the *prytanis* in office, Aurelius Kastorion, in order that it (i.e., the council) carry out an inspection, making an estimate of the states of the stream and which [*sic*] are covered with stones, situated in the village of Tanis of your nome, so that we too can benefit from the coming up of the flax and have drinkable water and sow the plain of our village and stay in our own *idia* and have benefit from our own possessions. Farewell.

Yet it is a petition from the previous third century that most clearly reveals the sense of local rights and state responsibility constituted by the Roman *penthēmeros*. In a petition addressed to a regional governor (*epistratēgos*) Antonius Colonianus, a collectivity of farmers from the northeastern Fayyūm village of Kerkesoucha claims that although they were eager to undertake the rural labors annually demanded of them, certain local officials known as the "supervisors of sowing" (*katasporeis*)[84] had failed to provide their customary material support for the maintenance of a local *emblēma*, a transverse dike that built up a head of water in a canal either to create a reservoir or to divert water into a local branch canal.[85] As a result of this alleged negligence, the village's land may go dry, thus placing at risk the revenues due to the fisc. The only solu-

84. Bonneau, *Le régime administratif*, 168–73.
85. Bonneau, *Le régime administratif*, 39–44.

tion was immediate intervention on behalf of the dependent farmers of the village:

> Ἀντωνίωι Κολωνιανῶι τῶι κρατίστωι ἐπιστρατ<ήγ>ωι παρὰ Γελλίου Σερήνου καὶ Γεμέλλου Ὡρίωνος καὶ τῶν λοιπῶν γεούχων καὶ δημοσίων γεωργῶν κώμης Κερκεσούχων. πρὸ τῶν ὅλων, ἐπιτρόπων μέγιστε, οἱ κατὰ κερὸν (*l.* καιρὸν) ἡγεμόνες τὴν πρόνοιαν ποιούμενοι τῆς γῆς γράφουσι περὶ τῆς τῶν χωμάτων καὶ διωρύχων τὴν ἀπεργασίαν γίνεσθαι. καὶ ἡμῖς (*l.* ἡμεῖς) βουλόμενοι ὡ̣ς καὶ ἀεὶ προθυμότατα συντελῖν (*l.* συντελεῖν) τὰ ἀνήκοντα τ̣ῇ γ̣ῇ ἔργα, οἱ οὖν κατασπορηες (*l.* κατασπορεῖς) τῆς Ἀργαίτιδος φιάλης τ[ο]ῦ ἐνεστῶτος ιθ (ἔτους) οὐκ οἶδ' ὅπως ἤτοι κατ' ἀμελίαν (*l.* ἀμελείαν) οὐ παρήνεγκαν τὰ καθ' (*l.* κατ') ἔτος παραφερόμενα διὰ αὐτῶν ξύλα καὶ ὕλην εἰς ἀνοικοδομὴν φρυγανικοῦ ἐμ̣βλήματος περὶ τὴν αὐτὴν κώμην καλουμένου Κορμοῦ οὐδὲ τὴν ἀπεργασίαν ἐποίησαν ο̣ὐδέν, \ὡς ε̣ἰ . . . . ου ὑπονοίας . . . . γυλ( ) τοῦ διώροχας (*l.* διώρυχας) τὴν δ̣ιαφορὰν ποιῆσαι./ τῆς γῆς κινδυνευούσης διὰ τοῦτο ἀβροχῆσαι καὶ τὸ ἱερώτατον ταμῖον (*l.* ταμεῖον) βλάψαι τὰ ὑπὲρ τούτων μετρούμενα δημόσια οὐκ ἐν ὀλίγαις μυριά\σι/ ὄντα, τοῦ ἱερωτάτου Νίλου προθυμότατα ἑαυτὸν ἐπ' ἀγαθοῖς ἡμῖν ἐπινενευκοτα (*l.* ἐπινενευκότος), ἀξιοῦμεν ἐὰν σου τῇ εὐμενεστάτῃ τύχῃ δόξῃ κελεῦσαι διὰ τῆς σῆς εὐτονίας τὸ ἔργ[ο]ν γενέσθαι <ὥστε> καὶ δυνηθῆναι ἡμᾶς τὴν ἐπ' ἀγαθοῖς ἐσομένην κατασπορὰν ἀμέμπτω[ς ε]πὶ τὸ πλέον αὐξῆσαι καὶ τῷ ἱερωτάτῳ ταμ̣ίῳ (*l.* ταμιείῳ) μηδὲν παραπολέσθαι.

To Antonius Colonianus, the most noble *epistratēgos*, from Gellius Serenus and Gemellus Horion and the rest of the landowners and public cultivators of the village of Kerkesoucha. First of all, most excellent of governors, those who serve as prefect [of Egypt] from time to time, devoting forethought to the land, issue written orders concerning the accomplishment of the labor of maintenance on the dikes and canals.[86] And although we were most zealously prepared to perform fully the labors that pertain to the land, the supervisors of sowing for the Argaitis bowl in the present nineteenth year, for some unknown reason or through simple negligence, did not produce the wood and materials which are annually provided by them for the reconstruction of the wattled weir in the vicinity of the same village called the "Log," nor did they in any way provide for maintenance, as if [sc. without the least?] suspicion of the fact that

86. Cf. Egerton Papyrus 13 above.

> canals make the difference [i.e., between prosperity and dearth?]. Since the land runs the risk of going dry on this account and inflicting injury on the most sacred treasury in the matter of the dues paid for these [*arourai*], which come to not a few myriads [of *artabai* of grain], at a time when the most sacred Nile has shown himself most favorable to us for a good issue, we request, if it seems best to your beneficent genius, that you order with your vigor that the work be done, so that we may be able blamelessly to maximize the imminent sowing—may it be for the good!—and nothing be lost to the most sacred treasury.[87]

In demanding the aid of a Roman high official with works that Roman high officialdom in turn demanded of them, the farmers of Kerkesoucha enacted and reinforced the bonds of subjectivity that tied them to their rural village of record and its public irrigation infrastructure. Yet theirs was also a leveling discourse (not unlike that of the Theadelphian Ptolemais), for it cast both these local farmers themselves and the local officials who allegedly failed them in a particularly close fiscal and labor relationship with the *epistratēgos* and the Roman state: after all, if they were willing to play their parts as subjects of this imperial environmental regime, so too should the local *katasporeis* (likewise Ptolemais' local antagonist, the *aigialophylax*). The petition from Kerkesoucha thus encapsulates the distinct duality of Roman coordinated localism in the Fayyūm: at once internally communal and externally coercive in character, local and imperial in scope, top-down and bottom-up in its practical administration.[88] This duality—its simultaneous embodiment of spontaneous communal agency and compulsory individual obligation—surely contributed to its durability. Indeed, even after the *penthēmeros* itself had disappeared, Rome continued enforcing dike- and canal-work in the Fayyūm, seemingly assimilating it to the *naubion*-corvée elsewhere in Egypt. Villagers at Karanis are thus attested moving *naubia* of earth in the late third or early fourth century,[89] as are residents of Narmouthis, Theadelphia, and Euhemeria on the opposite rim of the depression.[90] Yet however the annual corvée was administered, it was altogether reliant on a network of stable, vibrant, and self-perpetuating communities, whose populations were personally invested in the maintenance of their own irrigation infrastructure. As I will describe in the following chapter, the

87. *SB* 14.11478 (210–11 CE). Trans. Youtie, "P. Mich. Inv. 2920 = Sammelbuch 4.7361," 152.
88. See the petition *P.Oxy.* 12.1469 (298 CE) in which *naubion* laborers of the Oxyrhynchite village of Païmis complain in much the same terms of being exploited by local officials who allegedly imposed an unfair workload and did not properly credit them for work accomplished.
89. *O.Mich.* 266–69 (mid-third–early fourth century CE), 284, 292, 294 (third century CE).
90. *P.Sakaon* 53 (fourth century CE).

dissolution of independent village communities therefore eroded the human infrastructure upon which Roman coordinated localism was founded. I nonetheless close the present chapter with a glimpse at al-Nābulusī's Fayyūm, an era in which the Egyptian central government had largely retreated from the Fayyūm's countryside and left the management of the canal system in the hands of local villagers. The social practices of communal irrigation discussed in this section thus endured, albeit amid the advanced degradation of the physical infrastructure of Fayyūm irrigation. This snapshot accordingly illuminates the practical and material significance of the forms of intervention practiced by the earlier Ptolemaic and Roman states.

Coda: Durability and Decay in al-Nābulusī's Fayyūm

Near the beginning of his village survey al-Nābulusī makes a sharp accusation:

وتواترب مطالعات عبيد دولته وارقاء مملكته باحوال الفيوم وانه ربما فترب الهمة في عمارته واستمر اهمال المباشرين له حتى تغير عن حالته.

> Reports about the affairs of the Fayyūm by the servants of his state [i.e., the Sultan al-Mālik al-Ṣāliḥ, r. 1240–48 CE] and by the slaves of his kingdom have been copious. Indeed, attention to its cultivation has often slackened, and the negligence of its supervisors persisted until its situation had deteriorated.[91]

Upon visiting the Fayyūm, the sultan al-Mālik al-Ṣāliḥ himself allegedly told al-Nābulusī that "the local officials have been careless with this region until its neglect has become apparent."[92] Al-Nābulusī's dispatch to the province was a direct response to this supposed neglect and his subsequent survey was intended as a guide to its future rehabilitation.

The thirteenth-century Fayyūm was indeed much reduced following the political infighting, low floods, famine, and rampant starvation that marked the early years of the second Fatimid century (1068–1074 CE) and contributed to the social and economic decline of the province.[93] By the time the Fayyūm again becomes historiographically visible in al-Nābulusī's survey, the state had all but retreated. Al-Nābulusī records that more than 90 percent

91. Al-Nābulusī, *VF*, 31.
92. Al-Nabulusi, *VF*, 31–32. هذه البلاد قد غفل عنها عمالها حتى ظهر اهمالها
93. Rapoport, *Rural Economy*, 47–51.

of the Fayyūm's villages were distributed to members of the military as grants (*iqṭāʿ*), an institution that entitled recipients to the entirety of their revenues.[94] Compared to Rome's fiscal interest in the agricultural produce of every Fayyūm village, the Ayyūbid sultan's minimal 10 percent stake in the province provided little incentive for state intervention in the work of irrigation. As described in an important article by Yossef Rapoport and Ido Shahar, the near-complete absence of the state meant that the management of the thirteenth-century Fayyūm's canal system as well as the organization of water distribution and water rights fell to the various tribal units that made up the Fayyūm's population.[95] But whereas Rapoport and Shahar highlight the positive aspects of Ayyūbid localism—the continued upkeep of select infrastructure, noncoercive local mobilization, egalitarian water sharing, and so on—attention must also be paid to al-Nābulusī's discussion of the deleterious siltation of both the al-Lāhūn inlet and the mouth of the Baḥr Yūsuf al-Manhā in southern Egypt at Dayrūṭ, two critical pieces of infrastructure upon which the integrity of the entire canal system depended. This infrastructural decay, coupled with limited and irregular manpower mobilization, stands in stark contrast to Ptolemaic central organization and Rome's more thoroughgoing system of coordination cum coercion, thereby underscoring what could and could not be accomplished by local agency alone.

In al-Nābulusī's telling, the Fayyūm's troubles began at al-Lāhūn and extended upstream to the head (*raʾs*) of the Baḥr Yūsuf/al-Manhā at Dayrūṭ. Beginning in 1223–24 CE, the *iqṭāʿ*-holder of the Fayyūm, one Fakhr al-Dīn ʿUthmān, undertook to remedy these problems beginning with the al-Lāhūn inlet, which had become badly silted up and was no longer flowing at optimal capacity.[96] First, he removed vegetation from the banks of the inlet in an attempt to widen it. Though unsuccessful, this did no harm. Fakhr al-Dīn then raised the dam slightly, inadvertently causing a buildup of silt in the channel, which had to be removed by hand in May and June. Having been unsuccessful here as well, he turned to the head of the channel at Dayrūṭ (Darwat Sarabām), which was allegedly so impacted with sediment that it no longer delivered the amount of water it had carried in the past. The reason, al-Nābulusī claims elsewhere, was neglect in its maintenance:

94. Revenues: Rapoport, *Rural Economy*, 10.
95. Rapoport and Shahar, "Irrigation," 27–28.
96. The following depends on Rapoport and Shahar, "Irrigation," 11–14.

وقد كان راس هذا البحر المنهى فيما سلف يجف كل سنة رُبعة اشهر ويمد الفيوم ماؤه الحاصل فيه والنزر بقية السنة وهي ثمانية اشهر فانعكس الحال فيه لا همال حفوه وعدم العناية به فصار يجف الان ثمانية اشهر ويمده النيل رُبعة اشهر ومن دليل اهماله انه لم يوجد للاهتمام بحفره ذكر في الديوان ولا حساب يشهد به من مدة تزيد على مائة سنة.

In the past the head of this al-Manhā canal would only lie dry for four months every year. During the rest of the year, that is, for eight months, its water—that which flows into it [i.e., floodwater] as well as seepage water (*al-nazr*)—used to reach the Fayyūm. The situation has now reversed due to negligence in digging it and to lack of maintenance, and it now lies dry for eight months and the Nile flows into it for only four months. An indication of its neglect is that there is no account in the Treasury of any attention being given to it, and there is no receipt attesting to it for a period of over one hundred years.[97]

Yet Fakhr al-Dīn succeeded only in worsening the problem. According to al-Nābulusī, he had brought with him to Dayrūṭ a number of Fayyūmī "engineers" to design a solution to the channel's low water levels. These engineers—in reality simply men from al-Lāhūn who oversaw the operation of the dam and whose technical expertise al-Nābulusī accordingly scorned—suggested cutting a new head for the channel some 390 m below the old opening. The experiment was a failure, al-Nābulusī alleges, and actually reduced the amount of water that entered the channel. By divine providence, he claims, it soon silted up and returned to its previous condition. A second attempt was made, this time by sinking a number of boats at the head of the channel in the hopes that they would rapidly be silted over and form an artificial island that would divert water into the al-Manhā. Once again, the plan failed and water was diverted away from the head of the al-Manhā rather than into it, contributing to its eight months of dryness.[98] These physical manifestations of long-term state neglect, al-Nābulusī writes, were starkly visible along the ancient desert canals of Tanabṭawayh and Waradān. Both were choked with silt and abandoned, their banks lined with the ruins of ancient settlements.[99] Their total reclamation, he suggests, could be effected only with considerable time and expense. Although local efforts had resulted in the reoccupation of a handful of abandoned villages, little more could be achieved without state intervention, a

97. Al-Nābulusī, *VF*, 35.
98. Al-Nābulusī, *VF*, 46.
99. Al-Nābulusī, *VF*, 46–47.

prospect that would inevitably entail coercion and risked the potential flight of the Fayyūm's inhabitants.

Al-Nābulusī was of course incorrect to blame the desertions of marginal settlements on recent governmental neglect, since a number of these ancient sites were already deserted by the later Roman period. Al-Nābulusī's assertions of the Fayyūm's ruinous condition also rhetorically magnified the importance of his own assignment and simultaneously accentuated the sultan's paternal care for a sadly marginalized province. Still, the apparently poor condition of the al-Lāhūn inlet alone speaks to the indispensable role of state-directed labor mobilization in maintaining the canal system at its original extent. Indeed, the 181 laborers sent to Ptolemais Hormou from Tebtynis in 193 CE might have represented 2 to 3 percent of the village's total population.[100] If levies of similar scale were raised from all of the Fayyūm's rural settlements, anywhere between 4,000 and 6,700 men could have descended upon the headworks of the canal system each year to prepare it for the arrival of the flood.[101] While this is nothing more than a thought experiment, it nevertheless highlights a precipitous decline in manpower mobilization between the Roman and late-Ayyūbid periods. Absent this coercive authority from above, local initiative alone seems to have been insufficient to ensure that the inlet remained dredged and clear of obstruction.

While it would not be altogether unwarranted to regard such problems as evidence of unadulterated decline, Rapoport and Shahar depict the survival of irrigated agriculture in the province as a demonstration of successful localism in Egyptian irrigation.[102] As I have already argued in the first chapter, the al-Lāhūn dam had been simplified since the early Islamic period and transformed into a simple spillway that could be operated and maintained by a collection of locals dubbed "engineers" (*muhandisūn*) by virtue of their long experience with the structure.[103] Chapter 2 further demonstrated that the central Fayyūm remained highly productive after antiquity, the retreat of the state notwithstanding. The Ayyūbid Fayyūm had in other words established a new socioenvironmental equilibrium, one adapted to the absence of an extractive and thus highly interventionist central state. It may not have been as large or as wealthy

100. Assuming a population of between 5,400 and 7,400: Langelloti, *Village Life*, 58–59.

101. Assuming a Roman-era population of 170,000 to 200,000. See the Introduction above, "From Reclamation to Retrenchment."

102. Rapoport and Shahar, "Irrigation in the Medieval Islamic Fayyūm."

103. Al-Nābulusī, *VF*, 41.

as its ancient predecessor, but it remained inhabited, fruitful, and internally self-regulating. This durability notwithstanding, the decay of several vital pieces of infrastructure serves as a potent reminder of the limits of localism within this state-created and (formerly) state-maintained space. We must accordingly regard the flow of water through the much larger ancient Fayyūm as a product of unique entanglements between the water, infrastructure, rural society, and activist states—entanglements that had extensively unraveled by the mid-thirteenth century CE.

Conclusion

This brief case study of Fayyūm water governance suggests that the current valorization of localism in premodern Egyptian irrigation needs refinement. Although al-Nābulusī demonstrates that local agency could indeed maintain a functional Fayyūm irrigation system in the absence of intensive state penetration, state power was nonetheless deeply entangled with irrigation during the Ptolemaic and Roman periods and proved integral to the maintenance of the canal system's original extent. Yet even these earlier, more intrusive states were neither despotic nor totalizing and instead governed Fayyūm water flow through a series of contractual, cooperative, coordinating, and coercive relationships with the canal system's local beneficiaries. State-rural society relationships in the Fayyūm were thus marked by considerable diversity between the three historical periods discussed here. The perennial debate between statism and localism in premodern Egyptian irrigation must therefore eschew abstractions and generalizations and instead pay keen attention to state particularity, that is, the differing fiscal and political goals of Egypt's numerous state governments as well as their disparate institutional structures. In so doing it will become clear that the history of premodern Egyptian water governance was marked by no less change, rupture, and discontinuity than the country's political history.

But beyond these programmatic arguments, this chapter has placed particular emphasis on Roman coordinated localism, arguing that it constituted a form of rural subjectivity for the perpetuation of extant patterns of Fayyūm water flow. Yet by fashioning a subjectivity inseparable from place (*idia*), Roman coordinated localism remained coherent only as long as the irrigation communities resident in those places were healthy, cohesive, and thus collectively invested in annually (re)producing local water flows. The following chapter thus explores the internal constitution of Fayyūm irrigation communities and the methods by which they collectively produced and distributed the waters upon which they depended. This fuller understanding of the human social infrastructure of Fayyūm irrigation will in turn allow us to better establish the problems in the fourth century that plagued the western Fayyūm village of Theadelphia, whose decline and death are the subject of the final chapter.

Chapter 4
Communities of Flow

We smash them . . . We smash his face in!
—*P.Haun.* 3.58 (439 CE) and anonymous Fayyūmī farmers (2000s CE)[1]

Water Rights, Water Fights

ʿAbd al-Hādī was anxious. Water was scarce, tempers flaring, and conflict had become all but inevitable. The trouble had started a short time ago when Egypt's Ministry of Irrigation halved his village's statutory irrigation period from ten days per month to five. Under the previous ten-day regime, villagers had used a single animal-powered waterwheel (*sāqiya*) to transfer water from the public feeder canal to their village's canal network. Although the wheel lay on ʿAbd al-Hādī's land and was nominally owned by him, the other villagers possessed fixed time-shares in it and used it by turns to raise water for their own fields—a small irrigation cooperative known as a *sāqiya* ring, of which ʿAbd al-Hādī was the head, the *shaykh al-sāqiya*.[2] These fixed shares had been assigned in proportion to the amount of money that each villager had contributed to finance the wheel's construction and then further calibrated to the ten-day irrigation period. Elegant and equitable, this finely tuned system nonetheless immediately became obsolete when the Ministry reduced the village's water supply.

After a period of ratcheting tension, communal solidarity was suddenly and violently shattered during ʿAbd al-Hādī's own turn at the wheel. As he was

1. Luyendijk, *People Like Us*, 55–56.
2. Mehanna, Huntington, and Antonius 19–21; Gouda, *Social Capital*, 59.

walking alongside the ditch leading to his fields and monitoring the flow of the water he had raised, the steady stream suddenly slowed to a trickle. Alarmed, he quickly discovered that his neighbor Diāb had breached the earthen dike that separated their plots in order to divert the water onto his own fields and those of Diāb's own neighbor Mas'ūd Abū Qāsim. Angered by such brazen theft, 'Abd al-Hādī immediately confronted Diāb, who spit back in anger:

> اسمع ياعبد الهادى لما اقول لك!! . . . أنا ليه في الساقية يوم وجارى مسعود أبو قاسم يوم!! آخذ ميه على كيفى! . . . وللا علشان ما اسمها ساقيتك؟ ساقيتك قال! احنا لنافيها يوم ومحمد أبو سويلم له يوم، ومسعود أبو قاسم والناحية الشرقية يومين، وأنت بقيت العشرة ايام! أنا حاخذ يومنا في الساقية النهاردة. ياللا حل بهيمتك وأهى مرات مسعود أبو قاسم جايه أهى ومعاها البهيمة!

> Listen to what I'm saying to you 'Abd al-Hādī . . . I have one day at the wheel and my neighbor Mas'ūd Abū Qāsim has a day. I'll take water as I like . . . You say it's your wheel, do you? Yours!? We have a day of it, Muḥammad Abū Suweilim has a day, as does Mas'ūd Abū Qāsim, and the eastern sector of the village has two days, and you have the rest of the ten days. I'll take our day at the wheel today! Now c'mon, take your animal off because Mas'ūd Abū Qāsim's wife is coming with their animal![3]

The simultaneous arrival of both Mas'ūd Abū Qāsim's wife and a number of farmers from the village's eastern sector only inflamed the situation further, since each demanded to begin taking their respective shares of the wheel immediately. In his role as *shaykh al-sāqiya*, 'Abd al-Hādī attempted to mediate, reminding his fellow irrigators that their established water-sharing regime was now defunct:

> وحاول عبد الهادي أن يغير عزمهم، فقد كان لهم يومان عندما كانت أيام الرى عشرة أما الآن فلو انهم تمسكوا بيومين فلن يجد بقية الشركاء في الساقية ما يكفي لرى الأرض العطشانة! . . . الناحية الشرقية كان لها يومان من عشرة وأيام الرى الآن خمسة فلها يوم واحد. واختلطت اصوات الرجال والنساء في رفض لما يقول عبد الهادى . . . وارتفعت الشمس قليلا والمناقشة تحمى بين أهل الناحية الشرقية وبعضهم، وبينهم وبين عبد الهادى، وبين عبد الهادى ودياب. وأحس كل واحد من الواقفين كأنما الآخر يريد أن يسلبه الحياة نفسها! . . . وباسم الدفاع عن حياة الأرض—عن الحياة نفسها—مضى كل فلاح يضرب ويضرب بلا توقف كل من يريد أن يناقش حق الأرض في الماء!

> 'Abd al-Hādī tried to change their minds: yes, they [the eastern sector] may have had two days when the irrigation period was ten, but if they clung fast to

3. Al-Sharqāwī, *Al-Arḍ*, 153–54.

> these two days, there wouldn't be enough shares in the wheel left over to irrigate other thirsty land . . . The eastern sector previously had two days out of the ten but since the days of irrigation were now five, it should have one day. But both the men and the women raised their voices together in rejection of ʿAbd al-Hādī's words . . . As the sun began to rise, the conflict—between the people of the eastern sector and the others, between them and ʿAbd al-Hādī, and between ʿAbd al-Hādī and Diāb—grew heated. Each felt that the others were trying to deprive him of his very life! . . . In the name of defending the life of his land—indeed of life itself!—each farmer exchanged blows without ceasing, every one of them proclaiming his own land's right to water.[4]

The village elders soon intervened in an attempt to stop the violence but their pleas went unheard. Only when it was discovered that Masʿūd Abū Qāsim's frightened buffalo had fallen into the *sāqiya*'s well did the brawl come to an end. After all, such a disaster required the villagers to put aside their enmity, if only temporarily, and work together to drag the poor beast out.

• • •

This vignette is an excerpt from the Egyptian novelist ʿAbd al-Rahmān al-Sharqāwī's 1953 social-realist classic *Al-Arḍ* (*The Land*), which draws on the author's own rural upbringing in the Nile Delta to recount the dissolution of a fictional Delta village community in the 1930s due to water shortages forced upon them by a corrupt Irrigation Ministry.[5] I begin with this piece of modern Egyptian popular fiction because it encapsulates more vividly than any historical source the twofold argument of this chapter, namely, that water flows within rural irrigation communities are the products of both social and environmental factors, and that disruptions to established patterns of flow are significant drivers of conflict. As al-Sharqāwī's fictional conflict illustrates, water rights are fundamental to the structuring of water flow within and between irrigation communities. Water rights are here understood as the entitlement of an individual or a group to make use of a portion of the flow of a shared public water source, all backed by a legitimate authority, whether internal or external to the irrigation community, with power of enforcement.[6] By establishing an agreed-upon distribution of the common resources, water rights shape irrigators'

4. Al-Sharqāwī, *Al-Arḍ*, selections from pp. 154–56.
5. The abridged English translation by Desmond Stewart is published under the title *Egyptian Earth* (London: Saqi, 1990).
6. Drawing on language in Beccar, Boelens, and Hoogendam, "Water Rights and Collective Action," 3.

expectations and thus help to mitigate, if not eliminate, conflicts. Broadly speaking, water rights can be divided into two principal categories: property-based rights and land-based rights.[7] The former construes the right to water as private property. At least in theory, property rights in water eliminate competition and conflict by entitling each irrigator to a clearly defined portion of the common-pool water resources—for example, a certain volume of water or a fixed period of time during which a farmer may draw upon the common water supply, such as the time-shares of al-Sharqāwī's imagined *sāqiya* ring. This approach stands in stark contrast to land-based methods of apportionment in which the right to benefit from the flow of a public watercourse is generated simply by possessing land abutting or irrigated by it—for example, the common law doctrine of riparian rights.[8] Although the contemporary international development community argues that property rights in water more effectively manage tensions between irrigators and stimulate more efficient and environmentally sustainable water management practices, this enthusiasm is somewhat tempered by the acknowledgment that water supplies in many parts of the world are variable and uncertain, making it impossible to guarantee that every user's fixed water entitlement can be fulfilled indefinitely by the available resources.[9] Such is the case in al-Sharqāwī's fictional village. The seeming stability of its water-sharing regime was illusory, for it was founded upon the water supply of a canal whose flow could be altered at will by the Ministry of Irrigation. Rigid and inflexible, the system of time-shares was unable to accommodate a sudden diminution of the water supply.

To be able to accommodate change and uncertainty, durable and sustainable water-rights regimes must therefore be constructed in dialogue with local environments. Scholarship on contemporary irrigation communities accordingly shows that water-sharing practices tend to be deeply informed by the hydrology of the surrounding waterscape. For example, communities served by predictably unpredictable water supplies—waters whose arrival or volume regularly varies and cannot be predicted in advance—tend to share water in ways designed to accommodate such uncertainty. In spate- and flood-canal irrigation systems, where the timing and volume of the annual water supply fluctuate from year to year, water is generally apportioned not

---

7. For traditional land-based vs. modern property-based water rights see Hodgson, *Modern Water Rights*, 4–30.
8. Scott, *Evolution of Resource Property Rights*, 63–66.
9. Hodgson, *Modern Water Rights*, 98–99.

by fixed property-based entitlements but through a flexible, adaptable, and responsive set of guidelines that allows irrigators to distribute unpredictable floodwaters in a manner tailored to the unique character of each annual flood.[10] In these systems, rights *in* water are not assigned from without; rather, a right of access *to* the common-pool water resources is generated from within the community by maintaining amiable, reciprocal relationships with other canal-sharers, particularly through voluntary participation in the communal work of building and maintaining shared irrigation infrastructure, a labor ritual that serves as a "reaffirmation and reproduction of the relationship" among the members of an irrigation community.[11] Hence, water rights in such communities are best regarded as a concrete manifestation of social relationships between irrigators. Far from abstract legal entitlements assigned and adjudicated by an external authority, they "are inseparable from the way water management is organized . . . [and] part of a bundle of responsibilities to the entire group."[12] Emerging from within the community, such water rights are moreover enforced and defended by the community, both internally and against external encroachment.

This perspective will be brought to bear on the problem of communal water sharing in the premodern Fayyūm, both among the villages of the Graeco-Roman margins and those of the Islamic-era center. It is the former that seemingly present the more perplexing case. Although chapter 2 has demonstrated that the papyri shed light on the shape of the border canal system and its punctuated flow, they provide no evidence that a system of entitlements or quotas ever governed water sharing within or between marginal villages throughout the entire Graeco-Roman millennium. This silence is striking, particularly since we possess scattered papyrus records of inter- and intravillage water conflicts from the Ptolemaic, Roman, and early Islamic periods.[13] In an important article on water rights in the Roman Empire, historian Dennis Kehoe has critiqued this apparent absence of externally regu-

---

10. Mehari, van Steenbergen, and Schultz, "Water Rights and Rules."
11. Beccar, Boelens, and Hoogendam, "Water Rights and Collection Action," 6.
12. Mehari, Steenbergen, and Schultz, "Water Rights and Rules," 115. See also Ostrom, *Governing the Commons*, 69–81; Van Steenbergen, "Understanding the Sociology of Spate Irrigation, 359-61; Meinzen-Dick and Nkonya, "Understanding Legal Pluralism"; Clarke and Malcolm, "The Role of Property in Water Regulation."
13. For disputes in the Roman and Ptolemaic periods see, respectively, Anagnostou-Canas, "Les différends"; and Anagnostou-Canas, "Litiges en rapport." For the eighth-century CE Islamic Fayyūm see texts 16 and 17 (*P.Mich.inv.* 5613 and 5625) in Sijpesteijn, *Shaping a Muslim State*, 347–56. Modern comparative perspective in Price, "Water Theft."

lated water sharing, arguing that Rome's inadequate governance of the Fayyūm's commons created perverse incentives for upstream irrigators to use as much as water as possible thereby depriving downstream farmers and contributing to the collapse of marginal villages like Aurelius Sakaon's Theadelphia in the fourth century CE. Although Kehoe does not phrase it in such terms, his argument in some respects exemplifies the so-called "tragedies of the commons" famously described in 1968 by ecologist Garret Hardin, in which the absence of both externally imposed rules and internal communal self-regulation allegedly incentivizes the self-interested overexploitation and subsequent degradation or depletion of common resources.[14]

Yet such perverse incentives did not suddenly materialize in the early fourth century CE to the detriment of marginal villages. The durability of internally regulated water sharing along the Fayyūm's margins between the Hellenistic and early Islamic period must therefore be explained. Inspired by the late economist Elinor Ostrom, whose work demonstrates that farmer-managed irrigation systems do not inevitably result in tragedies of the commons,[15] I suggest that the absence of externally regulated water entitlements along the ancient margins was not a failure but a feature of irrigation practices in this portion of the depression: a practical adaptation to the predictably unpredictable water supplies of the flood-irrigated agricultural landscape described earlier in this book. This seasonal unpredictability is to be contrasted with the markedly different environment of the Fayyūm's central plain, which was characterized by greater water availability and easier irrigation. Here, at least in the thirteenth century, villages were assigned a defined water quota (*ʿibra*) from a specific canal, information recorded by al-Nābulusī.[16] According to al-Nābulusī, however, only villages with predictable water supplies drawn from gravity-fed perennial canals possessed such a quota. Settlements located on higher grounds—areas out of reach of or

14. Kehoe, "Economics and the Law of Water Rights." Garrett Hardin, "The Tragedy of the Commons," *Science*, 162.3859 (1968): 1243–48.
15. Ostrom, *Governing the Commons*; Ostrom, "Coping with Tragedies"; Ostrom, "Collective Action"; and Ostrom, "How Farmer Managed Irrigation Systems Build Social Capital." See also Ostrom and Gardner, "Coping with Asymmetries." For a primer on social capital and small-scale irrigation communities see Jeff Dayton-Thompson, "Social Capital, Social Cohesion, Community: A Microeconomic Analysis," in *The Economic Implications of Social Cohesion*, ed. Lars Osberg (Toronto: University of Toronto Press, 2003), 43–78, esp. 50–67.
16. The translation of عبرة as "quota" is not immediately obvious since the word has shades of meaning from "admonition" and "warning" to "advice" and "rule." In the Egyptian dialect, however, the related verb عبّر means "to measure." See El-Sayid Badawi and Martin Hinds, *A Dictionary of Egyptian Arabic* (Beirut: Libraire du Liban, 1986), 559. Regardless, "quota" in the sense of "measurement" or "share" is the only rendering of the word that makes sense of its use in this context, and it is so translated by Rapoport and Shahar in their edition of al-Nābulusī. On the nature of these quotas see further below under "Environment and Apportionment."

poorly served by gravity-fed perennial canals and thus watered only by the annual flood—were not assigned an *ʿibra*. Where water was plentiful and predictable, in other words, more explicit rights in water could be assigned and enforced, at least at the level of the village. Where water supplies were variable, unpredictable, and not infrequently scarce, water sharing was less tightly structured. The latter scenario obtained along margins of the Fayyūm in antiquity. On higher grounds or near the tail end of a long canal system, the water supply to these settlements was susceptible to significant interannual variation influenced by the height of the Nile's flood, changing patterns of upstream water consumption, and even simple conveyance loss through seepage and evaporation.[17] Given the persistent and acknowledged risk of unflooded land (*abrochos*) in all periods, the water supply to these tail end settlements was indeed always uncertain.[18] Since this unpredictable hydrological regime was ill-suited to apportionment via property rights in water, it is unsurprising that the handful of irrigation conflicts recorded in papyri—most but not all in the form of petitions—do not revolve around alleged violations of irrigators' private rights in water. Complainants instead focused their frustrations on physical modifications to the public irrigation infrastructure that altered its flow and thus effectively denied them their accustomed access to the commons. Sadly, if predictably, the information provided by petitions is incomplete, for they illustrate only instances in which water-sharing practices had broken down. Moreover, since the purpose of such documents was to solicit state intervention on one's own behalf, petitioners necessarily always portray themselves as the victims of an injustice. These limitations notwithstanding, such texts enable us to reconstruct the normative expectations of rural irrigators along with the sorts of behaviors they regarded as deviant. In rare cases—and despite their better efforts—petitioners also offer glimpses into the mundane if sometimes rough encounters between irrigators that characterize the everyday life in water-sharing communities—moments in which our petitioners found themselves in a weaker position.

Above and Below

The Fayyūm's canal system introduced the problem of endemic upstream/downstream water conflict to premodern Egypt. As Karl Butzer recognized

17. For modern Fayyūmī comparanda see Price, "The Cultural Effects of Conveyance Loss."
18. See chapter 2, pp. 95–96 at nn. 89–91.

decades ago, upstream/downstream competition for and conflict over water resources had never been a systemic or indeed existential problem in the Nile Valley and Delta since settlements within the floodplain had essentially equal access to the waters of the annual flood.[19] This is not to suggest that conflicts over water between neighboring communities never occurred. The complex tangle of basins and canals recorded from the northwestern Delta province of Buḥayra/Ḥauf Ramsīs in the eleventh and twelfth centuries was tightly regulated, the breaching and sealing of dikes and canals being performed according to an exacting schedule in order to maintain predictable water levels in feeder canals and forestall conflict between neighboring settlements.[20] The Fayyūm's irrigated environment, however, presented greater possibilities for intercommunal conflict. Lying distant from the river itself, its villages were directly linked to one another by canals that delivered water to their dependent communities in stages, settlement by settlement, from the head of each canal to its tail. Because of the fragility as well as the comparably small size and volume of canals, upstream settlements had the ability to deprive their downstream neighbors of water by failing to perform regular maintenance on their portion of the shared waterway, purposefully obstructing its flow, or by drawing from it so heavily that downstream settlements were left with insufficient resources.[21] As al-Nābulusī memorably writes, downstream users might also suffer from poor-quality water. Those at the end of the Baḥr Yūsuf, he claims, suffered greatly from the refuse and corpses discarded into the canal upstream, which befouled its waters and promoted ill health among its downstream users, particularly during the low-water season when the level of the channel was low.[22] These sorts of upstream/downstream tensions remain a fixture of Fayyūm agricultural life in the present day and exert a centripetal pull on the farmers at the furthest extremities of the canal system. The downstream position (Arabic *taḥt*, "below"), where water is most scarce, is known to be disadvantageous and "everyone wants to be what they call *fowq*, 'above,' meaning upstream."[23] The phenomenon, seemingly universal, is still more evocatively described by American writer and garlic farmer Stanley Crawford. According to the folk wisdom of the New Mexico *acequia* (irrigation canal, from Arabic *al-sāqiya*) community of which Crawford is a member:

19. Butzer, *Early Hydraulic Civilization*, 109.
20. Borsch and Sabraa, "Qānūn al-Riyy."
21. Bagnall, *Egypt in Late Antiquity*, 141–42.
22. Al-Nābulusī, *VF*, 39.
23. Barnes, *Cultivating the Nile*, 124.

> You can argue that the character of a man or woman can be as much formed by genetic and cultural material as by the location of their garden or chile patch along the length of a ditch, toward the beginning where water is plentiful or at the tail where it will always be fitful and scarce. "He's that way because he lives at the bottom of a ditch and never gets any water," is an accepted explanation for even the most aberrant behavior in this valley. The man who lives at the bottom of a ditch is forever expectant, forever disappointed.[24]

Given the realities of life along an irrigation canal, especially at or near the disadvantaged tail end, it is hardly surprising that upstream/downstream conflicts appear in the Fayyūm papyri as early as the third century BCE.[25] The most familiar and detailed instances, however, are contained in the fourth-century CE archive of Aurelius Sakaon of Theadelphia—texts we will revisit in the final chapter. Among these papyri the *locus classicus* is *P.Sakaon* 35 (ca. 332 CE), the beginnings of a courtroom *narratio* outlining a complaint levied by the Theadelphians against several villages upstream who shared the same waterway(s) with downstream Theadelphia:

> κατὰ τὸν προπέρυσι ἐνιαυτὸν καὶ πέρυσι τῶν ἐδαφῶν τῆς κώμης ἡμῶν ἐν ὑψηλοῖς τόποις ὄντων καὶ τῶν ἔγγιστα κωμῶν, Ναρμούθεως καὶ Ἑρμουπόλεως κώμης καὶ Θεοξενίδος, ὑποκλεπτόντων ἡμῶν τὰ ὕδατα καὶ οὐκ ἐπιτρεπόντων ἀρδεύεσθαι ἡμῶν τῆν γῆν, διὰ τὸ ἀρχὴν αὐτῶν εἶναι τῶν πάγων καὶ ἡμᾶς ὑστέρους εἶναι τοῦ πάγου, ἔρημον κώμην οἰκοῦντας ἡ γạ[ρ] φορολογία τῆς κώμης ἡμῶν συνάγ[ε]ι εἰς πεντακοσίας ἀρούρας τὰς [ἀεὶ] ἀβροχίας τυγχανούσας.

> The year before last, as well as last year, since the fields of our village are situated on high grounds (*en hypsēlois topois*) and the nearest villages, Narmouthis, Hermoupolis Village, and Theoxenis, steal (*hypokleptontōn*) our water and prevent our land from being irrigated because they are at the front of the districts and we are the last in the district and inhabit a deserted village, the tribute of our village amounts to 500 *arourai*, consistently unwatered (*abrochias*).[26]

On paper the situation is straightforward: the downstream village of Theadelphia lay at the rear of its rural district at the end of a public canal and was

24. Crawford, *Mayordomo*, 23–24.
25. *P.Lond.* 7.1967 (225 BCE).
26. *P.Sakaon* 35 (ca. 332 CE), ll. 4–15.

being deprived of water, allegedly illicitly, by several canal-sharing neighbors upstream. Yet the accusation of theft introduced here is anomalous, a rhetorical flourish rather than a claim of property rights violated. Nowhere else in Fayyūm papyri, including the remainder of the Sakaon archive, are conflicts over water explicitly described as the result of theft. Rather, complainants consistently highlight alleged disruptions to customary patterns of canal flow. A paradigmatic example is the late-Ptolemaic petition *P.Tebt.* 1.50 (112/111 BCE) in which a tenant farmer on public land (*basilikos geōrgos*) in the southern village of Kerkeosiris complains that his upstream neighbor has blocked up (*synechōsen*) their shared public canal (a "royal waterway," *basilikou hydragōgou*) for several years in a row, thereby denying him his right to irrigate from it in accordance with custom (*akolouthōs tois ethismois*). At the other end of antiquity Aurelius Sakaon and his fellow villagers make nearly identical complaints in *P.Sakaon* 33 (ca. 320 CE), a fragmentary record of proceedings before one Valerius Ziper, governor (*praeses*) of the province of Aegyptus Herculia.[27] Here the Theadelphians voice two separate but nearly identical grievances. First, they allege that the inhabitants of Andromachis, a village located somewhere upstream, "dam the channel (*proschōnousin* [*l. proschōnnyousin*] *to rhithron*) and do not allow the waters to flow in easily." In the second complaint the claim that a certain Manos, his associates, and his brothers, all of whom lie upstream (*hyperkathēmenoi hēmin*[28]) from Theadelphia and possess some twenty *arourai* in the Fayyūm's plain (*pedion*), "block up the channel (*apophrassousiv to rhithron*) and do not allow the waters to be sent to us" (*kai ouk eōsin eis ēmas pempesthai ta hydata*). Again, at issue is not the theft of Theadelphian property but the allegedly improper obstruction of public waterways.[29] The *praeses* rendered his judgments in Latin, Greek translations of which were included at the end of the document. While the Latin of the first judgment is badly fragmented, the Greek translation simply orders those who had blocked up the channel to clean it out so that the water might have its "customary influx" (*tēn synēthē eisrhoian*). In the case of Manos and associates, he

27. On the official's name see Ast, "Tziper, not Q. Iper."

28. Parássoglou's translation of ὑπερκαθήμενοι ἡμῖν in *P.Sakaon*—"Manos' associates and his brothers *are situated on higher ground than ours*"—misconstrues the relationship between the two villages. The translation in the DDBDP renders the phrase as "our superiors" based upon a piece of Russian scholarship to which I do not have access (see *BL* 8.300).

29. As discussed in the following chapter, the reference to the small size of the alleged offenders' holdings—"only twenty *arourai* (εἴκοσι ἀρούρας μόνας)—is also of significance and is subsequently reflected in Ziper's judgment. See p. 207.

simply orders (here the Latin is preserved) that the upstream users draw only the water they need, allowing the remainder to flow downstream (*superfluam in terris susceptorum tuorum tradant*). Nowhere, at least in the preserved text, do the Theadelphians claim any entitlement to a specified share of the canal's water. We see much the same in the fourth-century petition from Philadelphia, already discussed in the previous chapter under "Environmental Subjectivity," in which the village's chief administrators blame the upstream village of Tanis for their longstanding debt to the treasury. They accordingly request that the upstream portions of the channel's stone-lined bed (*rheithron*) be inspected so that they might "benefit from the coming up of the flax and have drinkable water and sow the plain of our village and stay in our own *idia* and have benefit from our own possessions." The Philadelphians do not allege any specific malfeasance on the part of Tanis, however, and the issue seems once again to concern only the disruption to a canal's downstream flow.[30]

In contrast to this Philadelphian entreaty, the *P.Sakaon* 33 clearly alleges the deliberate physical obstruction of public canals, although the methods and motivations of the upstream irrigators is not described. Indeed, no violations of local norms may have been intended, since the temporary damming-up of canals, complete or in part, by means of earthwork transverse dikes (*emblēmata*) was the standard method by which irrigators diverted the water from public canals onto their own fields. Given that this was a normal and largely unproblematic practice, it is attested only occasionally in the papyri, though it can still be observed in the contemporary Fayyūm.[31] In *P.Prag.Varcl.* 2.52, an undated third-century letter from the Theadelphian estate of Aurelius Appianus, a certain Kopres is ordered to dam up (*emblēmatisai*) a canal for a day and release the water (*apolythēnai*) on the following day.[32] Perhaps unsurprisingly, such *emblēmata* might become the object of downstream frustrations. Such is the case in the petition *P.Sakaon* 45 (334 CE), in which Sakaon claims that two men named Amies and Euporas along with their sons—"like tyrants and bandits" (*tyrannia chrōmenoi kai lēstrikō tropō*)—had installed an *emblēma* in a canal during the flood ("the time of the waters," *ton kairon tōn hydatōn*), an action that must have denied Sakaon or Theadelphia itself access to some por-

30. *P.Wisc.* 1.34 (305 CE).

31. On *emblēmata* see Bonneau, *Le régime administratif*, 39–44. The technique remains in use in the form of the "ad hoc mud dams" that contemporary Fayyūm farmers use to direct water to the desired location. Mehanna, Huntington, and Antonius, *Irrigation and Society*, 92 and 122-24.

32. Rathbone, *Economic Rationalism*, 222.

tion of the flood. For comparison, the perspective of a beneficiary of an *emblēma* is preserved in a first-century CE petition from the western village of Euhemeria, whose sender Penneis claims that a certain Onnophris has assailed a named *emblēma* and damaged it. Although the relationship between the two individuals is not specified, it is not implausible that Onnophris was downstream of the dike and felt disadvantaged by it:

> τῆι ιζ τοῦ ἐνεστῶτος μηνὸ(ς) Νέου Σεβαστοῦ τοῦ κ (ἔτους) Τιβερίου Καίσαρος Σεβαστοῦ ἐπιβαλὼν Ὀννῶφρις Ὀννώφριος εἰς τὸ λεγόμενον Ταορβελλείους ἔμβλημ(α) οἰκοδομήμενον μετὰ δαπάνης οὐκ ὀλίγων κεφαλαίων ἀργυρικῶν αὐθάδως κατέσπασεν ἀπὸ μέρους, ἐξ οὗ κινδυνεύει τῷ ὅλωι ἐξαρθῆνα[ι] καὶ τὰ ὑποκείμενα τούτῳ ἐδάφη οὐκ ὀλίγα εἰς ἄσπορον ἐκτραπῆν[α]ι.
>
> On the seventeenth of the present month of Neos Sebastos of the twentieth year of Tiberius Caesar Augustus, Onnophris son of Onnophris, having attacked the dike called Of Taorbelles, built at the expense of no small amounts of money, presumptuously pulled it down in part, for which reason the entirety is in danger of being carried away and the fields beneath it, not few, becoming unsown.[33]

Apart from this ambiguous case, the remainder of the papyri described above clearly narrate conflicts generated by the modification of public irrigation infrastructure, which altered prevailing patterns of water flow to the apparent detriment of downstream irrigators. That even Roman law recognized the dangers of altering shared waterways is also hinted at in *P.Sakaon* 45, where Sakaon claims that the actions of Amies and Euporas were explicitly illegal ("contrary to what is permitted," *para to mē exon*), claiming that "imperial law established that the *emblēma* not be built up."[34] The law, if any, to which Sakaon here refers is unclear, although Egyptian provincial law does seem to have explicitly prohibited damaging Nile embankments as well as reducing water flows by erecting dikes or breaching embankments.[35] Indeed, already by the

---

33. *P.Ryl.* 2.133 (33 CE), ll. 6–22.
34. *P.Sakaon* 45a, ll. 11–12: θῖος (*l.* θεῖος) νόμος ἐφύτ̣ευ̣σεν μὴ βαλέσθ[α]ι ἔμβλημα.
35. Ulpian, *De officio proconsulis* at *Dig.* 47.11.10, describes the penalties in Egypt for directly damaging the embankments (*chomata*) or the sycamores alongside that help firm them up. Diminutions in water flow produced by dikes and breaches are also proscribed (*deminutiones aeque coercentur: chomata etiam et diacopi, qui in aggeribus fiunt, plecti efficiunt eos, qui id* [ ] *admiserint*). The terminology—χῶματα and διακόποι—is clearly official in origin, for which see e.g., *P.Oxy.* 12.1409 (278 CE). Cited in Anagnostou-Canas, "Les différends," 26–27.

second century CE, Roman law had come to recognize the need to ensure that waterways shared by multiple rural irrigators remained unobstructed. The *lex rivi Hiberiensis* (ca. 117–38 CE), a statute that consolidated a single irrigation community from three fractious groups of irrigators along a canal from the River Ebro in Spain, established a fine of 250 silver *denarii* for blocking the canal, compared to a mere 25 *denarii* for failing to contribute to the annual maintenance of the waterway.[36]

Yet obstructing shared canals was only the most direct method by which water flow might be disrupted and conflicts ignited. Changes in cultivation patterns, including agricultural intensification or the planting of more water-intensive crops, necessarily altered patterns of water consumption upstream with potentially significant downstream ramifications.[37] Given the largely downstream provenance of the Graeco-Roman papyri, however, it is all but impossible to determine to what extent, if any, documented instances of water shortages in marginal Fayyūm villages were produced by changes in upstream water use (on which see further below, under "Collective Action.") The phenomenon is unambiguously attested papyrologically only in an Arabic document from the archive of one ʿAbd Allāh b. Asʿad, a mid-eighth-century administrator of a district that encompassed the southwestern portion of the Fayyūm. The text is fragmentary, but it concerns a water conflict between Narmūda (Greek Narmouthis) and the upstream settlement of Nuwayra, which had apparently begun to divert additional water to a grove of sycamore:[38]

[بسم الل]ه الرحمن الرحيم [من] ناجد بن مسلم الى عبد الله بن اسع[د سلم] عليك فانى
احمد [الي]ك الله لا اله [الا هو] [اما بعد] . . . ه بن ابى مسلم [يذكر ان] من ترضاه من [
ا[هل] . . . . [منذ يكون على حـ] . . . [ما ترسا نرمودة] . . . [ا كم]..[. . .] . . . [كتب لى
] . . . [من اهل نرمودة [. . . .] ونى باب خليج تبر[ون؟ . . .] [يسق]ون الما ارضهم التى
ع[ليها] الجميزة وليس لهم ذلك ولا نعمة عين لهم فامنعهم ما ترسا] . . . [ا . ا . لاحـ] . .

In the name of God the Merciful, the Compassionate. From Nājid bin Muslim to ʿAbd Allāh bin Asʿad. Peace be upon you and I praise for you God besides Whom there is no god but He. Further . . . bin Abī Muslim came to me remind-

36. Bannon, "Ancient Roman Water Rights," 23.

37. For the effects on canal flow of a transition from field crops to orchards in the modern Fayyūm see Mehanna, Huntington, and Antonius, *Irrigation and Society*, 101–3.

38. Nuwayra is perhaps identical with modern Nawwāra, ca. 8 km northeast of the site of Narmouthis/Medīnat Māḍī.

ing me that] the one you endorsed of the people of . . .] . . [. . .] water that you intended (?) for Narmūda. [. . .] of the people of Narmūda . . . me the sluice to the canal of Nuwayra they watered their land which has on it the sycamore trees which they are not allowed to do and have no permission for (?). So do not allow them the water you intended (?) for Narmūda[39]

The links between cropping patterns, canal flow, and landscape change become clearer still in al-Nābulusī's survey. Thanks to the recent increase in sugarcane, a notoriously water-intensive crop, he writes that a large settlement southwest of the capital by the name of Dahmā had been forced to abandon cotton cultivation for lack of water:[40]

كانت تزرع الاقطان قبل ان تصرف المياه الى الاقصاب. فلما كثرت الاقصاب استوعبت جميع المياه فتعطلت الناحية عن زراعة القطن.

Cotton was sown in it before the water was diverted to the sugarcane (*al-aqṣāb*). When the [cultivation of] sugarcane increased, it absorbed the entire amount of water, and the sowing of cotton in the village was discontinued.[41]

Similar situations obtained in the large northeastern village of Dhāt al-Ṣafā᾽ (*VF* 155) and in southwestern Shidmūh (*VF* 180). The former had recently abandoned rice in order to allocate water to sugarcane (*li-tawfīr al-mā᾽ ʿalā-l-aqṣāb*) while the latter had entirely abandoned summer field crops (*zaraʿ*) after sugarcane cultivation increased. Sugarcane produced still more significant disruptions at Shāna, a village near the eastern extremity of the Fayyūm at the foot of the limestone ridge that rings the depression (*fī dhayl al-jabal*, "at the base of the mountain"). The inhabitants of the village, al-Nābulusī relates, had recently migrated north and inward toward the Fayyūm's plain to found the village of New Shāna. Among the several reasons given for this settlement relocation was a "lack of water because of the increase of sugar cane in the Fayyūm" (*qillat al-mā᾽ lammā katharat al-aqṣāb bi-l-Fayyūm*).[42]

All of these instances of disruption demonstrate that human intervention

39. *P.Mich.inv.* 5613b (ca. 730–50 CE). Trans. Sijpesteijn, *Shaping a Muslim State*, 348–49.
40. Sugar consumes more water per *feddan* than any other Egyptian crop: Ibrahim and Ibrahim, *Egypt*, 123.
41. Al-Nābulusī, *VF*, 154.
42. Al-Nābulusī, *VF*, 178.

was essential to the production of water flow throughout the canal system. Upstream choices as to where and how water should flow invariably affected water availability downstream, provoking everything from minor local spats, to changes in downstream cultivation patterns, and, in the case of Shāna, even to the relocation of entire settlements.[43] The villages documented in the Graeco-Roman papyri were almost invariably located at or near the tail end position along their respective canals. As a result, their water supplies can be regarded as the product not only of the Nile flood and the slopes of their gravity-driven canals but also of every decision made by irrigators upstream. Yet when downstream irrigators entered into conflicts with their upstream neighbors, they never complained about violations of their legal rights *in* water. Rather, they protested actions that denied them unobstructed access to the flow of common-pool water resources. The next section suggests that this commons-based approach to water apportionment was not a shortcoming or a sign of insufficient state oversight but rather a productive adaptation to the unpredictability of the water supplies at the end of the canal system.

Environment and Apportionment

Unambiguous property rights in water were already a feature of Egyptian irrigation before the Graeco-Roman period. A small collection of fifth-century BCE demotic ostraka from ʿAin Manāwir in the Kharga Oasis preserves evidence for entitlements in the form of time-based allotments. The documents refer in various contexts to so-called "days of water," lengths of time marked in full days and fractions of days during which irrigators were permitted to draw upon a specific and occasionally named water source. These "days of water" were inextricably bound to the plot of land they irrigated; when the land was ceded the attached water rights likewise passed to the new owner.[44] The practice seems to have endured throughout the Roman period as suggested by two recently published fourth-century CE Greek ostraka from the city of Trimithis in the Dakhla Oasis that contain irrigation schedules measured in days (*hēmerai*) and fractions of days (*O.Trim* 1.39 and 2.466). This practice, unat-

43. For contemporary examples see Price, "Water Theft in Egypt's Fayoum Oasis"; and Barnes, *Cultivating the Nile*, 122–25.

44. "Days of Water": Chauveau, "Les *qanāts* dans les ostraca de Manāwīr." Kharga hydrology and irrigation in general: Wuttmann, "Les *qanāts* de ʿAyn-Manāwīr."

tested elsewhere in Egypt, was enabled by the unique hydrology of the oasis, where the perennial discharge of numerous spring mounds could be apportioned fairly among its dependents.[45] The same was true of the third-century CE irrigation community at Lamasba (mod. Marwāna, northern Algeria), where the water of a spring dubbed the *aqua Claudiana* (mod. *ʿAīn Marwāna*) was distributed to its dependents in time-allotments measured in hours and calibrated to the size of each landholding.[46] Water collected in large public cisterns can also be distributed according to fixed shares, since the volume of water available can be calculated with relative ease beforehand. A modern example obtains in the Yemenī highland valley of al-Aḥjūr, where reliable spring flows are collected in cisterns or in dammed basins for later distribution on a rotation cycle throughout the year. A single turn in the cycle is "measured at the cistern according to either a defined time unit or a measure of volume." One studied cistern had a 17-day rotation cycle, the smallest unit of irrigation time in which is the *rubʿ* or "quarter" of a 12-hour day. In other parts of the region, water is distributed only after the volume of a cistern had been measured in "hand widths" (*kufūf*). The irrigator's turn is then allotted in "hand widths," converted into a time unit. Few disputes occur since the aggregate water supply is quantified in advance and every irrigator knows the amount to which he is entitled.[47]

In the contemporary Fayyūm, water rights at the level of the *mesqa*—the small irrigation canals used and maintained by groups of farmers—are also structured on a time-based rotation system. Under the modern perennial irrigation regime, water flows into each *mesqa* continuously and its users decide among themselves how to allot the 10,080 minutes (i.e., 7 days or 168 hours) of water available each week. Every user is allotted a fixed number of minutes of the *mesqa*'s weekly flow based upon the size of their landholding and the water

45. On the hydrology of the region see Bravard, "Water Resources and Irrigation."

46. Shaw, "Lamasba," esp. 72–73. See also Shaw, "Water and Society"; and, briefly, Lloris, "Irrigation Infrastructures," 129–30. For corrections and updates to Shaw's studies see Leone, "Water Management in Late Antique North Africa." Cf. Pliny, *HN* 18.188 on the time-based distribution of spring water in Tacape (mod. Gabès, Tunisia): *ternis fere milibus passuum in omnem partem fons abundat, largus quidem, sed certis horarum spatiis dispensatur inter incolas* ("There is a spring that distributes water over a space of about three miles in every direction, giving a generous supply, but nevertheless it is distributed among the population only at special fixed periods of the day." Trans. Harris Rackham).

47. Varisco, "*Sayl* and *Ghayl*."

needs of the crops grown on it—for example, field crops or more water-intensive orchards. These schedules are nowhere written down and are popularly regarded as being of great age.[48]

Although we lack the evidence necessary to reconstruct premodern Fayyūm water-sharing practices at such a fine level of granularity, Arabic sources offer abundant testimony to a system of precisely articulated water rights by which irrigation water was apportioned between canal-sharing village communities in the central Fayyūm. Already well established by the ninth century CE if not earlier, these practices were, according to our sources, founded on and sustained by the unique hydrology of this portion of the depression. Our earliest testimony, Ibn ʿAbd al-Ḥakam's *Futūḥ Miṣr*, situates the origins of this system, unparalleled elsewhere in Egypt, in pre-Islamic antiquity by attributing it to the patriarch Joseph. In his account, after Joseph had reclaimed the Fayyūm he next established a system of water rights that assigned each village a fixed amount of water measured in "fist-lengths" (*qabaḍāt*) so that "none falls short of its right (*ḥaqq*) nor exceeds its measure (*qadr*)."[49] A similar notice appears in Ibn Mammātī's twelfth-century CE administrative handbook *Kitāb qawānīn al-dawāwīn*. As quoted in chapter 2, Ibn Mammātī claims that the Fayyūm "contains many renowned (*mashhūra*) canals, filled to the brim and overflowing (*ʿāmira wa ghāmira*)," all of which provide every village with a fixed water-right (*shurb maʿlūm*) at a known time (*waqt mafhūm*).[50]

The underlying veracity of these brief notices is confirmed by al-Nābulusī, who records the water quota (*ʿibra*) assigned to most central Fayyūm villages as well as the name of the canal(s) from which this water quota was drawn. These rights are indeed measured by the "fist length" (*qabḍa*, pl. *qubaḍ*), a linear measurement of roughly 10.9 cm.[51] From the evidence of al-Nābulusī alone, Rapoport and Shahar have argued that these measurements refer to the width of the opening on the weir governing the head of each village's principal

48. Barnes, *Cultivating the Nile*, 23 with note 28. For more detail see Mehanna, Huntington, and Antonius, *Irrigation and Society*, 100–108.

49. Ibn ʿAbd al-Ḥakam, *Futūḥ*, 16. Rapoport and Shahar, "Irrigation," 19.

50. Ibn Mammātī, *Qawānīn al-Dawāwīn*, 229.

51. The size of the *qabḍa* seems to have varied over time. El-Sayid Badawi and Martin Hinds, *A Dictionary of Egyptian Arabic* (Beirut: Libraire du Liban, 1986) put it at 12.5 cm, while P. S. Girard in the *Description de l'Égypte* states, probably wrongly, that the unit in which Fayyūm water rights were assigned, his *palme*, was equivalent to 18 cm (Girard, "Mémoire sur les irrigations," 334). The actual size of the unit nonetheless has no bearing on the argument here.

public canal.[52] This suggestion is entirely correct, since the practice of proportional water sharing by means of precisely calibrated openings at the heads of canals has been in continuous use until the present day. As described by P. D. Martin in the *Description de l'Égypte*, echoed by Linant de Bellefonds in the 1870s, the nine canals that branched from the terminus of the Baḥr Yūsuf on the western edge of the Fayyūm's capital were governed at their heads by weirs whose heights were determined by "the length of the ground to be traversed [sc. by the canal] and by the area of the land which it must water," that is, the canal's command area.[53] Also in the *Description*, P. S. Girard states that the chief administrator of the Fayyūm maintained a written register containing "the number of villages, and the quantity of water which was to be distributed to each." Girard claims to have examined just such a register dating to 1014 A.H. (1605 CE), which gave a total Fayyūm water supply of "598 palms" (*palmes*), that is, *qubaḍ*.[54] In the contemporary Fayyūm, the head of each canal that branches from one of the Fayyūm's main Baḥr Yūsuf–fed waterways is governed by a weir (*haddār*), the width of whose opening is correlated to the canal's command area. Each of these branch canals in turn feeds numerous small *mesqa*s. *Mesqa*s on the same branch canal receive their water through a cluster of weirs (*nasba*), whose openings are likewise correlated to the size of the *mesqa*'s command area and the cropping patterns therein—for example, 1 mm of width per *feddān* of field crops and 2 mm per *feddān* of orchards.[55]

In further support of this interpretation of the water rights reported by al-Nābulusī, Rapoport and Shahar note that rights tend to increase toward the downstream ends of canals, presumably representing larger and larger weir-openings designed to compensate for the progressive diminishment of canal volume owing to upstream use and conveyance loss (primarily evaporation and seepage). The pattern appears clearly in evidence along the western Minyat Aqnā canal, to which al-Nābulusī assigns four settlements, here listed in upstream-downstream order along with their shares of the canal's total flow of 54 *qubaḍ*:

The pattern is equally clear along an unnamed canal branching from the

52. So suggested by Rapoport and Shahar, "Irrigation," 18.
53. Martin, "Description hydrographique," 23; Linant de Bellefonds, *Mémoires*, 16.
54. Girard, "Mémoire sur les irrigations," 333–34. This is significantly less than the roughly 812 *qubaḍ* reported in al-Nābulusī's survey and may represent the still-further diminution of the Fayyūm's cultivated area after the troubles of the eighteenth century. For a list of Fayyūm villages and their water rights see Yossef Rapoport and Ido Shahar's database of al-Nābulusī at https://projects.history.qmul.ac.uk/ruralsocietyislam/database.
55. Gouda, *Social Capital*, 60–61 and 111–16; Mehanna, Huntington, and Antonius, *Irrigation and Society*, 94–100; and Wolters et al., "Division of Irrigation Water," 162.

Table 4.1. Minyat Aqnā Canal Rights

Village	Water Allotment (*qabḍa*)
Dinfāras of Jardū and Ihrīt	4
Babīj Anqāsh	4.5
Al-Ḥanbūshiyya	14
Minyat Aqnā	50

Table 4.2. Ṭubhār-Ibshāyat al-Rummān Canal Rights

Village	Water Allotment (*qabḍa*)
Ṭubhār	8
Jardū	8
Babīj Unshū	9
Abū Ksā	13
Ibshāyat al-Rummān	16

Table 4.3. Dhāt al-Ṣafā' Canal Rights

Village	Water Allotment (*qabḍa*)
Miṭr Ṭāris	18
Akhṣāṣ al-Ḥallāq	6
Furqus	5.5
Ibrīziyā and al-Zarbī	10
Sirsinā	16.5
Dhāt al-Ṣafā' and Akhṣāṣ al-Najjār	31
Minyat al-Baṭs	13.666

terminus of the Baḥr Yūsuf in the Fayyūm that brought, again, 54 *qubaḍ* to five still-extant settlements. While the precise route of the canal is not known, the *ʿibra* of each village nonetheless increased downstream. At least in the Napoleonic period, the two tail end villages lay at the *termini* of their respective canals and possessed a reservoir:

This interpretation of water rights in al-Nābulusī is less obvious, though still broadly defensible, along the northeastern Dhāt al-Ṣafā' canal:

Yet Rapoport and Shahar also remark that this pattern breaks down along the Dilya and Tanabṭawayh canals, since certain villages toward the heads of these waterways had particularly large weir openings.[56] This is a strong hint that the water quotas recorded by al-Nābulusī likely took into account not only

56. Rapoport and Shahar, "Irrigation," 21.

progressive reduction in canal flow but also the size of each village's cultivated area and its cropping patterns, not unlike the contemporary Fayyūmī practices described above. From this perspective, the variability of water rights along the Dhāt al-Ṣafā' canal becomes more intelligible. Toward the head of the canal, the large village Miṭr Ṭāris (*VF* 217) was "a bride among the brides" (*'arūs min 'arā'is*) of the Fayyūm with its abundant orchards, streams, fruit trees, and other produce. So too Akhṣāṣ al-Ḥallāq (*VF* 78), another of the Fayyūm's "brides," with its orchards, streams, plantations, fruit trees, vineyards, and flowers. Farther downstream, yet with a far smaller quota, Furqus (*VF* 196) cultivated only field crops, dates, and figs. Farther along still, Dhāt al-Ṣafā' itself and its hamlet Akhṣāṣ al-Najjār (*VF* 155) were replete with gardens, vineyards, diverse fruits, date palms, and all varieties of field crops, along with a water-powered mill, all hardly suggestive of significant downstream reduction in canal flow. Finally, Minyat al-Baṭs (*VF* 226) at the tail end had a far smaller water quota than Dhāt al-Ṣafā', which irrigated only date palm shoots, acacia, and both winter and summer field crops. Its quota was nonetheless significantly greater than Akhṣāṣ al-Ḥallāq, despite the latter's apparently abundant water and greater productivity.

Water rights in al-Nābulusī's Fayyūm were therefore not simply mechanistic functions of village position, command area, or cropping patterns, but rather as an amalgam of each of these variables. Yet by whatever means these quotas were assigned, they provided a firm foundation upon which complainants in intervillage water conflicts could base their claims against neighboring communities. Wakako Kumakura describes just such a dispute in the mid-sixteenth century CE between the northeastern villages of Ṭāmiya and its upstream neighbor al-Rawḍa. Although both villages were entitled to 5 *qubaḍ* from their shared canal, an additional ditch leading to Ṭāmiya had been excavated, denying upstream al-Rawḍa its full entitlement. The offending ditch was therefore stopped up and judgment was rendered that Ṭāmiya was to take only its five-*qubaḍ* share.[57]

As already mentioned above, however, precise water quotas were not universal in the later central Fayyūm since villages on higher ground were ill-served by gravity-fed perennial canals and thus could not easily be assigned a water quota. Al-Nābulusī explicitly states that multiple villages were "without

57. Kumakura, "The Early Ottoman Rural Government System," 103.

Table 4.4. Villages without water quota due to elevation

Village Name	Water Supply	Reason for Lack of Quota
Dimūh al-Dāthir (*VF* 152)	Canal (*khalīj*) without quota for winter crops (*al-shatawī*).	Lies on high ground (*min ʿuluww*)
Ṣunūfar (*VF* 182)	Canal (*khalīj*) only for winter crops in the form of a ditch (*misqā*) branching from the northern bank of the Baḥr Yūsuf. Two additional ditches (*masāqī*) without weirs or quota.	"Because of the elevation of the land" (*bi-ḥukmi ʿuluww al-arḍ*)
Ghābat Baja (*VF* 189)	Two canals (*khalījayn*) and one ditch (*misqā*) without weir or quota.	One canal has a lower quota "because of the elevation of its lands" (*bi-ḥukmi ʿuluww arāḍihā*).
Qushūsh (*VF* 202)	Three ditches (*masaqī*) branching from the northern bank of the Baḥr Yūsuf for winter cultivation only.	"Because of the elevation of its land" (*li-ʿuluww arḍihā*)
Abū ʿUṣayya (*VF* 205)	Ditch (*misqā*) without quota branching from the Baḥr Yūsuf.	"Because of the elevation of its surrounding area" (*li-ʿuluww nāhīyatihi*)
Minyat al-Dīk, Banū Majnūn, Shalmas (joint settlements, *VF* 228)	Banū Majnūn Canal, one sluice gate, without weir or quota, for winter crops; ʿAnz Shalmaṣ Canal for winter crops; ditch (*misqā*) without quota for summer crops (*al-ṣayfī*), ditch without quota for orchards; occasional ditch, ca. one *qabḍa*.	Banū Majnūn Canal: "the elevation of its lands" (*ʿuluww arāḍīhā*) Summer ditch: "high [elevation]" (*ʿāl*)

a quota" (*min ghayr ʿibra*) due to the elevation of their lands.[58] He also records no water quotas for villages (or parts of villages) irrigated only by the flood (see table 2.4), though only in two cases—Bandīq and Dimūh al-Dāthir—does he explicitly state that such settlements drew water from canals "without quota" (*bi-ghayr ʿibra*).

His entry for the village of Ṭubhār, still extant some 13 km west of the capital, most clearly describes the difficulties faced by settlements on higher ground. Although possessing a water quota of eight *qubaḍ*, Ṭubhār was still faced with shortages after the flood due to its elevation:

58. Rapoport and Shahar, "Irrigation," 19.

Table 4.5. Flood-irrigated villages without water quota

Village Name
Al-Ḥammām (*VF* 97)[1]
Al-Lāhūn and Umm al-Nakhārīr (joint settlements, *VF* 95)
Bāja (*VF* 109)
Bandīq (*VF* 129)
Babīj Ghaylān and Kawm al-Raml (joint settlements, *VF* 130)
Dumūshiyya (*VF* 145)
Dimūh al-Dāthir (*VF* 152)
Sidmant (*VF* 173)
Ṭimā (*VF* 183)
Minyat al-Usquf (*VF* 205)
Nāmūsatayn (*VF* 234)

[1] The village is exterior to the Fayyūm proper, lying northeast of the al-Lāhūn inlet.

تزرع الشتوي لا غير. اراضيها عالية لا يصل اليها الماء الا بكلفة ويقل عنها عند احتراق الماء وربما عدمته بالكلية فيشرب اهلها من الابار. وتطلق لها المياه عند حاجة الكروم اليها في زمن الاحتراق.

Winter crops are sown but nothing else. Its lands are elevated (*ʿāliya*) and water does not reach it except with difficulty (*illā bi-kulfa*). [Water] lessens at the burning-up of the water (*iḥtirāq al-māʾ*, i.e., the low-water season) and its disappearance may be complete. At this time its people drink from wells (*al-ʾābār*) and water is discharged for it [sc. from alternate sources?] for the need of the vines (*al-kurūm*) during the time of the burning-up.[59]

Each of these settlements lacked easy access to the gravity-driven perennial canal system that supported the regime of water-quotas throughout the rest of the thirteenth-century central Fayyūm. They may be regarded, I suggest, as analogous to the settlements of the Fayyūm's ancient margins. As described in the Theadelphian trial *narratio* preserved in *P.Sakaon* 35, such villages lay on higher ground toward the ends of the canal system. These marginal ancient settlements did not possess a reliable, perennial water supply and were instead irrigated with whatever floodwaters they received.[60] It is this unique topo-

59. Al-Nābulusī, *VF*, 185.
60. High lands are described as unflooded (ἄβροχος) in *P.Bagnall* 9 (199–175 BCE); *SB* 20.14179 (185/4 BCE); *P.Bagnall* 45 (= *P.Tebt.* 4.1117c, 119 BCE). The account published as *P.Wisc.* 2.77 (254 BCE) also references the digging of a canal to irrigate high ground (τὰ ὑψηλὰ, l. 37). Cattle-breeders are said to retreat to the highest places (τοὺς ὑψηλοτάτους τόπους, l. 172) during the flood in *P.Tebt.* 3.703 (210 BCE).

graphical and hydrological situation—the predictable unpredictability of their annual water supply—rather than any failures of ancient water governance that explains the lack of clearly defined water entitlements or quotas.

Irrigation practices along the Graeco-Roman margins of the Fayyūm should therefore not be regarded as dangerously underregulated but as adaptations to unique environmental circumstances. Harder to irrigate by virtue of their location and thus ill-suited to schemes of apportionment by entitlement, water rights on higher ground and at the tail ends of the ancient canal network were simply the right to access the water of a shared canal during the flood, whatever its quantity, without hindrance or obstruction. Yet this internally regulated, farmer-managed system required communities to pool their energies not only to maintain the local infrastructure that delivered and distributed their water supply but also at times to assert and defend their access to the commons. It is ultimately these spontaneous and internal rituals of communal labor and collective action—the shared production and shared defense of shared waters—that constitute and define an irrigation community. The final section describes these communal aspects of irrigation—the social production of flow at ground level. By highlighting the fundamental importance of stable and self-reproducing communities, this discussion will also throw into sharp relief the severe handicaps suffered by the largely depopulated village of Theadelphia in the fourth century CE, whose final days are the subject of the concluding chapter.

Collective Action and the Social Production of Flow

Irrigation via shared public canals is inherently and necessarily communal. As described by geographer Jessica Barnes,

> it is the community that makes the water flow; a community that is generated not by turning up at a meeting but by everyday practices of blocking, unblocking, digging, and weeding an irrigation ditch. It is a community generated not through a collective imagination but through the shared work of maintaining the flow of water.[61]

61. Barnes, *Cultivating the Nile*, 87. Cf. Anwar al-Sādāt's nostalgic remembrance of irrigation in his native Delta village of Mīt Abū-l-Kaum: "We had to leave at dawn for the special canal that filled to overflowing for no more than two weeks, our 'statutory' irrigation period, during which all land in

Human social infrastructure is thus intimately entwined with the physical infrastructure of a canal irrigation system, an entanglement evocatively described by Stanley Crawford in whose own *acequia* community "the arteries of ditches and bloodlines cut across each other in patterns of astounding complexity."[62] This is to say that, while gravity may power the flow of water through a canal system, the collective labor of the system's dependents gives shape to these flows—fluid, material manifestations of communal relations continually reinscribed on the landscape.

Yet the work of producing water flow in concert with fellow irrigators was by and large mundane and uneventful; for this reason it made little impact on the papyrological record, save for the proxy evidence generated by the annual dike- and canal-work corvée. What evidence there is for everyday aspects of Fayyūm water flow was exploited earlier in this study to illustrate the hydrology and hydraulics of the border canals. Still, several surviving petitions generated by instances of local conflict offer glimpses into the normative behavioral expectations of rural irrigators, along with the means by which communities produced and defended their water supply. The production of drainage-water flow appears to have been particularly contentious, since wastewaters released from field basins (*perichōmata*) posed significant potential threats to nearby fields. Petitioners accordingly complained of instances in which they were harmed—accidentally or purposefully—by their neighbors' drainage. In a petition of 218 BCE, a farmer from the southern village of Kaminoi complains that after sowing his field in grass pea (*arakos*), two neighbors flooded (*kateklysan*) his field with drainage so that his seed was utterly ruined (*P.Enteux* 60). In a petition of 113 BCE addressed to Menches, the village scribe of southern Kerkeosiris, one Apollophanes likewise complained that his neighbor released (*eklyontos*) the water from his land thus flooding (*katakeklyken*) Apollophanes' own two and one-fourth *arourai*, which were at that time being plowed (*P.Tebt.* 1.49). A similar situation is attested in a petition of early 86 BCE (*P.Tebt.* 1.54). Here, one Melas of Theogonis complains to a local official that the ten *arourai* he farmed had been flooded when three brothers from the

the village had to be watered. It was obviously necessary to do it quickly and collectively . . . The main thing was to ensure that at the end of the 'statutory' period all the land in the village was irrigated. That kind of collective work—with and for other men, with no profit or any kind of individual reward in prospect—made me feel that I belonged not merely to my immediate family at home, or even to the big family of the village, but to something vaster and more significant: the land." Anwār al-Sādāt, *In Search of Identity* (New York: Harper and Row, 1979), 2–3.

62. Crawford, *Mayordomo*, 23.

neighboring town of Kerkeosiris had entered his fields by night (*tēi nykti*) and released (*eklelykan*) the water of their land into his own, inundating it (*kataklysthēnai*) just as it had been made ready for sowing. A Tebtynis farmer claims much the same in *PSI* 15.1529 (169–72 CE), complaining that a freedman (*apeleutheros*) named Neilos flooded his field (*kataklyston epoiēsen*)—this time, however, through sheer stupidity and stubbornness.

Needless to say, these petitions indicate that Fayyūm farmers expected their neighbors to drain their own wastewater responsibly, lest nearby plots be ruined. Likewise, they expected their neighbors to contribute to the annual work of canal- and dike-maintenance, work that ensured the unobstructed flow of water to and through the community. This expectation of communal labor participation underlies the second-century BCE petition *SB* 18.13735, discussed in the previous chapter, in which the author Protomachos describes the customary maintenance undertaken each year on a local canal by its beneficiaries. A much later lease of a palm-garden (*phoinikōn*) in the northeastern village of Hephaistias dating to 208 CE also alludes to these shared rituals (*P.Ryl.* 2.172). Here the lessee promises, in addition to the regular duties of embanking and irrigating the plot, to do a third share (*triton meros*) of the embanking of local canals (*anabolēs diorygōn*) and a half share of their cleaning (*hymisou* [*l. hēmisy*] *tēs katharseōs*), the other shares of this work presumably belonging to the neighboring landowners or lessees who all drew from the same waterway.

Comparative evidence further hints at the relative ubiquity of such practices in small-scale irrigation communities. Surely drawing on extant local practices, the *lex rivi Hiberiensis* stipulates that the responsibility for canal cleaning and maintenance lay with the canal's beneficiaries, each of whom was responsible for the share of the canal they used. "To the clearing and repairing of the channel Hiberiensis Capitonianus from its uppermost part as far as the bottom dam," the *lex* reads, "all the rural residents (*pagani*) must contribute each in proportion to his share."[63] The practice continued in medieval Valencia, where each member (*hereter*) of a community of irrigators (*comu* or *comuna*) dependent upon the same canal (*cequia*) was responsible for the annual maintenance of the canal during the month of April, before the planning of the

63. From § 2b.21–26: *ad rivom Hiberiensem Capitonianum purgandum reficiendumve ab summo usque ad molem . . . omnes pagani pro parte* (*vacat* 4) *sua quisque praestare debeant.* Text in Lloris, "An Irrigation Decree," 154 with English translation at 173. On local practice and the *lex* see Bannon, "Rivers, Rights, and 'Romanization.'"

spring crop.[64] So too in eighteenth-century Ottoman Egypt, where Alan Mikhail notes that canal workers are often described simply as *ahālī al-nāḥiya* ("the people of the village") or more precisely as *man yasta ʿīnūnuhu bihi* ("the one who benefits from it," i.e., the farmer who draws from the canal).[65] Contemporary farmer-managed irrigation systems display much the same forms of spontaneous internal self-organization. In the contemporary Fayyūm, the cleaning of a shared local canal (*mesqa*) is overseen by a local authority, who coordinates the labor of the canal's beneficiaries and collects money from them for expenses related to cleaning and maintenance.[66] During the annual cleaning periods his principal task is to secure the participation of all the beneficiaries of the *mesqa* by "harangu[ing] everybody into going to the *mesqa* site, as the absence of some farmers causes the others to refuse to work. Why should they work for the absentees?"[67] Writing of modern Andean irrigation communities, anthropologist Paul Trawick similarly remarks that participation in collective labor on behalf of the community is intimately connected to the right to irrigate and free riding is ill-tolerated. In these communities, even the work of irrigating is a public act and "larger landowners in particular, who get more water than most people, must assert and protect their rights personally, and quite publicly, with a shovel in the act of irrigating."[68] Among *acequia* irrigators in modern Mexico and the American Southwest—inheritors of Hispano-Roman and medieval Spanish irrigation practices—each community is presided over by an elected *mayordomo*. The annual cleaning (*limpia*) of the main feeder canal from which the community draws water (*acequia madre*) is coordinated by the *mayordomo* and performed by the canal's dependents. Their labor, in turn, is directly tied to water rights. The written 1919 code of the acequia of Corrales, NM, entitled the *mayordomo* to prosecute any community member who attempted to draw water from the canal if he/she has failed to provide an annual day of labor or a cash equivalent. All those with property bordering the ditch are responsible for those sections that traverse their property and must keep them in a state of good repair for the good of the community as a whole.[69]

Following Elinor Ostrom, we may conclude that face-to-face relationships

64. Glick, *Irrigation and Society*, 48–49.
65. Mikhail, *Nature and Empire*, 175.
66. Gouda, *Social Capital*, 116.
67. Mehanna, Huntington, and Antonius, *Irrigation and Society*, 114.
68. Trawick, "Successfully Governing the Commons." See also Trawick, "The Moral Economy of Water," 368.
69. Rivera, *Acequia Culture*, 95.

are critical to the work of water sharing in locally managed irrigation communities. Such personal relationships facilitate trust, foster a sense of collective identity and solidarity, and reinforce the unwritten social norms that govern the use of common resources.[70] It follows, then, that irrigators must maintain a continual presence on the ground and be personally involved, along with their neighbors, in the work of irrigation. A late second-century BCE petition gives us a glimpse, albeit oblique, of such social realities in antiquity. Writing to Menches, the village scribe of Kerkeosiris, a farmer named Pasis claims that he has been wronged while he was away from the village:

> [Μεγχε]ῖ κωμογραμ[ματ]εῖ Κερκεοσίρεως[παρὰ] Πασῖτος το[ῦ Πε]τεσούχου βασιλικοῦ γεωργοῦ τῶν [ἐκ τ]ῆς αὐτῆς. ἔτ[ι ἐκ τ]ῶν ἔμπροσθεν χρόνων [ἐθισ] μοῦ ὄντος π[οτί]ζεσθαι τὴν ὑπάρχουσάν [μοι] περὶ τὴν αὐτὴν κώμην βασιλικὴν γῆν [δι]ὰ τοῦ διείργοντος διὰ τῆς ἐπικειμένης Λύκου τοῦ Ζωπυρίωνος καὶ ἑτέρων βασιλικῆς τε καὶ ἱερᾶς γῆς βασιλικοῦ ὑδραγωγοῦ, ἐν δὲ τῶι β (ἔτει) χωρισθέντος μου εἰς ἀλλοδημίαν περὶ ἀναγκαίων πραγμάτων τῶν Ἀσκ[λ]ηπιάδου τοῦ συγγενοῦς ὁ προγεγραμμένος Λύκος νομίσας καιρὸν εὐφυηι ἔχειν ἐπιβαλὼν συνέχωσεν τὰ ἐν τῆι ἑαυτοῦ γῆι μέρη τοῦ σημαινομένου ὑδραγωγοῦ.
>
> To Menches, village scribe of Kerkeosiris, from Pasis son of Petesouchos, crown cultivator from the same village. It has long been customary for me to water the royal land belonging to me near the same village with the royal canal that passes through the adjacent crown and temple land of Lykos son of Zopyrion and others. But in the second year while I was abroad on urgent business for Asklepiades kinsman [of the king], the aforementioned Lykos, having thought that he had an ideal opportunity, dammed up the portions of the aforementioned canal [that lie] on his own land.[71]

The complainant Pasis' absence (*eis allodēmian*) from the village is the critical issue here. As I have already claimed above, the marginal Fayyūm's environment lacked a system of water rights assigned and enforced by an external authority. Hence, personal contribution to the annual work of main-

70. Ostrom, "A Behavioral Approach to the Rational Choice Theory"; and Ostrom, "Coping with Tragedies of the Commons." See also Castiglione, "Social Learning and the Bonds of Self-Governing Communities."

71. *P.Tebt.* 1.50 (112–111 BCE), ll. 1–14.

tenance as well as presence on the ground in the fields during irrigation periods was an essential component of successful irrigation. Pasis' absence therefore left him vulnerable to water theft. Even in the modern Fayyūm where perennial irrigation and water rights are assigned and enforced by a central water bureaucracy, it falls first upon farmers themselves to defend their water rights in the field:

> Everyone gets his fair share of water (more or less) but at the cost of watching not to be cheated, which means that when someone is irrigating, someone else must stay at the branching point on the canal to make sure nobody comes along and diverts the water.[72]

Yet beyond simply monitoring one's neighbors to enforce social norms of water sharing passively, internal management of the commons could also require more aggressive, at times violent, confrontations. In the modern Fayyūm most conflicts are small-scale, occurring between users of the same *mesqa*. If, like the fictional ʿAbd al-Hādī above, one user of the *mesqa* discovers that water intended for his fields is being diverted into those of a neighbor, he follows the stolen water to its destination, whereupon "an argument follows and perhaps a small fight." If it comes to blows, the conflict can quickly spiral out of control, "as friends and relatives step in, half of them ending in hospital and the other half in jail." Physical altercations are nonetheless rare, since the parties to a dispute usually settle the matter quickly among themselves or, failing that, through the intercession of a local authority figure.[73] This was likely the case in antiquity as well, hence the scarcity of papyri documenting small-scale water conflicts between neighbors. This modern perspective further suggests that the attack upon the *emblēma* in *P.Ryl.* 2.133 above, for which the petitioner posits no motive beyond the offender's alleged presumptuousness, might be interpreted not as inexplicable malice but as one farmer's blunt, physical assertion of his right to the water of a shared canal. But groups of irrigators could also band together to take aggressive collective action in defense of their water supply, a phenomenon alluded to in the fragmentary petition *P.Merton* 1.11 (39–40 CE). The text is badly damaged and cannot be reconstructed in full, but its outlines are reasonably clear. Harpaesis, a farmer of one hundred *arou-*

72. Mehanna, Huntington, and Antonius, *Irrigation and Society*, 109.
73. Mehanna, Huntington, and Antonius, *Irrigation and Society*, 133.

rai of royal (*basilikēs*, i.e., public) land in Philadelphia, complains that he had been prevented (*ekōlythēn*) from irrigating by a certain Sambas and a crowd (*synodou*) of fifty men. Although the text is lacunose at this crucial point, it seems that Sambas thereafter diverted the water to his own plot, which bordered that of Harpaesis (*hautou klēron geitniōnta emou prokimenou*).[74] While the reasons behind the conflict are unstated and irrecoverable, it should be noted that Harpaesis' one hundred *arourai* will have consumed a substantial amount of the available water, a fact that might have drawn the ire of smaller farmers, especially in years of scarcity. In such times, even if Harpaesis were to draw only such water as he needed, the remainder of the community might have to do with less. Speculation aside, it remains clear that whatever their motivation, the farmers of Philadelphia were capable of mobilizing in numbers for the purpose of aggressively redirecting water flow within the community.

Offensive collective action is also described in *P.Sakaon* 32 (254–68 CE). The papyrus preserves the minutes of a hearing held before an Arsinoite *stratēgos* to adjudicate an irrigation conflict between Theadelphia and the village of Philagris, which was located just over 8 km upstream (south) along the western border canal.[75] Although the text is (again) badly fragmented and the details of the conflict consequently irrecoverable, it seems to have concerned purposeful damage to the stone-lined mouth (*stomion*) of a shared canal, which was somewhere in the vicinity of Philagris. According to a written report presented by irrigation inspectors (*hydrophylakes*) at the hearing, a number of Theadelphians had cut away at (*laxeusai*) the mouth of the canal, from which stone had also been carried off, albeit possibly by villagers of Philagris. According to one Hermias, a resident of Philagris present at the hearing, he and his fellow villagers had not removed any stone from the mouth. Rather, he claims that the main bone of contention was a lock (*kleidion*) of a sluicegate on the canal's mouth, on account of which the Theadelphians viewed Philagris with malicious ill-will (*phthonos*).[76] While no more can be ascertained with certainty, what is clear is that a group of Theadelphians had unilaterally—and presumably self-interestedly—modified the mouth of a shared canal, actions that were subsequently contested by Philagris. The motive of the Theadel-

74. Ll. 20–24: ἀναλαβόντος . . . ϲτ̣ . [ἀν]θρ̣ώ̣[πους]. . . .τ̣ο̣δ̣ . . [-ca.?- ὕ]δω̣ρ. . . .[. . . εἰς τὸν] αὐτοῦ κλῆρον γ[ειτνιῶ(ντα)] ἐμοῦ προκιμένου.

75. TM Geo 1766. Now the village of al-Ḥāmūli.

76. *P.Sakaon* 32, l. 33–4: Ἑρμίας ἀπὸ Φιλαγρί[δος εἶ(πεν)]· οὐδὲν βεβάστακται. οὗτοι φθόνῳ ἡμᾶς περὶ κλ̣ε̣ιδείου (*l.* κλειδίου) κατα[("Hermias from Philagris said 'Nothing has been carried off. These ones [the Theadelphians] [sc. 'regard us,' *vel sim.*] with envy on account of the lock'").

phians is unstated, though it is plausible that their intent was to increase the downstream flow of the channel. This would have the virtue of explaining their alleged ill-will toward the *kleidion* at Philagris, which would have modified or restricted the downstream flow of the shared canal.[77]

The frankest evidence for collective action in defense of water rights is *P.Haun.* 3.58 (439 CE), the latest dated papyrus from Karanis. Peppered with unusual vocabulary and orthography, the text has eluded accurate translation and has been regarded by some as evidence of the collapse of the village's agricultural regime prior to its abandonment.[78] When correctly translated and placed in the context outlined here, however—a social environment in which collective, at times aggressive, action was a viable method by which to assert and defend water rights—its meaning becomes clear. The text is a declaration (dubbed both a *cheir* and an *apodeixis*) in which a group of men, their own residence uncertain, announce to the priests, deacons, and rest of the population of Karanis that no one from the village shall draw water from a source identified by the obscure Egyptian toponym Thanesamēn—perhaps referring to a reservoir and its attendant well—nor shall they assert any claim over the lands that the source irrigated.[79] Should any resident of Karanis be discovered in violation of the declarants' water rights, the consequences will be grave:

77. On the term see Bonneau, *Le régime administratif*, 73–75. Cf. the similar vocabulary for the locks (κλεῖθρα, κατάκλειδες) at Ptolemais Hormou in chapter 1 under "Capturing the Flood." Cf. also a conflict accidentally incited on 29 August 1886 by British engineers in the town of Desūq in the western Delta. In order to accomplish unspecified irrigation works in the town, the British engineers had erected a dike across the town's main feeder canal. This had the effect of cutting off the water supply to area farmers during the irrigation and sowing season. After their appeals were rebuffed, "the natives took the law into their own hands, and a force of some 400 or 500 armed men attacked the dike and attempted to cut it through. The dam was defended by twenty native policemen, who had been collected there by the local authorities in consequence of the threatening aspect of the inhabitants. The police succeeded in beating off the attacking party, one of whom was killed." Although additional police were sent and the leaders of the riot were arrested, orders were simultaneously sent to breach the dike and allow water through. Gerald H. Portal to the Earl of Iddesleigh, 4 September 1886, *Further Correspondence Respecting the Affairs of Egypt: In Continuation of "Egypt no. 5, 1886."* Great Britain, Parliament (London: Harrison and Sons, 1887), 98.

78. Van Minnen, "Deserted Villages," 50–52; Hengstl, "Wasser in den Urkunden," 224–25.

79. Thanesamēn: TM Geo 2342. The proposed etymology of the *ed. pr.* (Bülow-Jacobsen and Ebbesen, "Five Copenhagen Papyri," at 27) ⲑⲁ-ⲛ-ⲥⲁⲙⲧ, "at the pools," i.e., referring to a location with some sort of reservoir or cistern, has not found acceptance, nor have the several suggestions of Bonneau, "Un règlement de l'usage," 8 with note 5. ⲑⲁ- points to *Tȝ-ḥw.t-* "the mansion" (Katelijn Vandorpe, "Egyptische geografische elementen in Griekse transcriptie" [PhD diss., KU Leuven, 1988], 130–37). The following ⲛ is resolvable as the plural article *nȝ*. The terminal -σαμην remains unclear. A well (πμουν, Egyptian, "the water of") by the name *Tsamen* (Τσαμεν) also appears in a third- or fourth-century CE ostrakon from the Dakhla Oasis, *O.Trim.* 98, though the etymology is likewise unknown. On the nature of the source, Bonneau suggests that it may designate a large reservoir and well for the irrigation of its attached territory, for which she cites the 190 *arourai* and their ὕδρευμα in *P.Col.* 7.174 (342 CE?). This is plausible, particularly in view of the proscribed activity of drawing water (ἀναλαβεῖν, lit. "take up").

μηδὶς (*l.* μηδεὶς) τῆς κώμης ἐξουσεύσει ἀναλαβι (*l.* ἀναλαβεῖ<ν>) νειρῶν (*l.* νηρὸν) εἰς τὼ (τὸ) Θανεσάμην μ̣η δ̣έτινα τῆς αὐτῆς κώμης ἐξουσεύσει κλήρων τῶν ἐμπροστὰ τ̣ῆς α̣ὐ̣τ̣ῆς Θανεσάμην, διὰ τοῦτω (*l.* τοῦτο) πεποιήμεθα τήνδε τὴν χῖραν (*l.* χεῖρα) ὀμνύοντες Θεὸν παντωκράτωρ (*l.* παντοκράτορα) καὶ Νίκην τῶν δεσποτῶν τῆς οἰκουμ̣έ̣νης Θεοδοσίου Οὐαλεντιανοῦ (*l.* Οὐαλεντι<νι>ανοῦ) τῶν αἰωνίον (*l.* αἰωνίων) Αὐγούστων εἴ τι[80] τινα εὑρήσκομεν (*l.* εὑρίσκομεν) κάτινον (*l.* κάτινα)[81] τῆς αὐτῆς κώμης Καρανίδος ἀναλαμβάνον (*l.* ἀναλαμβάνοντα) νειρῶν (*l.* νηρὸν) εἰς Θανεσάμην καὶ συνκλ̣ά̣σομεν αὐτοὺς οὐκ ἔχομεν μέμψιν παρά τινος τῆς κώμης καὶ διὰ τοῦτω (*l.* τοῦτο) πεποιήμεθα τήνδε τὴν χῖραν (*l.* χεῖρα) πρὸς ἡμᾶς (*l.* ἡμῶν?) ἀσφάλειαν.

Nobody from the village shall have the authority to draw water at Thanesamēn, nor shall any from the same village have authority over the plots in front of the same Thanesamēn. For this reason we have made this declaration, swearing by Almighty God and the Victory of the masters of the world [the emperors] Theodosius and Valentinian, the eternal Augusti. If we discover anyone whatsoever of the same village of Karanis drawing water at Thanesamēn and we smash them (*synklasomen autous*), we incur no blame from any person from the village, and for this reason we have made this declaration for our security.[82]

---

80. Editions before that of Rea ("*P.Haun.* III 58") bracketed the τι as dittography. Rea rightly eliminated the brackets yet incorrectly hypothesized that τι refers forward to νειρών in l. 13. It is instead to be taken together with εἴ as a single unit equivalent to εἴ alone, of which there are well over two hundred examples in published papyri. Krikē, "Μορφοσυντακτικοί νεωτερισμοί," 116–20.

81. The interpretation of this clause has been disputed, given the obscurity of the word *katinon* (κατινον). The *ed. pr.* regarded it as a gendered antecedent of modern Greek *kati* (κάτι, "something, anything"). Uncomfortable with the idea that a declaration signed and witnessed by a representative of the state (a *numerarius* in ll. 19–21) could endorse acts of physical violence against private persons, John Rea proposed a derivation from Latin *catinus*, referring to a basin or cistern (Rea, "*P.Haun.* III 58," 93–94), thereby transferring the violence of *synklasomen* ("break; smash") to physical infrastructure (cf. *P.Ryl.* 2.133 and *P.Sakaon* 32). So also Bonneau, "Un règlement de l'usage," 20–22, whose circuitous and unconvincing solution likewise attempted to read violence out of the text by interpreting *synklasomen* as a fiscal term referring to the return of derelict land to cultivation and taxation. The original solution was closer to the mark, since *katinon* is indeed an early variant of the medieval Greek indefinite *katis* (κάτις, "somebody/anybody"), whose case structure has here been assimilated to adjectives in -ος (cf. similar instances in *Chr.Wilck.* 110A, ll. 15–17 [110 BCE] and *SB* 14.12144 [198–99 CE]). *Katinon* accordingly reinforces the previous *tina* (τινα), as already recognized by Larsen and Bülow-Jacobsen in *P.Haun.* 3.58. note to ll. 12–13. The pair is thus most accurately rendered as "anyone . . . whatsoever." For this analysis see Krikē, "Μορφοσυντακτικοί νεωτερισμοί," 121–22. For κάτις see David Holton, Geoffrey Horrocks, Marjolijne Janssen, Stamatina Lentari, Io Manolessou, and Panagiotis Toufexis, *The Cambridge Grammar of Medieval and Early Modern Greek* (Cambridge: Cambridge University Press 2019), 1070–75.

82. *P.Haun.* 3.58, ll. 8–10, 12–16. Trans. modified from Rea, "*P.Haun.* III 58." Cf. Palladius' *Lausiac History* on the life of the holy virgin Piamoun, who writes that the virgin's village was at one time set upon during the flood by its spear- and club-wielding neighbors, "for they fight about the distri-

While this declaration concerns a localized and seemingly private water source rather than the common waters of the Fayyūm, the behavior the declarants describe is, if more graphic, much the same as that glimpsed in the previous examples: the collective (and here unapologetically violent) defense of water rights. That such collective action might at times leave others smashed and broken—the literal sense of *synklasomen*—seems to be taken entirely for granted. It bears mentioning that the only other clear attestation of the verb *synklaō* in the documentary papyri is a roughly contemporaneous Oxyrhynchite report of a robbery and assault that left one victim half-dead, his every limb broken (*synklasthentōn*).[83] We would therefore do well to take the declarants of *P.Haun.* 3.58 at their word and rightly understand what they mean to do: break and batter anyone who lays hands on their water. Such remains the informal rough justice of the contemporary Fayyūm. When asked in the early 2000s by Dutch journalist Joris Luyendijk how Fayyūmīs deal with a water thief, two unnamed farmers quickly and laughingly responded "We smash his face in!" The elder of the two nonetheless quickly added that thanks to the local water-user associations recently created in partnership with Dutch NGOs, at least there were now "fewer stabbings."[84]

Conclusion

Although powered by gravity, the flow of the Fayyūm's water was ultimately shaped by human decision-making. Water rights were a fundamental element of this socioenvironmental system, for they established the patterns by which shared water flows were apportioned within and between communities. This human element nonetheless ensured that water flows were in a perpetual state of flux in response to changing patterns of water consumption or modifications to the physical infrastructure of the canal system. Downstream irrigators consequently occupied the most vulnerable position in the canal network, since they were exposed to the cumulative results of every decision made upstream

bution of water" (μάχονται γὰρ εἰς τὰς ὑδρομερισίας). *HL* 31.1. Cited in Bagnall, *Egypt in Late Antiquity*, 138.

83. *SB* 24.15970 (455 CE): πληγῶν δὲ ἐ̣[π]ε̣νεχθεισῶν τῷ ἐμοὶ διαφ̣έροντι Μουσαίῳ ἡμιθανη (*l.* ἡμιθανὴς) ⟦αὐτὸν⟧ κατ̣ε̣σ̣τάθη, συνκλασθέντων (*l.* συγκλασθέντων) ὅλων αὐτ̣ο̣ῦ τῶν μελῶν ("and after blows had been inflicted on my servant Mousaios he was left half-dead, all his limbs broken").

84. Luyendijk, *People Like Us*, 55–56. On the overall failure of Dutch-created WUAs see Gouda, *Social Capital*.

along a shared canal. While these upstream-downstream relationships at times provoked conflicts, their relative rarity in the papyri at least suggests that most were resolved locally before they spiraled out of control and generated documentation designed to solicit state intervention. Indeed, although direct evidence is thin, the papyri hint that ancient Fayyūm irrigation communities were largely capable of producing, contesting, and defending their own water supply, work that might entail assaulting not only offending infrastructure but also, at extremes, other irrigators. It was through this shared production of water flow, moreover, that Fayyūm irrigation communities were internally constituted and bound to their own local stretch of the canal system. Fayyūm irrigation must therefore be regarded as an entangled coproduction not simply of nature and human agency but of nature and the *collective* agency of robust and internally cohesive communities willing and able both to maintain the infrastructure that delivered their water and to defend those waters against outside competitors. This communalism-in-place was in turn the socioenvironmental bedrock upon which Roman coordinated localism was founded, an administrative system that sought to perpetuate extant patterns of water flow by binding rural subjects not simply to their villages of record but to the canals that served them. Communal decline therefore represented a critical threat to this system, since population loss undermined an irrigation community's ability to take collective action for the production and defense of their water supply. At such moments, the bonds that tied the community's remaining members to one another and to their portion of the canal system began to weaken. This understanding of the Fayyūm's "communities of flow" helps to clarify the plight of the struggling village of Theadelphia in the fourth century CE, a badly diminished community still bound by the state to its canals and expected to irrigate, yet increasingly unable to accomplish the collective work of making water flow. As the final chapter will demonstrate, this clash between rural realities and state expectations proved irresoluble and was the defining feature of the community's final days.

Chapter 5
The Tail End

We are at the end. We are tired. The problem is the projects upstream that take all the water.
—Anonymous farmer, Thānia, 2000s CE[1]

Wise about Agriculture

Sometime in October or November of 257 BCE a group of Egyptian farmers came into conflict with a Greek. The setting was the new Fayyūm village of Philadelphia, specifically the 10,000-*aroura* gift-estate of Apollonios, the Ptolemaic chief of finances. The Egyptians in question were natives of the Heliopolite nome, a district near the apex of the Nile Delta just to the north of modern Cairo. Like many other Egyptians in this period, they had migrated to the recently reclaimed Fayyūm as agricultural laborers and had undertaken a lease of 1,000 *arourai* on the Apollonios estate.[2] Something, however, had gone wrong. As the farmers relate in a petition to Apollonios, shortly after they had sown the land, a Greek by the name of Damis suddenly took back some 200 *arourai* and thereafter arrested three of their elders in order to force them to sign a deed renouncing the lease entirely. They also claim that another Egyptian, perhaps the local village scribe, was chasing them away from the town and not letting them take up residence there. In closing they plead that Apollonios send for some of them and to listen to their story, for their resources were otherwise exhausted.

1. Quoted in Barnes, *Cultivating the Nile*, 123.
2. Westermann, "Egyptian Agricultural Labor"; Clarysse, "Toponymy of Fayum Villages"; and Clarysse, "Village Names."

The background to this conflict is unknown, although it has been plausibly interpreted as a confrontation between two cultural understandings of land rights.[3] Yet it is not the altercation itself that interests me here but rather the stinging accusation with which the Egyptians punctuate the narrative portion of their petition. "And there are no few failures (*hamartēmata*) in the 10,000 *arourai*," they assert, "for there is no man here wise about agriculture (*syneton peri geōrgian*)."[4] A cutting slight, to be sure, but the insult also doubles as a claim: a self-confident assertion of the farmers' own superior local knowledge of Egyptian agriculture. I use the term "local knowledge" here in a loosely Geertzian sense to refer to a way of knowing that is "practical, collective and strongly rooted in a particular place."[5] Dubbed *mētis* by the political scientist James C. Scott from the Greek for wisdom, cunning, or craft, local knowledge is nonetheless not systematic—an "organized body of considered thought," as Geertz would have it[6]—but rather a context-specific collection of acquired skills and expertise that enables individuals to respond productively to changes in their human or natural environments.[7] At once practical, experiential, and spatially circumscribed, local knowledge is therefore acquired and transmitted not by language or through study but by routine practice in place. This joint embeddedness and embodiedness is readily discernible in agricultural contexts where knowledge of the rural landscape emerges only from quotidian labor—the unique bundle of everyday practices required to cultivate a particular plot of earth. Nature, labor, routinized practice, and knowledge are thus intimately, indeed generatively, entwined, a relationship nowhere more evocatively expressed than by American historian Richard White, who writes that only when its demands are felt in bone and sinew is nature truly known.[8]

The word *local* thus carries significant conceptual weight here, for it speaks to the entanglement of local knowledge with the particulars and peculiarities of place, a sociospatial context from which it cannot be disentangled without los-

3. Manning, *Land and Power*, 114–16, discussing also the somewhat similar conflict *PSI* 5.502 (257 BCE). Cf. the Heliopolitan farmers' additional complaints in *P.Lond.* 7.1955 (257 BCE).
4. *P.Lond.* 7.1954 (257 BCE), ll. 7–8: καὶ οὐκ ὀλίγα δὲ ἁμαρτήματά ἐστιν ἐν [τ]αῖς μυρίαις ἀρούραις διὰ τὸ μὴ ὑπάρχειν ἄνθρωπον συνετὸν περὶ γεωργίαν.
5. Corburn, *Street Science*, 48.
6. Geertz, *Local Knowledge*, 75.
7. This formulation draws on both Scott, *Seeing Like a State*, 313; and Di Giminiani, *Sentient Lands*, 110. For the critique of Geertz's characterization of "indigenous agricultural knowledge systems" see Paul Richards, "Cultivation: Knowledge or Performance," in *An Anthropological Critique of Development: The Growth of Ignorance*, ed. M. Hobart (London: Routledge, 1993), 61–78.
8. White, *The Organic Machine*, 4, on the work of learning to navigate the Columbia River.

ing all integrity and meaning. Local agricultural knowledge thus stands in marked contrast with managerial, institutional, or administrative knowledge of an agricultural landscape, a systematizing and homogenizing way of knowing that is produced and transmitted at greater physical remove. Just such a perspective on the early Hellenistic Fayyūm's countryside survives in the masterplan of one of these very 10,000-*aroura* estates, probably that of Apollonios himself, preserved on a Greek papyrus of 259 BCE, two years before the altercation at Philadelphia. Enumerating the dimensions of the estate, the size and number of the embanked field-basins (*perichōmata*) it would contain, the total amount of earth to be moved, as well as the timing and cost of realizing the endeavor, the plan is accompanied by a rough sketch of the entire domain.[9] While this blueprint was undoubtedly instrumental in the planning and initial establishment of the estate, it nevertheless reflects a detached and schematic view of the landscape that effaced the surely heterogeneous character of more than 27.5 million square meters of Egyptian earth.[10] This heterogeneity—the unique demands that each field would come to make on bone and sinew—could be known only at ground level through the work of bringing these virgin *arourai* into productivity. Yet unlike their native counterparts at Philadelphia, the Greek newcomers initially lacked the acquired skills and expertise needed even to begin to work, let alone to know, the land they had reclaimed. In a letter to the *architektōn* Kleon of 11 October 257 BCE, roughly contemporaneous with the conflict at Philadelphia, the Apollonios estate's manager Panakestor demands Kleon's aid in diverting water into the estate from a public feeder canal, blaming the inexperience of him and his men (*apeiroi esmen*) for their total failure to irrigate.[11] Haplessly unable to produce the water flows their fields required, it seems the new Greek "farmers" at Philadelphia were anything but. So while it is merely speculation, we might imagine that the many failures to which the Egyptians allude in their petition to Apollonios reflect a troubled transition from plan to practice—the mistakes of those who knew the land from without but had too little dirt beneath their fingernails.

• • •

9. *P.Lille* 1.1 (259 BCE). See the translation and illustration in Thompson, "Irrigation and Drainage," 118–20.

10. Cf. the reference to sloping ground or hollows (ll. 28–29: τινες τόποι τοιοῦτοι καὶ κοῖλοι) that made up on portion of the 10,000 *arourai*.

11. *P.Petrie Kleon* 17, l. 4: μὴ βρεχομένην τὴν γῆν. See chapter 3 under "Central Coordination."

This final chapter returns to the moribund village of Theadelphia in the early fourth century CE, framing its troubles as another fraught encounter between two ways of knowing the Fayyūm's irrigated landscape—a bookend of sorts to the Philadelphian encounter almost six centuries earlier. The preceding chapters have established a firm context for Theadelphia's struggles by describing the hydrology of the Fayyūm, the form and function of its canal system, the administration that coordinated its annual maintenance, and the local rituals of communal labor, water sharing, and collective action that structured the flow of water through the system. All told, these discussions have demonstrated that the "irrigation system" of the Fayyūm, particularly during the Roman period, must be regarded as a dense entanglement between environment, local communities, and state authority, each element working in tandem to move water throughout the entirety of the ancient canal network. The virtual depopulation of Theadelphia had nonetheless shattered the integrity of this system along the western fringes of the depression by the early 300s CE, thereby triggering persistent water conflicts between the remaining Theadelphians and their canal-sharing neighbors upstream. Despite several interventions by Roman officials, these conflicts proved irresoluble and the site of the village was permanently abandoned sometime after midcentury.

Theadelphia's water conflicts are well known thanks to the robust documentation they generated, namely the petitions from Aurelius Sakaon and his companions, who were among the small handful of residents remaining.[12] Yet also preserved in Sakaon's archive is the transcript of a hearing before one Valerius Ziper (or Tziper), governor (*praeses*) of the province of Aegyptus Herculia, in which two of Theadelphia's ongoing conflicts were adjudicated (*P.Sakaon* 33, ca. 320 CE). It is this hearing that I characterize as an encounter between local and administrative knowledge or, more analytically, as a moment of apposition between two epistemic frameworks for understanding the irrigated landscape at Theadelphia. Termed "waterscapes" in contemporary water studies, such frameworks are informed not only by the sociocultural needs and expectations of water users but also by simple circumstance—that is, time, place, and perspective.[13] As we will see below in selections from Sakaon's archive, the expectations and perspective of the Theadelphians were those of small and struggling farmers at ground level. Their knowledge of the Fayyūm's

12. *P.Sakaon* 35 (ca. 332 CE?), 42 (ca. 323 CE), 44 (331–32 CE), and 45 (334 CE).
13. Bender, *Water Brings No Harm*, 8–10. Cf. Karpouzoglou and Vij, "Waterscape."

irrigated countryside was correspondingly narrow and shaped by the everyday work, increasingly futile, of producing the local water flows upon which they and their fields depended. This parochialism is duly reflected in the language of their complaints, which consistently stress the socioenvironmental particularity of their troubles—namely, the difficulty of watering an increasingly depopulated village situated on high ground near the tail end of a canal (or canals) shared with more fortunate and more populous settlements upstream. Closely circumscribed and deeply contextualized, theirs was a waterscape born of routine practice in place and the local knowledge it engendered. Language nevertheless proved a poor vehicle for the transmission of a knowledge hard felt in bone and sinew. The *praeses* consequently failed to perceive the peculiar local intersection of position, population, and relative power that underlay these conflicts and instead situated them within the universalizing framework of Roman water law, which made strong normative claims about the duty of rural irrigators to share water in proportion to need. Flattened and universalized, his was a legal-administrative waterscape in which irrigation figured not as an expression of locally embedded socioenvironmental relationships but as normative praxis.

This reading enriches our understanding of Theadelphia's final days in two principal ways. First, it humanizes its problems by emphasizing their singularity. Far from reflecting some larger Egyptian irrigation crisis or general late-Roman malaise, the village's terminal struggles were a localized human-scale tragedy, the product of a breakdown in the socioenvironmental infrastructure of water flow in a single place at a single point in time. Consequently, although other marginal Fayyūm villages would likewise be deserted in the ensuing years and centuries, we cannot casually assimilate their abandonment histories to the unique concatenation of circumstances at Theadelphia. Second, by centering the indispensable role of community in Fayyūm irrigation, this reading makes clear that the largely depopulated village's problems were at this stage altogether resistant to external resolution. While this fact does not absolve the Roman state of all blame, barring a large and coercive population transfer, provincial authorities could do little to make Theadelphia's water flow again. As I will suggest in closing, the encounter between Ziper and the Theadelphians also opens a window onto the management of the commons, specifically the perversity of attempts to fit the local and the particular into global and universalizing frameworks.

But how did things come to this point? While the economic and population decline of the Fayyūm's marginal villages seems to have begun already in the late second century CE, its cause(s) remain a subject of debate and a full accounting of the scholarship on the changing landscape, society, economy, and administration of later Roman Egypt lies outside the scope of this book. Instead, the following section extends the critique of the new scientism articulated in the Introduction to this book to recent contributions that identify exogenous environmental phenomena—epidemic disease or climate change—as the principal drivers of demographic and landscape change in the later Roman period. While the reservations enumerated below are neither intended nor should they be construed as a wholesale rejection of natural-environmental agency, they nonetheless underscore that arguments for environmental (mono) causality often rest on unproven assumptions, uncertain evidence, and unfortunate elisions of the complicating testimony of traditional sources, contemporary and comparative alike. Yet even if the causes of Theadelphia's initial second-century decline remain undefined—though perhaps to be sought in the nexus of environmental *and* socioeconomic change—the relative weakness of its irrigation community is obliquely yet clearly discernible in papyri of the following third century, a period in which the village otherwise seems to have enjoyed an economic efflorescence. When viewed through the lens of the socioenvironmental account of irrigation advanced in preceding chapters, however, it becomes apparent that Theadelphia *qua* irrigation community was already in precipitous decline decades before the period documented in Sakaon's archive.

Plague, Floods, Estates, and Population

As catalogued by Roger Bagnall in his prosopography of Aurelius Sakaon's Theadelphia, the population of the village had declined precipitously from the estimated 2,100–2,300 in the 120s and 130s CE to scarcely more than two dozen adult males between the years 313 and 336 CE. Although these numbers would rebound slightly in some years, in others the site was all but deserted.[14] The origins of this dramatic falloff are often located in the so-called Antonine Plague of the 160s to the 180s CE—an empire-wide epidemic usually diag-

14. Bagnall, "The Population of Theadelphia." Second-century population estimate in Sharp, "The Village of Theadelphia," 164 with note 22.

nosed as smallpox.[15] A common account holds that widespread mortality (ca. 20 percent of the population) left large amounts of Fayyūm land ownerless or, in the case of state-owned land, without its former tenant farmers. Wealthy absentee landowners subsequently acquired numerous parcels of abandoned land in villages scattered throughout the Fayyūm, thereby creating the decentralized great estates attested in papyri of the third century CE.[16] Theadelphia was no exception to these trends and by the third century it had become home to divisions of several aristocratic estates, most notably that of the Alexandrian councillor Aurelius Appianus. The documentation generated by the management of the Theadelphian division (*phrontis*) of his estate is vast and it dominates our view of the village in this period.[17] By the last decades of the third century, however, even the estate was gone and the rump community documented in Sakaon's archive begins to come into view.

While the dramatic impact of the Antonine epidemic on Egyptian demography—losses on the order of 20–30 percent—is considered settled by some,[18] among Romanists more broadly opinions on the epidemic remain mixed, running the gamut between arguments for mass mortality[19] and assertions that any quantitative conclusions are all but impossible.[20] The root of this disagreement is the lack of hard evidence for case fatalities from which estimates of overall mortality might be extrapolated. As a result, historians have taken to constructing what Romanist Keith Hopkins dubbed "wigwam" arguments: assemblages of multiple evidentiary "poles," none of which are sufficient on their own but which together support a larger argument, in this case for a mortality crisis in the late second century CE.[21] Parallel epidemics in China, statistics on Lydian burials, discharge *diplomata* from the Roman army, interruptions in documentary dating series, spikes in Palmyrene tomb-building, apparent rises in magical amulets against disease, together with changes in agricultural leases and wages visible in Egyptian papyri: when collectively assembled these disparate strands of evidence seem to point toward a significant demographic shock.[22] Focusing squarely on the papyri, Walter Scheidel

15. Classic account in Boak, "The Population of Roman and Byzantine Karanis."
16. E.g., Kehoe, "Property Rights over Land," 97; Rathbone, "Economic Rationalism," 263; Van Minnen, "Economic Growth," 259 and 265.
17. Rathbone, *Economic Rationalism*.
18. Rathbone, "Roman Egypt," 700.
19. E.g., Harper, "People, Plagues, and Prices"; and Harper, *The Fate of Rome*, 65–118.
20. Greenberg, "Plagued by Doubt"; Bruun, "The Antonine Plague in Rome and Ostia."
21. So Elliott, "The Antonine Plague," 4–5.
22. The classic examples of this approach are Duncan-Jones, "The Impact of the Antonine Plague"; and

sees an Egyptian reflection of these socioeconomic perturbations in the rise in the wages of laborers and a simultaneous decrease in the price of land—an outcome predicted both by the rules of neoclassical economics and by the later European experience of the Black Death.[23] Others have mined the papyri more deeply still for wheat and land prices as well as rents and wages in the attempt to plot the general trajectory of the Roman Egyptian economy before, during, and after the epidemic. Kyle Harper's recent quantitative analysis of this evidence concludes that falling rents and land prices as well as rises in real wages clearly reflect a sudden mass-mortality event.[24] Such conclusions buttress William Harris' recent categorical pronouncement that "no minimalist position [on the epidemic's death toll] . . . merits further consideration as far as Egypt is concerned."[25]

Yet for all the methodological sophistication of recent scholarship on the epidemic, it has been remarked that the ancient historiographical community as a whole has not managed to improve upon the situation described by Roman historian J. F. Gilliam in 1961: "There is not enough evidence to identify satisfactorily the disease or diseases responsible, to trace the epidemic's origin and spread with much accuracy, or to determine even approximately the number of those who died."[26] Economic historian Colin Elliott has also cautioned against placing too much confidence in conclusions arrived at through quantitative means, remarking that they "lend a sense of empirical plausibility to what are essentially speculative arguments."[27] Even when researchers rely on the same Egyptian "dataset," they can arrive at strikingly different conclusions. Elliott's own methodologically inquisitive analysis of the same data exploited by Harper is a case in point, since he concludes that the epidemic had a decidedly more moderate if not minor impact on Egypt. Such discrepancies can emerge from simple methodological preferences—variations in the way data of uncertain date are graphed—as well as one's confidence (or lack thereof) in the direct causal links between the epidemic and observed changes in wage and

---

Duncan-Jones, "The Antonine Plague Revisited." Quote from Elliott, "Disease Proxies," 297.

23. Scheidel, "A Model of Demographic and Economic Change"; and Scheidel, "Roman Wellbeing." See also Scheidel, "Demography," 60.

24. Harper, "People, Plagues, and Prices." Cf. Harper, *The Fate of Rome*, 65–118, which puts total deaths at 7–8 million out of a total Roman imperial population of 75 million. Cf. Elliott, "Disease Proxies."

25. Harris, "The Great Pestilence," 334.

26. Haldon et al., "Plagues, Climate Change, and the End of an Empire (2)," (citing Gilliam, "The Plague under Marcus Aurelius," 227).

27. Elliott, "Disease Proxies," 297.

price structures. For Elliott, these uncertainties indicate that scholars must consistently refine their quantitative models, not only by seeking more and better data but also by interrogating the assumptions upon which the models themselves are founded. Such work necessitates continued engagement with traditional historiographical sources and methodologies, since quantitative data become meaningful only by means of qualitative contextualization.[28]

Yet at least as far as Egypt is concerned, any attempt at qualitative contextualization immediately reveals the thinness of the evidence and the significant obstacles to reaching anything approaching robust conclusions. Of primary concern is the identity of the disease itself, which is commonly identified as smallpox (*Variola major*) or a closely related poxvirus, a retrospective diagnosis based on the descriptions of symptoms in ancient medical literature.[29] Smallpox is a well-known and still relatively recent killer, whose "impact on a virgin population can be catastrophic" with average mortality rates of 25–30 percent "rising to 40–50 percent in the very young and the very old."[30] As Walter Scheidel has summarized, a smallpox outbreak in the Roman Mediterranean "could have killed from 20 to 50% of those infected," with infection rates reaching "60 to 80% of the total population." Based on sophisticated computer modeling, Yan Zelener predicts that a smallpox outbreak in the Roman Empire would have produced an initial demographic contraction of some 20–25 percent.[31] Moreover, Scheidel has also suggested that Egypt's unique environmental conditions—the majority of its population confined to the narrow strip of the Nile Valley—and its comparably high level of urbanism may have contributed to mortality rates higher than those in other regions.[32]

The centrality of this retrospective diagnosis to arguments about the Antonine epidemic cannot be understated. Indeed, it effectively begs the question by assimilating the Antonine pathogen to a well-known and deadly modern virus, thereby establishing in advance the epidemiological parameters of the ancient disease. With the pathogenicity, virulence, transmissibility, and mortality of modern *V. major* as an empirical baseline, scholarship therefore assesses not *whether* the Antonine epidemic had a significant impact on Roman demog-

28. Elliott, "Disease Proxies," 321, cf. the methodological considerations at 300–302.
29. The diagnosis is made in Littman and Littman, "Galen and the Antonine Plague." It seems accepted by all contributors to Lo Cascio, *L'impatto della "peste antonina."* See esp. Zelener, "Genetic Evidence," 168–70.
30. Duncan-Jones, "The Antonine Plague Revisited," 44.
31. Zelener, "Genetic Evidence," 174–75.
32. Scheidel, "A Model," 99–100.

raphy but *how* significant this impact was.[33] Observed (or inferred) perturbations in the Egyptian or wider Roman economies as well as disruptions to other evidentiary streams accordingly become proofs of a conclusion already implicit in the diagnosis—death on a statistically and thus socioeconomically significant scale. While such circular reasoning[34] is worrisome enough, medical historian Rebecca Flemming has also recently argued against the smallpox diagnosis on literary, medical-historical, and genomic grounds. Not only is there no known poxvirus capable of producing the array of symptoms described in Roman medical literature, recent genomic analyses also prove that modern smallpox is entirely that—an evolutionary development of the eighteenth and nineteenth centuries.[35] The identity of the Antonine pathogen and the nature of its host interactions must therefore remain an open question.[36] Flemming's argument is consequential, since a return to agnosticism would effectively strip scholarship on the Antonine epidemic of the empirical framework derived from *V. major*, forcing estimates of its mortality to stand on their own. The burden of proof therefore rests not on the doubters but on those whose arguments are founded upon the characteristics and behavior of a virus that did not yet exist, at least not in its modern form.[37]

Second, attempts to ground arguments in the testimony of the papyri run hard up against the near-total lack of evidence. Papyrologist Paul Schubert has already summarized the situation, remarking that "the impact [of the epidemic], if it

---

33. This is the approach of Zelener, "Genetic Evidence."
34. On circularity in the identification of evidence for the epidemic see already Wolfgang Hameter, "The Afterlife of Some Inscriptions from Noricum: Modifications and Falsifications," in *The Afterlife of Inscriptions: Reusing, Rediscovering, Reinventing & Revitalizing Ancient Inscriptions*, ed. Alison E. Cooley (London: Institute of Classical Studies, School of Advanced Study, University of London), 37–46, esp. 44. See also Erdkamp's similar criticism of attempts to identify evidence for climate change ("War, Food, Climate Change," 432).
35. Citing A. T. Duggan et al., "17th Century Variola Virus Reveals the Recent History of Smallpox," *Current Biology* 26 (2016): 3407–12. P. Pajer et al., "Characterization of Two Historic Smallpox Specimens from a Czech Museum," *Viruses* 9 (2017): 200, https://doi.org/10.3390/v9080200, date the common ancestor of the modern variola virus to ca. 1350 CE, a date pushed forward to the late sixteenth or early seventeenth century CE by A. Porter et al., "Comment: Characterization of Two Historic Smallpox Specimens from a Czech Museum," *Viruses* 9 (10) (2017): 276, https://doi.org/10.3390/v9100276; and C. Smithson, J. Imbery, and C. Upton, "Re-Assembly and Analysis of an Ancient Variola Virus Genome," *Viruses* 9 (2017): 253, https://doi.org/10.3390/v9090253. On the remarkably rapid mutation and evolution of pox viruses through encounters with human immune systems see reportage on the 2022 global monkeypox outbreak, e.g., Ben Turner, "Monkeypox May Have Undergone 'Accelerated Evolution,' Scientists Say," *Live Science*, 25 June 2022, https://www.livescience.com/monkeypox-mutating-fast.
36. Flemming, "Galen and the Plague," 232–40.
37. *Pace* Zelener, "Genetic Evidence," 171.

existed, has left few unambiguously identifiable traces."[38] This evidentiary dearth is critical, since of the various "poles" composing most "wigwam" arguments about the epidemic, none is made to carry as much weight as Egyptian papyri.[39] Over and above any qualms regarding the cliometric analysis of fragmentary and poorly contextualized datasets drawn from papyri of a few agriculturally marginal regions, no unambiguous testimony to a widespread demographic contraction has been found. The oft-cited example of *P.Oxy.* 66.4527 (185 CE) is illustrative. The papyrus is a fragment of a grain-tax account from the *meris* of Herakleides, one of the three administrative divisions of the Graeco-Roman Fayyūm. Depending on how text lost in a lacuna is restored, the account can be read as evidence either of continuity in agricultural production after the epidemic or of dramatic decline in rural revenues—that is, proxy evidence for widespread socioeconomic disarray.[40] Any conclusions based upon one or the other restoration are therefore unsound—a "history from square brackets" in which scholarly conjecture becomes historical fact.[41] Yet apart from this single disputed case, the Egyptian papyri offer little more than hints of a mortality event during these years. Recently assembled and described by Isabella Andorlini, these few traces, some altogether uncertain, are listed in the table below:[42]

Scant and indirect, such testimony is at least suggestive of a sudden uptick in mortality. But death is only one explanation for the disappearance of Roman Egyptians from documentary sources. Indeed, the only published papyrus that seems to directly attest the epidemic simultaneously muddies the waters by confirming that its impact was entangled with other local sociopolitical developments.[43] As recorded in a carbonized roll of administrative papyri from the western Delta city of Thmouis, various taxes in arrears owed by a village named Kerkenouphis had been suspended in 168–69 CE in response to a cascading array of local calamities, particularly an outbreak of violence and raiding among the *Nikōchitai*, otherwise known as the *Boukoloi* ("herdsmen"), peoples who lived amid the swamps and marshes on the margins of the cultivated land of the Delta:

38. Schubert, *Philadelphie*, 156.
39. So Elliott, "The Antonine Plague," 4.
40. Stability: Bagnall, "*P.Oxy.* 4527 and the Antonine Plague in Egypt." Decline: Van Minnen, "P. Oxy. LXVI 4527." See also Van Minnen's response in "The Changing World of the Cities," 163n17. Summary of the dispute in Schubert, *Philadelphie*, 149–51.
41. Ernst Badian, "History from 'Square Brackets,'" *Zeitschrift für Papyrologie und Epigraphik* 79 (1989): 59–70.
42. Andorlini, "Considerazioni," 21–22."
43. Blouin, *Triangular Landscapes*, 255–56.

Table 5.1. Possible papyrological attestations of Antonine epidemic mortality

After Andorlini, "Considerazioni sulla 'peste antonina,'" 21–22.

- *Karanis*: tax rolls indicate a population decrease of 33–47 percent by 171 CE (*P.Mich.* 4.223–5).[1]
- *Ptolemais Euergetis/Arsinoe*: statement of simultaneous death of 3 persons in 175/6 CE (*BGU* 1.79 = *C.P.Gr.* 2.55).
- *Soknopaiou Nesos*: deaths of 78 of the 244 male taxpayers in 179 CE (*SB* 16.12816).
- *Theadelphia*: several documents suggest numerous deaths among local cultivators between 161 and 210 (*SB* 26.16675) and 216 CE (*P.Strasb.* 7.688, col. 1).[2]
- *Terenouthis*: Latin inscriptions, though of disputed date, attest multiple deaths on the same day (*SB* 3.6585; 8.9996, 10162).

[1] Boak, "The Population."
[2] Sharp, "The Village of Theadelphia," 185–89.

> ὁ αὐτὸς κωμογρα(μματεὺς) καὶ τὰ ἐπὶ κώμης Κερκενούφ(εως) ἀναλαμβ(ανόμενα) καὶ ἄλλα ἐκούφισεν φήσας τοὺς πλείστους τῶν ἀπὸ τῆς κώμης ἀνειρῆσθαι [*l.* ἀνῃρῆσθαι] ὑπὸ τῶν ἀνοσίων Νεικωκειτῶν, ἐπελθόντων τῇ κώμῃ καὶ ἐμπρησάντων αὐτήν, καὶ ἄλλους τῷ λ̣οιμικῷ καταστήματι τετελ(ευτηκέναι) καὶ τοὺς λοιποὺς ὀλιγίστους ὄντας ἀπο<πε>φευγέναι.
>
> The same village scribe has alleviated the [contributions] and the other taxes incumbent on the village of Kerkenouphis, having stated that the majority of those from the village had been killed by the impious Nikochitai, who descended on the village and burned it, while others had perished in the pestilential situation (*tō loimikō katastēmati*), and the very few remaining had fled.[44]

Even if this extract is interpreted as clear testimony to the Antonine epidemic,[45] it simultaneously demonstrates that its impact was mediated through local socioeconomic and political conditions.[46] We must therefore be prepared to accept that the effects of the epidemic would have varied widely across Egypt, a diversity that will likely remain forever invisible to historians whose view of Graeco-Roman countryside is largely confined to the marginal regions from which papyri survive in quantity. Consequently, it would be irresponsible both to assume catastrophic mortality from what little evidence we do have and to generalize from these marginal cases to the rest of Egypt.[47]

44. *P.Thmouis* 1, col. 104, ll. 9–18, with Blouin, *Triangular Landscapes*, 253–58. On the *Boukoloi* uprising and its causes see Blouin, *Triangular Landscapes*, 267–97.
45. Andorlini, "Considerazioni," 20, on the phrase τῷ λ̣οιμικῷ καταστήματι.
46. Drawing on language in Erdkamp, "War, Food, Climate Change."
47. See already Bagnall, "*P.Oxy.* 4527," 292.

Still, the near-complete absence of papyrological testimony has itself been regarded as evidence of the epidemic's severity. Following Richard Duncan-Jones, Andorlini suggests that the observable decline in papyri year-dated to the period of the epidemic is consistent with the context: a catastrophe of a scale and intensity sufficient to disrupt regular documentary practices, public and private alike.[48] While the scenario is outwardly plausible, James Greenberg notes that the effect of the epidemic on the papyri is only apparent, since the rate of documentary production had already begun to decline in the 140s and never dips below that of the first century CE. Comparative evidence from later European epidemics also indicates that rates of document production might remain stable or even increase during and after mass mortality events.[49] Comparative perspective from Egypt itself is also highly suggestive, since the country's later experience of the Black Death reveals the implications of mass mortality crises so starkly that the silence of the papyri becomes harder to explain away. As described by Stuart Borsch, outbreaks of the bubonic plague in the thirteenth and fourteenth centuries CE carried off a significant proportion of Egypt's rural inhabitants and drove large numbers of surviving farmers into nearby urban centers.[50] The subsequent inability of the remaining rural population to maintain public irrigation infrastructure was perhaps the most visible and significant socioeconomic casualty of the plague. Even villages whose populations were largely spared and which were therefore able to continue maintaining their own local dikes and canals eventually suffered due to breakdowns in public water infrastructure.[51] As Borsch writes, "well-maintained and ordered local systems now collapsed suddenly during years when decay [in the public infrastructure] led either to severe flooding or to equally devastating droughts caused by breaches in the larger network of dikes and canals."[52] This increasing vulnerability to both low and high floods reduced the overall resiliency of rural society and led to more frequent crop failures and food shortages.[53]

Borsch's observations recall the arguments of chapter 3 above, which described the mobilization of large numbers of human bodies for annual mainte-

48. Andorlini, "Considerazioni," 23. Cf. Duncan-Jones, "The Impact of the Antonine Plague," 124–30.
49. Greenberg, "Plagued by Doubt," 415–16.
50. On the methodologies for assessing mortality rates see Borsch and Sabraa, "Refugees of the Black Death."
51. On the distinctions between *sulṭānī* and *baladī* dikes and canals see Ibn Mammātī, *Qawānīn*, 232–33, trans. in Cooper, "Ibn Mammati's Rules," 76–77.
52. Borsch, *The Black Death*, 47.
53. Borsch, *The Black Death*, 47. See also Borsch, "Environment and Population"; and Borsch, "Plague Depopulation and Irrigation."

nance of the Fayyūm's public water infrastructure. Widespread disruption of this socioenvironmental system would necessarily have had a detrimental effect on water flow throughout the region. Perspective on this intimate entanglement between laboring bodies and flowing water emerges with microscopic clarity from a first-century CE Fayyūm letter in which a father plaintively begs his son for aid on a plot of land whose water supply had all but failed for want of labor:

> Ἑρμοκράτη[ς Χαιρᾷ] τῷ υἱῶι [χαίρειν]. πρ[ὸ] τῶ[ν ὅλων ἐρρωσθαί σ]ε εὔχο[μαι . . . . δέ]ομε (*l.* δέομαι) σε ε[. . . . . . .γ]ράφειν π[ερὶ] τῆς ὑγίας (*l.* ὑγείας) σου καὶ [ὅ]τι βούλι (*l.* βούλει), καὶ ἄλλοτέ σοι ἔγραψα περὶ τῆς τ[. .]ψυα καὶ οὔτε ἀντέγραψας οὔτε ἦλθας, καὶ νῦν, αἰὰν (*l.* ἐὰν) μὴ ἔλθῃς, κινδινεύω ἐκστῆναι οὗ ἔχω [και]ροῦ. ὁ κοινωνὸς ἡμῶν οὐ συνηργάσατο, ἀλλ' οὐδὲ μὴν τὸ ὕδρευμα ἀνεψήσθη, ἄλλως τε καὶ ὁ ὑδραγωγὸς συνεχώσθη ὑπὸ τῆς ἄμμου καὶ τὸ κτῆμα ἀγεώργητόν ἐστιν. οὐδεὶς τῶν γεωργῶν ἠθέλησεν γεωργεῖν αὐτό, μόνον διαγράφω τὰ δημόσια μηδὲν συνκομιζόμενος μόλις γὰρ μίαν πρασεὰν (*l.* πρασιὰν) ποτίζι (*l.* ποτίζει) τὸ ὕδωρ, ὅθεν ἀνανκαίως (*l.* ἀναγκαίως) ἐλθέ, ἐπὶ (*l.* ἐπεὶ) κινδυνεύει τὰ φυτὰ διαφωνῆσαι. ἀσπάζεταί σε ἡ ἀδελφή σου Ἑλένη καὶ ἡ μήτηρ σου μέμφεταί σε, ἐπὶ (*l.*ἐπεὶ) μὴ ἀντέγραψας αὐτῇ. ἄλλως τε καὶ ἀπαιτῖται (*l.* ἀπαιτεῖται) ὑπὸ τῶν πρακτόρων ἱκανὸν ὅτι οὐκ ἔπεμψας πρός σε τοὺς πράκτορες (*l.* πράκτορας), ἀλλὰ καὶ νῦν πέμψον αὐτῇ. ἐρρῶσθαί σε εὔχ[ομ]αι. Παοῖνι (*l.* Παῦνι) θ.

Hermokratēs to his son Chairas, greetings. Above [all] I hope that you [are well]. I ask you [. . .] to write about your health and what you desire, and at other times I wrote you about Tapsoia (?) but you neither replied nor came. And now, if you do not come, I risk abandoning the place I possess. Our partner was of no help and the reservoir (*hydreuma*) was not even cleaned out and the canal (*hydragōgos*) was also filled up with sand (*ammos*) and the holding is uncultivated. None of the tenants wanted to farm it. I am simply paying the taxes with no benefit, for the water scarcely irrigates one garden plot. So by all means come, for the plants risk coming to harm. Your sister Helen sends her greetings and your mother is angry because you didn't write back to her. Besides, she is being hounded well enough by the tax collectors because you didn't send them to yourself. So send to her right now (sc. what the tax collectors demand). Farewell. The ninth of Payni.[54]

---

54. *BGU* 2.530 (first century CE). For the proposed restoration of ll. 8–9 ἔγραψα περὶ τῆς τ[. .]ψυα see *BL* 7, 13.

Given the serious implications of labor shortages for Fayyūm irrigation, it is all but inevitable that a mortality event on the order of 20–30 percent would have magnified these sorts of disruptions across the whole of the region, even if only in the short term. The absence of evidence for such a crisis is therefore striking, all the more so when we recall that it is usually only in moments of disruption that water and irrigation make a mark on the papyrological record. Yet the only papyri thus far proposed as proxy evidence of systemic irrigation problems during and immediately following the epidemic are the so-called *abrochia* declarations. First appearing in the latter half of the second century CE, this document type enabled farmers to attest that their land was unflooded (*abrochos*), uncultivable, and thus eligible for a remission of yearly imposts. As Christer Bruun writes, the earliest *abrochia* declarations predate the arrival in Egypt of the Antonine epidemic (ca. 165 CE) and thus cannot be regarded as a documentary-administrative response to the disease. He nonetheless suggests that peaks in the number of surviving declarations in the 180s and 190s may indicate the cumulative effects of declining water infrastructure.[55] While this conclusion is not unreasonable, it must be remembered that the water shortages were a fact of life at the tail end of the Fayyūm's canal system and were thus accounted for in land leases of both the Ptolemaic and Roman date.[56] Moreover, chapter 2 has already suggested that a significant portion of the land in downstream villages might be unflooded even in normal years.[57] It is therefore likely that local flood-failures were a common occurrence in all periods, only becoming more visible through the introduction of a novel documentary instrument in the *abrochia* declaration. More important, however, is the fact that, as this study has demonstrated, Fayyūm water flow was at all times a socioenvironmental production. Therefore, the virtual invisibility to papyrologists of all but a handful of marginal settlements in antiquity means that we cannot uncritically link downstream water shortages here to upstream failures. Absent relevant evidence either way, it is no less plausible to suggest that increasing water shortages at the ends of the canal system were produced by thriving upstream conditions and subsequently greater water consumption, for instance by changing cropping patterns or agricultural intensification, phenomena described in

55. Bruun, "The Antonine Plague and the 'Third-Century Crisis,'" 205–6. On the declarations in general see Habermann, "Aspekte des Bewässerungswesens."
56. See chapter 2 above at n. 89.
57. See the discussion of *P.Cair.Isid.* 6 (300–305 CE) above at pp. 95–96. For irrigation problems in the southern Fayyūm in the late second century BCE see Thompson, *Kerkeosiris*, 117–21.

the previous chapter.[58] Since it is effectively impossible to validate either speculation, the testimony of the *abrochia* declarations is effectively ambiguous.

While we must remain open to the possibility that the epidemic had a serious effect on Egyptian demography, these reservations highlight the difficulty of reaching anything approaching empirical certainty.[59] Yet arguments for climate causality are more problematic still, a function of their relative novelty, the protean instability of their evidentiary base, and their critical misreading of Egyptian irrigation. Focusing on the Nile Delta, Colin Elliott, for instance, has suggested that shortfalls in the Nile flood during the late second century CE, rather than the Antonine epidemic, precipitated the disorder visible in the papyri from Roman Thmouis. This argument is premised on the hypothesis that global volcanism and subsequent atmospheric shrouding during the latter half of the second century CE disrupted the El Niño-Southern Oscillation (ENSO)—variations in winds and sea-surface temperatures in the tropical eastern Pacific—in turn suppressing the East African and Indian monsoons, whose rains become the Nile flood.[60] Such climate-forced reductions in the flood, Elliott argues, will have led to an increase in water shortages and crop failures, potentially sparking significant unrest.[61] In a contribution on climate and the decline of marginal Fayyūm villages, Sabine Huebner has likewise pointed to historic volcanism and other climate proxies such as African lake sediment cores that seem to indicate an overall trend toward drier conditions in later antiquity. Declining flood levels, she suggests, made irrigation increasingly difficult at the multiple tails of the Fayyūm's canal system.[62] So too Kyle Harper, who draws on "one sedimentation record from Ecuador" to suggest that a "quiescent ENSO" and relatively reliable Nile flooding gave way after 155 CE to more erratic flood levels, more frequent crop failures, and declining Egyptian grain production.[63]

The evidence upon which these studies are based poses significant difficul-

58. Cf. the modern *comparandum* in Barnes, *Cultivating the Nile*, 122–25.
59. Cf. the reservations of demographer M. Livi Bacci in "Note demografiche," 344–45: "I tempi, i luoghi, le circostanze, l'organizzazione della società non potrebbero essere più distanti. La scarsità delle testimonianze e di dati oggettivi rendono fragili le interpretazioni."
60. For earlier scholarship on ENSO-Nile flood connections see Alfatih A. B. Eltahir, "El Niño and the Natural Variability in the Flow of the Nile River," *Water Resources Research* 32.1 (1996): 131–37; and Guiling Wang and Alfatih A. B. Eltahir, "Use of ENSO Information in Medium- and Long-Range Forecasting of the Nile Floods," *Journal of Climate* 12.6 (1999): 1726–37.
61. Elliott, "The Antonine Plague."
62. Huebner, "Climate Change."
63. Harper, *The Fate of Rome*, 132–36, quote at 134.

ties from a historiographical perspective. Elliott's arguments, for instance, depend upon scientific catalogs of historical volcanic eruptions, an assemblage whose dating has been described by environmental historian Timothy Newfield and geologist Inga Labuhn as poorly resolved and prone to frequent adjustment in response to rapid improvements in dating methodologies. They further note that the links between Roman-era volcanism and the Nile are also at present ill understood. As a result, Elliott's hypothesis linking Roman-era volcanism to Nile flood failures and sociopolitical disorder in the Egyptian countryside is intriguing but shakily founded.[64] Of still greater methodological concern is climate scientist Sylvia G. Dee's recent claim that the impact of historical volcanism on the ENSO has been significantly overestimated. Dee and her team instead link most of the observed ENSO variations during the preindustrial portion of the last millennium to endogenous climate variability rather than to volcanism and atmospheric shrouding.[65] That even this argument has in turn been disputed should strongly caution historians against placing too much weight on science (broadly construed).[66] Fluid and ever in the process of change and refinement, scientific evidence makes an exceedingly unstable foundation for most if not all historiographical argumentation.[67]

A potential source of external confirmation, however, is the record of Nile floods assembled in the early 1970's by papyrologist and historian Danielle Bonneau, data that allegedly point to an abrupt downward trend in Nile flood levels beginning in the middle of the second century CE.[68] Drawn from literary, papyrological, and numismatic evidence and covering the years 261 BCE to 299 CE, Bonneau's flood list has also been utilized by historians Michael McCormick and Kyle Harper, who propose that the reliability of the flood

---

64. Newfield and Labuhn, "Realizing Consilience," 220–27.
65. Dee et al., "No Consistent ENSO Response."
66. Alan Robock, "Comment on 'No Consistent ENSO Response to Volcanic Forcing over the Last Millennium,'" *Science* 369, issue 6509 (2020), https://doi.org/10.1126/science.abc0502.
67. So Steak-umm (@steak_umm) in response to Neil deGrasse Tyson (@neiltyson): "Science is an ever refining [*sic*] process to find truth, not a dogma." Twitter, 12 April 2021. See also the brief exchange in response to Joseph R. McConnell et al., "Extreme Climate after Massive Eruption of Alaska's Okmok Volcano in 43 BCE and Effects on the Late Roman Republic and Ptolemaic Kingdom," in *PNAS*, July 7, 2020, 117 (27) 15443–15449. Response to McConnell et al.: Sebastian Strunz and Oliver Braeckel, "Did Volcano Eruptions Alter the Trajectories of the Roman Republic and the Ptolemaic Kingdom? Moving beyond Black-Box Determinism," *PNAS*, December 22, 2020, 117 (51) 32207–32208. Response to Strunz and Braeckel: Joseph R. McConnell et al., "Reply to Strunz and Braeckel: Agricultural Failures Logically Link Historical Events to Extreme Climate Following the 43 BCE Okmok Eruption," PNAS, December 22, 2020, 117 (51) 32209–32210.
68. Elliott, "The Antonine Plague," 25–28, drawing on the catalog in Bonneau, *Le fisc et le Nil*, 221–58.

decreased markedly after 155 CE and became more unreliable still between 261 and 299 CE.[69] The problems with Bonneau's often subjective and impressionistic "data" are legion and have been thoroughly described elsewhere.[70] Their dubious reliability notwithstanding, other scholars suggest that her list in fact shows relative continuity in flood levels rather than the sharp downward trend in the late second or early third centuries CE.[71] For lack of additional evidence, debate over the trajectory of Nile flood levels has therefore stalled with little sign of resolution in either direction.

While greater exploration of historical Nile flood patterns should be encouraged, the current conversation is in two respects deeply flawed, at least as far as the Fayyūm is concerned. In the first place, any discussion of flood levels, irrigation, and the landscape history of the Fayyūm must take into account the claim in Arabic sources that the Fayyūm then became irrigable at a rise of twelve cubits (7.9 m), as opposed to the sixteen or seventeen cubits (10.5–11.3 m) necessary for the irrigation of the Nile Valley.[72] This testimony should be treated with caution, of course, given the tendency of Arabic authors to hyperbolize the productivity of Fayyūm irrigation. Yet the fact that the figure of twelve cubits appears in the administrative texts of both Abū Isḥāq and al-Nābulusī gives reason for pause. Indeed, Abū Isḥāq explicitly describes the al-Lāhūn dam as being opened to admit the flood when it had reached the requisite height of twelve cubits, at which point water entered and began to flow toward the capital (see appendix below). As suggested in the first chapter, this unique feature of Fayyūm irrigation may have owed to the depression's unusual hydrology, namely, the fact that water directly entered the depression through the al-Lāhūn inlet rather than having to rise high enough to overtop the river's banks as in the Nile Valley. One might object that perhaps the Fayyūm's central alluvial plain, the only portion of the region still occupied in the Fāṭimid and Ayyūbid periods, was irrigable by floods that would otherwise have been too weak to water the tail ends of the much larger Graeco-Roman canal system. While this is plausible, there are at present no grounds for certainty and we must reserve judgment in the hope that future modeling of premodern Nile hydraulics will lend useful perspective. All the same, we must simultaneously

69. McCormick, "What Climate Science," 76–81; and Harper, *The Fate of Rome*, 133.
70. Bonneau assigned floods to one of six categories: *mauvaise*, *faible*, *mediocre*, *normale*, *bonne*, and *abondante*, often on exceedingly shaky evidence. For thorough critiques see Haldon et al., "Plagues, Climate Change, and the End of an Empire (2)," 5; and Huebner, "Climate Change," 507–10.
71. Haldon et al., "Plagues, Climate Change, and the End of an Empire (2)," 5.
72. See above at pp. 42–43.

remain open to the possibility that the flood levels required for a successful crop in the Nile Valley may be largely or entirely inapplicable to the unique irrigation system of the Fayyūm.

Second, it is conceptually misguided to propose direct causal links between Nile flood levels and agricultural outcomes in Egypt, for this necessarily reduces irrigation to a dehumanized natural-environmental phenomenon entirely governed by exogenous environmental stimuli. Such a reading effaces the agency of rural irrigators, transforming them into "automatons, responding mechanically to forces beyond their control."[73] As Stuart Borsch's work on irrigation in early Islamicate Egypt has consistently demonstrated, flood-control and irrigation works throughout Egypt were well-adapted to enable cultivation even in years of high or low flood. Egyptian authors like al-Maqrīzī therefore often blamed the crop failures and other problems extending from poor floods not on the water supply but on the Egyptian government, accusing officials of squandering or embezzling revenues dedicated to the maintenance of rural water infrastructure.[74] Likewise this study, which has depicted irrigation as a socioenvironmental coproduction of nature and human agency. This is not at all to dismiss the real and significant hardships that poor floods, whether low or high, could cause over the short term.[75] Rather, it is simply a reminder that the socioeconomic impacts of changes in the Nile flood were not ecologically determined but were instead mediated through an array of water-management practices and institutions at both the local and the state level. In consequence, if future scholarship incontrovertibly demonstrates that the reliability of the Nile flood declined after the mid-second century CE, we must then investigate the degree to which Egyptian practices and institutions did (or did not) facilitate resilience in the face of a changing environment.[76]

---

73. McHugh, "Inside, Outside," 74. Cf. the similar critique by Jan De Vries, Review of Parker, *Global Crisis* (2013), *Journal of Interdisciplinary History* 44 (2014): 369–77, quote on 375: "This approach reduces the economy to little more than a direct physical relationship between weather and harvest results, but seventeenth-century economies in most of Eurasia were not so simple." Cited from Erdkamp, "War, Food, Climate Change," 437n60.

74. In general: Borsch and Sabraa, "Qānūn al-Riyy"; and Borsch, *The Black Death*, 40–54, for specific failures, cf. 127–33.

75. See e.g., the description of the high flood of 1887—"the terror reigning over the whole country"—in William Willcocks, *The Assuân Reservoir and Lake Moeris* (Cairo: National Printing Department, 1904), 28–30. Willcocks' dramatic narrative nonetheless details the "steady, business-lake manner" with which rural cultivators immediately responded to dangerously high waters and worked to prevent significant damage.

76. Drawing on language in Erdkamp, "War, Food, Climate Change," esp. 253.

• • •

Reflecting on the use of quantitative methods in historiography, Colin Elliott writes that they are often used destructively "to falsify hypotheses rather than in support of new arguments."[77] The above critique does much the same work, albeit from a more conventional historiographical posture. To reiterate, these remarks are not intended as a rejection of natural-environmental causality *tout court* but rather as a call to integrate emerging model- and science-based approaches with traditional historiography. Indeed, although it seems to have escaped much notice,[78] there exists a competing model of the decline of marginal Fayyūm villages and the rise of large estates in the third century CE based solely upon papyrological sources. In his 2007 study of Philadelphia, papyrologist Paul Schubert proposed that gradual accumulation by wealthy absentee landowners was the primary cause of demographic and tenurial transformations in the village between the second and third centuries CE. While it is impossible on current evidence to clearly discern the genesis of the contemporary estate of Aurelius Appianus at Theadelphia, Schubert nonetheless suggests that the same dynamics may have been at play on the opposite edge of the Fayyūm.[79] The scenario is plausible. Papyrologist Michael Sharp has already remarked upon the large number of Roman-citizen and Alexandrian absentee landowners in Theadelphia in the middle of the second century CE. Given that these individuals possessed a disproportionately large share of vineyard and garden-land in the village, it is possible that these patterns of land tenure anticipate the socioeconomic developments that would culminate in estates like that of Appianus in the following century.[80] As Dominic Rathbone has likewise noted, at least some of the Theadelphian vineyards later included in the estate of Aurelius Appianus were already present in the village a century earlier.[81] While both Rathbone and Sharp also cite the Antonine epidemic as a potentially significant causal factor, it is nonetheless clear that the effects of the disease would have interacted with socioeconomic processes already underway, namely, the gradual accumulation of land in fewer and fewer hands and an attendant transition away from arable subsistence agriculture and toward

77. Elliott, "Disease Proxies," 320.
78. The work is not cited in Römer, "Why Did the Villages"; or Huebner, "Climate Change."
79. Schubert, *Philadelphie*, esp. 159–68.
80. Sharp, "The Village of Theadelphia," 174–85.
81. Rathbone, "Economic Rationalism," 263, citing the labor contract for the fertilization of a vineyard *P.Col.* 10.255 (131 CE).

the market-oriented production of cash crops like garden produce and wine—the latter being the primary focus of the Theadelphian *phrontis* of the Appianus estate in the following century.[82] At present, however, the relationship and mechanisms of interaction between these internal socioeconomic developments and external environmental phenomena are indeterminate. It is accordingly the work of future scholarship both to reveal and unravel this potentially consequential socioenvironmental entanglement.

On High Ground / at the End / We Are Three

Whatever the causal mechanism(s) at play, a fair portion of state and private land at Theadelphia had been subsumed by a division (*phrontis*) of the estate of Aurelius Appianus by the middle of the third century CE. There were multiple divisions of this estate throughout the Fayyūm in this period, all centrally administered from the nome capital by a certain Alypios. Each *phrontis* was also under the direct oversight of a local manager (*phrontistēs*). While there was also a division of the estate in nearby Euhēmeria, only the *phrontis* in Theadelphia is known in any detail, thanks to the survival of an archive kept by one Heroninos, its *phrontistēs* between 249 and 268 CE. The Theadelphian *phrontis* itself was substantial, occupying an estimated minimum of 400 *arourai* out of the 5,200 or fewer *arourai* of land in the village, although that at Euhēmeria may have been larger still. Appianus was also not the only large landholder in Theadelphia, which was likewise home to divisions of the estates of several landed proprietors, the sizes of whose holdings are nonetheless unknown.[83]

Tightly managed and oriented toward market production of wine from water-intensive vineyards, the Theadelphian *phrontis* of the Appianus estate was heavily invested in maintaining and improving irrigation infrastructure in the village itself and the surrounding territory.[84] In a letter of 257 CE, the central manager Alypios wrote to a number of estate employees to remind them to undertake their duty (*dosis*) on a local dike—the annual dike- and canal-work

82. See already Harris, "The Great Pestilence," on the entanglement between the epidemic and extant socioeconomic phenomena.
83. Rathbone, *Economic Rationalism*, 38–41.
84. The following citations are drawn from Rathbone, *Economic Rationalism*, 222–27.

owed to the state.[85] Likewise in 253 or 256 Appianus himself ordered Heroninos and other employees to contribute to unspecified local works at Theadelphia.[86] The estate also frequently hired wage laborers specializing in dike- and canal-maintenance, so-called "river-men" (*potamitai*). A large, albeit ill-defined, project is referenced in another letter of 257 CE in which Alypios informs Heroninos that he has dispatched a soldier to ensure the accomplishment of works being done on a "new canal" (*kainēi diōrychi*) as well as the primary canal that served the village of Theadelphia (*kōmētikēi* [sc. *diōrychi*]).[87] In 260 CE the estate again contributed both its own laborers and hired *potamitai* to another ill-defined but large-scale irrigation project involving a "new canal" (*kainēn diōryga*) that was of benefit both to Theadelphia and Taurinou, a nearby village of uncertain location.[88] Elsewhere in the archive the central manager Alypios is seen dispatching *potamitai* to the village of Theogonis south of Theadelphia, there employed in the removal of silt (*ammos*) from the bed of a canal.[89] So too Heroninos' accounts, which reveal cash expenditures for wage laborers working on local irrigation infrastructure.[90]

Although few and scattered, these notices hint at the deep involvement of the Appianus estate in the production of water flow in and around Theadelphia, work that likely ensured the continued reliability of local water infrastructure. Yet as argued in the previous chapter, the production of water flow through collective labor was central to the social life of independent irrigation communities. In consequence, having already reduced the size of the irrigation community at Theadelphia by displacing an unknown but significant number of smallholders, the Appianus estate—and perhaps also the estate units of other landowners in the village—further eroded the social infrastructure of the community by assuming a portion of the collective labors through which the community was formerly constituted and bound to its local section of the canal system. While this integration of estate and community provided temporary stability, the estate itself proved ephemeral, since both Appianus and other large landowners at Theadelphia had disappeared by the end of the third cen-

85. *P.Flor.* 2.133.
86. *P.Prag.Varcl.* 1.5.
87. *P.Rein.* 2.115.
88. *P.Flor.* 2.273.
89. *P.Flor.* 2.157 (249–68 CE).
90. *SB* 6 9408.1 (250 CE); *SB* 20.14645 (251 CE); *SB* 20.14197 (253 CE); *P.Flor.* 3.321 (254–60 CE); *P.Lond.* 3.1170v (259 CE); *P.Prag.* 3.236r (259 CE).

tury—an event that still defies a universally accepted explanation.[91] In the aftermath, the rump community left behind was ill equipped to compete with upstream settlements for a share of the common water supply.[92] Lacking the bodies necessary for regular collective action, Sakaon and his companions instead adopted a model of water management by complaint in which state authorities at both the nome and provincial level were drawn by petition into Theadelphia's affairs. The brief era of the Appianus estate thus represents a critical stage in the life of Theadelphia's irrigation community: a transitional interlude between independent agency and the passive dependency visible in the archive of Sakaon.

Our final glimpse of the earlier, more self-reliant irrigation community at Theadelphia comes in a papyrus dating to the life of the Appianus estate: *P.Sakaon* 32 (254–68 CE).[93] Already described in the previous chapter, the text is the lacunose transcript of a hearing convened before the nome *stratēgos* to adjudicate a conflict between Theadelphia and its upstream neighbor Philagris. As previously noted, the poor preservation of the text obscures the origins and nature of the conflict. All that can be recovered with certainty is that an unknown number of Theadelphians (*tōn apo Theadelpheias*) were accused of physically altering a canal (*diōrycha*) and a stone-lined water-conduit (*stomion*) at Philagris, allegedly because of their frustrations with a sluicegate (*kleidion*) somewhere in the vicinity of the latter village. Regardless, it is not the details of the conflict that are important here but rather the fact that in the middle of the third century, the village's population still possessed one of the most salient characteristics of an independent irrigation community: the ability to take collective (and here seemingly aggressive) action to secure its water supply. *P.Sakaon* 32 is, moreover, an outlier to the archive of Sakaon proper, which principally dates to the years 280 to 342 CE.[94] Consequently, while the text may well testify to the beginnings of Theadelphia's later water struggles, it may also simply be evidence of an irrigation community unilaterally asserting its rights to the common water supply, confrontational though such action

91. Van Minnen, "The Changing World of the Cities," suggests that inflation after 275 made the market-oriented vineyard sector less profitable.

92. Rathbone, *Economic Rationalism*, 408. Cf. Ermatinger, "The Economic Death of Theadelphia," 8–9.

93. On the date see D. Hagedorn, "Zum Amt des διοικητής im römischen Aegypten," *Yale Classical Studies* 28 (1985): 167–210 at 205 (*BL* 8, 299).

94. On the text as outlier see Rathbone, "Villages and Patronage," 192 with note 10. On the date of papyrus see *BL* 7, p. 299.

might be. Whatever the case, collective action is emphatically not in evidence elsewhere in Sakaon's archive, which instead reveals a community that had become entirely dependent on outside intervention, a strategy that proved altogether ineffective.

The rest of this chapter is concerned with these later papyri, mostly petitions, which date to the third and fourth decades of the fourth century. As suggested above, I read this documentation as an attempt to communicate a fundamentally local knowledge of irrigation at Theadelphia, that is, an understanding, born of routine practice, of the unique environment of the village, of the work required to produce its water supply, and, most importantly, of the significant socioenvironmental impediments to the continued production of these water flows. This interpretation reflects a sympathetic reading of the archive, here meaning only that I take the Theadelphians' complaints seriously and attempt to ground them in their socioenvironmental context.[95] When read in this light, Sakaon's papyri, although often clumsy and naïve, present a coherent narrative of the incapacity and failure of an irrigation community—a narrative, moreover, that cannot be decoupled from the particulars and peculiarities of place.

Central to this narrative was Theadelphia's location, both in absolute and relational terms. In *P.Sakaon* 42 (ca. 323 CE), a petition to the *praeses* of Aegyptus Mercuria,[96] Sakaon and his companions describe their water shortages as the result of their location "very far back" (*to poly hysteran*). At least once but perhaps twice, a clear distinction is drawn between Theadelphia's topography and a settlement on the "plain" (*pedion*). In *P.Sakaon* 33 (ca. 320), discussed in greater detail below, the Theadelphians claim that a group of allegedly water-thieving individuals, their precise location not specified, lie upstream (*hyperkathēmenoi*, lit. "sitting above") and possess agricultural land "on the plain" (*epi tou pediou*). While the term *pedion* can refer simply to a village's embanked and floodable agricultural area,[97] in this context it marks a contrast between the location of Theadelphia's marginal fields on high ground above the level of local canals and villages whose fields were below the level of their canals, which were therefore able to be irrigated by gravity. In *P.Sakaon* 42 as well, the "weaker" (*asthenesteras*) Theadelphia is contrasted with "fortunate" (*eupothmousēi*) Hermoupolis, which is perhaps said to be located on

95. For this approach to petitions see Bryen, *Violence in Roman Egypt*, 4–5.
96. Later Arcadia, a late Roman province roughly corresponding to the old Heptanomia, i.e., from somewhat south of Oxyrhynchus to the apex of the Delta.
97. E.g., *P.Wisc.* 1.32 (305 CE), l. 19. Bonneau, *Régime administratif*, 46.

the plain, although the reading is restored.[98] The absolute and relational topography of the village is described most clearly in *P.Sakaon* 35 (ca. 332?), the opening lines of a legal advocate's *narratio*, seemingly unused, on behalf of the Theadelphians:[99]

n(arratio). λέγεις ὑπὲρ Ζακ[α]ῶνος καὶ Ἥρωνος καὶ Καναοῦγ τῶν ὑπολειφθέντων ἐρήμου κώμης Θεαδελφίας η πάγου Ἀρσινοΐτου νομοῦ. κατὰ τὸν προπέρυσι ἐνιαυτὸν καὶ πέρυσι τῶν ἐδαφῶν ⟦τῆς ἐδαφῶν⟧ τῆς κώμης ἡμῶν ἐν ὑψηλοῖς τόποις ὄντων καὶ τῶν ἔγγιστα κωμῶν, Ναρμούθεως καὶ Ἑρμουπόλεως κώμης καὶ Θεοξενίδος, ὑποκλεπτόντων ἡμῶν τὰ ὕδατα καὶ οὐκ ἐπιτρεπόντων ἀρδεύεσθαι ἡμῶν τὴν γῆν, διὰ τὸ ἀρχὴν αὐτῶν εἶναι τῶν πάγων καὶ ἡμᾶς ὑστέρους εἶναι τοῦ πάγου, ἔρημον κώμην οἰκοῦντας

Narratio. You speak on behalf of Zakaōn (*l.* Sakaon) and Herōn and Kanaoug, those left behind in the deserted village of Theadelphia in the eighth *pagus* of the Arsinoite nome. Both the year before last as well as last year, since the fields of our village lie on high ground (*en hypsēlois topois*) and the nearest villages, Narmouthis, the village of Hermoupolis, and Theoxenis, steal our water and do not allow our fields to be irrigated, because they are at the front of the *pagi* and we are at the back of the *pagus*,[100] we inhabit a deserted village (*erēmon kōmēn*)

These issues mattered greatly. As described in chapter 2, water shortages were endemic in downstream communities even in normal times, a problem surely exacerbated by the increasing inability of a depopulated village like Theadelphia to maintain its local infrastructure. So too the previous chapter, which described in greater detail the ways in which these disruptions to water flow were produced. As for the differences between fields on high ground versus those on the plain, we need only recall the circular sent by Alypios to all the *phrontistai* of the Appianus estate in the Fayyūm, which clearly distinguishes

98. The restored reading in ll. 17–18—τῇ εὐποθμουσῃ (*l.* εὐποτμούσῃ) κώμ[ῃ τοῦ πεδί]ου ("the more fortunate village on the plain"—is not implausible but is not secure. Although not clear, the black and white image online also seems to show a portion of the ι, suggesting that the text might be revised as κώμ[ῃ τοῦ πεδ]ίου.

99. Bagnall, "The Population of Theadelphia," 57.

100. Beyond the clear upstream/downstream relationship it is not clear how the text understands the disposition of rural *pagi*. Derda, ΑΡΣΙΝΟΙΤΗΣ ΝΟΜΟΣ, 271–72 with the map on 273. Narmouthis (TM Geo 1421) is modern Madīnat Maḍī. Hermoupolis (TM Geo 813) and Theoxenis (TM Geo 2386) cannot securely be linked to any known locale.

between vineyards in marginal regions like Theadelphia, which were irrigated by water-lifting (*antlētika ktēmata*), and vineyards "of the plain" (*ta de epipeda* [*ktēmata*]), which required no such mechanical assistance (*P.Flor.* 2.148 [266–67 CE]). The evidence from al-Nābulusī collated in the previous chapter likewise demonstrates the significant agroenvironmental distinctions between low-lying villages and those on high ground, which "water does not reach . . . except with difficulty."[101] Indeed, throughout the central plain of the contemporary Fayyūm, water is still largely delivered by gravity, while marginal villages must raise water from canals mechanically.[102]

Yet these inherent locational disadvantages became crippling only in the context of Theadelphia's minimal population, which in three instances is said to consist of only Sakaon and two other men (*P.Sakaon* 33, l. 3; 35, l. 22; 44, l. 3–4), of whom at least Sakaon himself remained encumbered by multiple liturgical obligations to the state and was thus less able to sever ties and abandon the village.[103] To be sure, the claim of near-absolute desertion is rhetorical exaggeration as well as a plaintive reminder that the entirety of the village's tax burden now fell on but a few heads (e.g., *P.Sakaon* 35, l. 18–19), a burden made all the heavier by the late Roman state's perverse imposition of taxation even on lands that not been flooded (*abrochos*).[104] All the same, the claim has critical resonance in a socioenvironmental context in which water flow was the product not only of gravity but also of communal labor. Moreover, that this residual population was dwindling still further is also made plain in *P.Sakaon* 44 (331/2 CE), a petition to the prefect Flavius Hyginus. Here Sakaon narrates, somewhat pathetically, his failed attempts to return a handful of Theadelphian absconders to their *idia*. He claims to have located five erstwhile Theadelphians on a hamlet in the Oxyrhynchite nome but was repulsed "with violence" (*meth' hybreōn*) when trying to apprehend them. A further three migrants, he adds, were found in the nearby Kynopolite nome, there farming more than 100 *arourai* of public land. They too, it seems, showed no inclination to return. Similarly, in the badly fragmented petition to a *praeses* of Aegyptus Herculia

101. See chapter 4 above, pp. 165–66. The quote is from al-Nābulusī's description of Ṭubhār at *VF*, 185: لا يصل اليها الماء الا بكلفة

102. See chapter 2 above, pp. 88–93. The few attestations of places dubbed "high" (ὑψηλός) in the papyri are almost invariably described as unwatered: *P.Tebt.* 3.703, l. 172 (ca. 210 BCE); *P.Bagnall* 9, l. 35 (200–176 BCE); and *P.Bagnall* 46, 1. 221 (119 BCE).

103. See the list of liturgical obligations in fourth-century Theadelphia in Karolien Geens' description of the Sakaon archive at Trismegistos.org: https://www.trismegistos.org/arch/archives/pdf/206.pdf.

104. Rowlandson, *Landowners and Tenants*, 64–65.

published as *P.Sakaon* 93 (314–23 CE), the author, whose name is lost, claims that the rest of Theadelphia's former inhabitants had abandoned the site, perhaps a year before, and had taken up residence in another location (*kai en heterōi topōi tēn oikēsin*). He and his wife alone remain and are living in the temple (*en tōi hierōi*) for the purpose of guarding it, yet they are continually harassed by local officials on account of taxes owed by others, presumably their absconded fellow Theadelphians.[105]

Even if not yet technically the "deserted village" dramatically described in *P.Sakaon* 35,[106] this was scarcely a community capable of the sorts of collective agency described decades earlier in *P.Sakaon* 32. Sakaon's own incapacity is nowhere clearer than in *P.Sakaon* 45 (334 CE), a petition addressed to a nome law-enforcement official (*eirēnarchēs*). Here he complains that several men upstream had installed an *emblēma* in a shared canal during the flood (*ton kairon tōn hydatōn*, "the time of the waters"). Seemingly unable to solve the problem either by himself or in concert with others, he instead alludes to an unspecified imperial law (*theios nomos*) against the installation of *emblēmata* and begs that the matter be referred to the court of the prefect. Yet it is the earlier *P.Sakaon* 42 that most clearly illustrates the dependency of fourth-century Theadelphia as a whole, since the text not only complains of present problems but also alludes to a previous and apparently ineffectual round of petition and administrative intervention. Indeed, apart from attracting the attention of local officials and exhibiting their barren fields, there is no suggestion here that the community had the ability to take action of its own accord. Moreover, the solution the Theadelphians here propose—the administrative linkage of their village to the "fortunate" upstream settlement of Hermoupolis—is all but an admission of total failure. The proposal might also be read, if only metaphorically, as an expressed wish to abandon this unviable downstream settlement and move to better lands upstream—from *taḥt* to *fouq* in contemporary Fayyūmī parlance:

105. *P.Sakaon* 93, ll, 3–5: τῆς προειρημέ[νης κώμης -ca.?-]. . [. . . . .]. μ. . ου γεγενημένης, οἱ ἀπὸ τῆς κώμης ἀνκαταλ[ε]ιπόντες (*l.* ἐγκαταλιπόντες) [-ca.?-]. . τ. ἐν [τῇ] κώμῃ καὶ ἐν ἑτέ[ρῳ τ]όπῳ τὴν οἴκησιν ἐσχηκότες πρὸ ὅλου [ἐνιαυτοῦ(?)-ca.?-]. . . . ου. . [.]. μετὰ τῆς συμβίου κ[α]ταλειπόντες ἐν τῷ ἱερῷ τοῦτο φυλάττιν (*l.* φυλάττειν). "Since the aforementioned [village] having become [abandoned?], those from the village, having left behind . . . in the village and having taken up residence in another place an entire [sc. year ago?] . . . with [my] wife, staying behind in the temple to guard it."

106. *P.Sakaon* 33 restores l. 3–4 παν[ερήμου οὔσης τῆς κώμης ("the village being completely deserted"), likely after the model of πανερήμου [οὔσης τῆς κώμ]ης in *P.Sakaon* 38 (312 CE).

[Σαβινια]νῷ τῷ διασημοτάτωι ἡγεμόνι Μερκουριανῆς Αἰγύπτου [παρὰ Σ]ακαῶνος καὶ Ἐσούρεως καὶ Ἀρείωνος ἀπὸ κώμης Θεαδελ[φίας.] ἐνετύχαμεν τῇ σῇ ἀρετῇ, κύριε, ἐξ ἐπιπέδου, ὡς μὴ μετα[λαμβαν]όντων ὕδωρ πρὸς τὴν ἀρδίαν (*l.* ἀρδείαν) τῶν ἐδαφῶν—διὰ τὸ καὶ [ὀλιγ]οϋδρῖν (*l.* [ὀλιγ]οϋδρεῖν) τὴν κώμην ὡς ἐκ τούτου ἐξεσθενηκέναι (*l.* ἐξησθενηκέναι)—τε[λούντ]ων ὑπὲρ τῶν τοσούτων ἀβροχούντων ἱκανῷ χρό[νῳ ε]ἰς πενίαν ἐλθόντες, ἀξιοῦντες εὐεργεσίας τυχεῖν. [ἐκέλε]υσας τοῖς χωματεπίκταις (*l.* χωματεπείκταις) μετὰ ὀφικιαλίου (*l.* ὀφφικιαλίου) καὶ τοῦ [πραι]ποσίτου τοῦ πάγου, ὥστε ἐπ[ὶ] τῶν τόπων γενέσθαι [καὶ ἐ]πιθεωρῆσαι τὴν ὄψιν, οἵτινες κατελθόντες εὗραν [ἡμᾶς] μὴ μεταλαβόντας τοῦ ὕδατος οὐ μόνον ἐφέτος [ἀλλὰ] ἱκανο[ῖς] χρόνοις διὰ τὸ πολὺ ὑστέραν εἶν[αι τὴν κώμ]ην. ἀξιοῦμεν τοίνυν τὴν σὴν ἀνδρίαν (*l.* ἀνδρείαν) εὐεργε[τῆσαι ἡ]μᾶς κα[τ]ὰ τοὺς νόμους καὶ τὰς κελεύσις (*l.* κελεύσεις), ἡγεμών, [σοῦ καὶ] ἄ[λ]λων ἀρχόντων, τὰς ἀσθενεστέρας κώμα[ς ταῖς εὐ]ποθμ[ο]ύσαις (*l.* [εὐ]ποτμ[ο]ύσαις) κώμαις συνάπτεσθαι, καὶ ἡμᾶς [κοινωνη]θῆναι τοὺς μερισμοὺς τῇ εὐποθμούσῃ (*l.* εὐποτμούσῃ) κώμ[ῃ τοῦ πεδί]ου, λέγομεν τῇ Ἑρμουπόλει, ᾗ καὶ ἔτι πρότ[ερον . . . .] τοῖς δεκαπρώτοις ἐπενεμήθημ[ε]ν αὐτ[ῇ. ἐπειδήπ]ερ κα[ὶ] τῶν ὑδάτων πρωτοτύπως με.[. . . . .]οι, κ[α]ὶ παρ' αὐτῶν τὴν εἰσφορὰν γενέσθαι [δεῖ τῶν τ]ελεσμάτων τῶν γῶν τῆς κώμης ἡμῶν, ἵνα τούτου τυ[χόντες δυ]νηθῶμεν ἐν τῇ ἰδίᾳ συνεστάναι, καὶ διὰ παντός σοι εὐχαριστῆσαι. διευτύχει.

To Sabinianus, the most eminent *praeses* of Aegyptus Mercuria, from Sakaon and Esouris and Ariōn from the village of Theadelphia. We petitioned your excellency, lord, asking to partake of your beneficence, since we are not receiving water for the irrigation of our fields—and on account of this water shortage the village has become debilitated—and have been reduced to poverty by paying taxes on so many unwatered [fields] for so long. You commanded that the dike-inspectors along with a member of your staff and the chief of the district come to the place to make an inspection of the situation, and when they came down they discovered that we were not receiving water, not only in the present year but for a long time, because the village is too far back. We therefore ask your virility to show beneficence to us and that in accordance with the laws and the commands, lord, both of yourself and other governors, that weaker villages be attached to fortunate villages, and that we share our assessments with the fortunate village of the plain (?), we mean Hermoupolis, to which even previously we were allotted . . . the *dekaprōtoi*. Since they [are the beneficiaries?] of the waters in the first instance, they also should contribute to the dues on the lands of our village, so that, after benefiting from this we may be able to stay in our *idia* and above all be grateful to you. Farewell.[107]

The hearing before the *praeses* Valerius Ziper in *P.Sakaon* 33 must be read in this context—the total collapse of the local socioenvironmental relationships that constituted a functioning irrigation community—since it better enables us to perceive the sheer powerlessness of Roman administration to resolve Theadelphia's problems. The two conflicts adjudicated at the hearing were briefly discussed in the previous chapter (see Above, "Above and Below"). In the first complaint, whose text is fragmentary, residents of a village named Andromachis (not securely located but somewhere just to the south/upstream of Theadelphia), are accused of damming up a canal and not letting the water flow down.[108] Local memory of the contributions of the Appianus estate also seem to have survived, since reference is made to one Alypios, who had in the past (*palai*) done something to "a canal leading to the plain (*pedion*) [sc. of Theadelphia?]" and equipped it with a stone-lined conduit (*rheithron*), all with a view to the "improvement of irrigation" (*paramythian tēs ardeias*).[109] The judgment in this complaint, here delivered in Latin, is likewise lacunose, although a full Greek translation is appended to the transcript:

> ὁ πραιπόσιṭος τοῦ πάγου καὶ οἱ χωματεπίκται τῶν τόπωṿ τούτους οὓς ἂν καταμάθοιεν χώμασιν τὸν τόπον ἀποπεφρακέναι μετὰ πάσης εὐτονίας διὰ ταχαίωṿ (*l.* ταχέων) καταναγκασιουσιν (*l.* καταναγκάσουσιν) τὸν αὐτὸν τόπον ἀνακαθᾶραι, ἵνα τὸ υὕδωρ (*l.* ὕδωρ) τὴν συνήθη εἴσροιαν ἔχειν δυνηθῇ.

> The *praepositus pagi* and the dike inspectors of the area will discover those who have blocked up the place with [earthwork] embankments (*chōmasin*) and will vigorously and swiftly compel [them] to clean out the place, so that the water may be able to have its customary influx.

The Theadelphians' second complaint is essentially identical in substance. According to the Theadelphians' advocate, a group of farmers somewhere

107. This and following translations are modified from Parássoglou.
108. For the village and its relative location see TM Geo 172.
109. *P.Sakaon* 33, ll. 4–6: ἔτι τοίνυν πάλαι Ἀλύπιος διώρυχα φέρουσαν εἰς τὸ πεδίον [- ca.25 -] καὶ λιθίνην κατεσκεύασεν, τὸ καλούμενον ῥῖθρον (*l.* ῥεῖθρον) ἵνα δηλαδὴ διὰ τὴν [- ca.25 -]ε̣ισ̣ι παραμυθίαν τῆς ἀρδίας (*l.* ἀρδείας). Rathbone, *Economic Rationalism*, 227.

upstream from the village were likewise blocking up a shared canal and not allowing water to flow downstream. Here, however, attention is drawn to the relatively small amount of land farmed by the accused—"only twenty *arourai*" (*eikosi arouras monas*)—suggesting that the Theadelphians objected here not only to the blockage of the canal but their upstream competitors' disproportionate water consumption:[110]

> οἱ περὶ Μάνον καὶ οἱ τούτου ἀδελφοί, ὑπερκαθήμενοι ἡμῖν καὶ εἴκοσι ἀρούρας μόνας κεκτημένοι ἐπὶ τοῦ πεδίου, ἀποφράσ[σο]υσιν τὸ ῥῖθρον (*l.* ῥεῖθρον) καὶ οὐκ ἐῶσιν εἰς ἡμᾶς πέμπεσθαι τὰ ὕδατα. ἐπεὶ τ<ούτ>οις οὖν, προσήκει τούτους ἢ συγχωρῖν (*l.* συγχωρεῖν) ἡμῖν τὰ ὕδατα, ἡμῶν ἡδέως [[εχον]] ἀναδεχομένων τῶν εἴκο[σ]ι ἀρουρῶν τὰ τελέσματα, ἢ δηλαδὴ ἐκείνων ἀποφρασσόντων τὰ ὕδατα ἔχειν τὴν ἡμετέραν γῆν—καὶ ἐξιστάμεθα αὐτοῖς ἀπεντεῦθεν.

> The associates of Manos and his brothers, who are above us (*hyperkathēmenoi hēmin*) and possess only twenty *arourai* on the plain, are blocking up the channel and preventing the water from being sent on to us. Now inasmuch as it is only proper that they either allow us to have the water, in which case we gladly undertake to discharge the dues on the twenty *arourai*, or else, if they continue blocking the water, take over our land—and we cede it to them henceforth.

Ziper's Latin judgment accordingly reflects the specifics of the complaint, for he orders that the alleged offenders take only the water they required (*sufficiente aqua iuxta terram*) and allow the rest to flow downstream:

> *Tziper v(ir) p(erfectissimus) praes(es) Aeg(ypti) Herc(uliae) d(ixit): praepositus pagi providebit quatenus hi adversus quos postulatur percepta sufficiente aqua iuxta terram quam posside\n/t superfluam in terris susceptorum tuorum tradaṇṭ, quo idem quoque possint terras ad se pertinentes inrigare.*

> The most eminent Ziper, *praeses* of Aegyptus Herculia, said: "The *praepositus pagi* will see to it that they against whom this charge is brought after having drawn sufficient water in proportion to the land they possess will send on the remaining water to your clients' land, so that they too may be able to irrigate their own land."

---

110. Trans. modified from Parássoglou.

Ziper's judgments would seem a victory for the Theadelphians: clear and authoritative declarations of their right to access the commons. Practically, however, they were obtuse, for they rested on the erroneous assumption that in the absence of illicit interference the flow of water downstream was a given, rather than a product of the sorts of communal labors that Theadelphia's minuscule remaining population could no longer undertake. The verdicts thus reflect a decontextualized and mechanistic understanding of Fayyūm irrigation, which recognized neither the need for healthy communities capable of collective action, nor the socioenvironmental disparities between Theadelphia and its upstream neighbors, nor also the central role of these disparities in precipitating and sustaining its conflicts—entirely local issues of which, I suggest, the Theadelphians themselves were keenly aware. This fraught encounter between a moribund Egyptian irrigation community and a Roman *praeses* therefore represents a failure to communicate between two ways of knowing of the Fayyūm's waterscape—one rooted in a local knowledge that had emerged from the work of producing and sharing local water flows, the other in the abstract normative principles that ostensibly governed such practices. That the hearing was conducted in Greek and its verdicts delivered in Latin, all before a small audience of presumably Egyptian-speaking farmers, whose predecessors had required a translator (*hermēneus*) in such settings, further hints at the mutual incomprehensibility of the two sides.[111] This inability to communicate was consequential, for beyond the liturgical obligations that continued to tie Sakaon himself in place, it only reinforced the bonds of rural subjectivity that kept Sakaon and his companions administratively bound to their *idia*—the barely irrigable, underproductive, and hence increasingly untaxable territory of an irrigation community that was already effectively defunct. From this perspective the remaining years of the village's life seem a tragic pantomime: the continued reenaction of communal irrigation practices in the absence of a functioning irrigation community.

Lastly, as a piece of environmental history, the death of Theadelphia offers us more than a glimpse of the unraveling of socionature in one tiny and ultimately insignificant corner of the later Roman Empire. It also underscores the importance of bridging between local and administrative knowledge in the management of the commons.[112] In our case, the failure of the Roman state to

111. In *P.Sakaon* 32, l. 23 and 33.
112. Rathwell, Armitage, and Berkes, "Bridging Knowledge Systems."

comprehend the Theadelphians' problems on their own terms advanced neither side's interests, since state intervention succeeded only in perpetuating a vision of rural subjectivity that prioritized the fixity and permanence of an irrigated landscape that had already irrevocably changed. At a deeper level, however Valerius Ziper's judgments highlight the difficulty, if not the futility, of governing locally contingent socioenvironmental relationships in accordance with universalizing principles, even when those principles are themselves inspired by the very practices they presume to govern. Indeed, the ostensibly Roman principle of proportional water sharing cited by Ziper in *P.Sakaon* 33 did not originate in Roman law but had entered it via the experience of governing irrigation-dependent societies elsewhere in the empire, from Spain to North Africa, where proportional water sharing was practiced.[113] Embodied in a rescript of the emperors Marcus Aurelius and Lucius Verus (161–69 CE), the principle is preserved in *Dig.* 8.3.17:

> *Imperatores Antoninus et Verus Augusti rescripserunt aquam de flumine publico pro modo possessionum ad irrigandos agros dividi oportere, nisi proprio iure quis plus sibi datum ostenderit. item rescripserunt aquam ita demum permitti duci, si sine iniuria alterius id fiat.*
>
> The emperors Antoninus and Verus Augustus replied that water from a public river for irrigating fields should be divided in proportion to the property holdings, unless someone show that more has been given to him by an individual right. Likewise, they wrote that water is only allowed to be channeled if it can be done without harm to another.[114]

As the previous chapter argued, however, water-sharing regimes in small-scale, farmer-managed irrigation communities are deeply informed by hydrology of local environments. Consequently, each manifestation of proportional water sharing encountered on the ground by Roman imperial administrators was unique, the product of local traditions and local socioenvironmental relationships, all rooted in local knowledge of local natures. Proportional water-sharing regimes are accordingly best regarded as internal expressions rather

113. Bannon, "Rivers, Rights, and 'Romanization.'" For examples of proportionality see the *Lex rivi Hiberiensis* in Lloris, "An Irrigation Decree," 170–71; and the Lamasba irrigation schedule in Shaw, "Lamasba," 82–83.

114. Trans. Bannon, "Rivers, Rights, and 'Romanization,'" 66.

than external causes of communalism in irrigation societies—emergent products of a thriving community's collective dialectic with its surrounding environment. The absorption of proportionality into Roman law thus represents the abstraction of a universal prescription—fairness—from numerous deeply embedded particulars. Unmoored from these generative contexts, however, the principle alone was meaningless and without substance. As a result, its bare prescription could not regenerate a functioning water-sharing community at Theadelphia.[115] What had once been generated from within, in other words, could not be reimposed from without.

Conclusion

As at Philadelphia in the third century BCE, fourth-century CE Theadelphia was host to a unique encounter between local Egyptian farmers and outsiders who did not know the land like those who farmed it. By adopting the ground-level perspective of the Theadelphians, we are better able to understand the waterscape of the village on the terms of those who farmed it: a village perched high near the end of a canal and all but unirrigated for lack of the bodies necessary to produce a local water supply. Drawing on the previous chapter's socio-environmental reading of Fayyūm irrigation communities, these arguments have reemphasized that moving water is work—the work of individual bodies, yes, but more importantly of cohesive communities constituted through the reproduction of the local water flows by which the community is in turn bound and sustained. This socioenvironmental contextualization sheds new light on Theadelphia's terminal period, most importantly by calling into question whether the village can still rightly be described as a community in this period. From the perspective of the late Roman state it clearly was, for it continued to owe taxes on five hundred largely unirrigated *arourai*[116] as well as yearly contributions to the work of repairing public irrigation infrastructure.[117] However, if my earlier reading of Fayyūm villages as "communities of flow" is accepted, then this community was already functionally extinct, since it was no longer able to undertake the collective actions through which Fayyūm irrigation com-

115. Paraphrasing Scott, *Seeing Like a State*, 310: "By themselves, the simplified rules can never generate a functioning community, city, or economy."
116. *P.Sakaon* 35 (ca. 332 CE), ll. 18–22.
117. *P.Sakaon* 53 (fourth century CE).

munities were internally constituted. The failure or, more charitably, the inability of provincial authorities to understand the socioenvironmental roots of Theadelphia's problems is what accounts for the village's continued administrative half-life in this period as a unit of fiscal and liturgical account, to which Sakaon and a few others remained bound even after most of their fellow inhabitants had moved on. This representation of Theadelphia's demise as an internal phenomenon long in the making also makes clear that any external obstructions to its water supply were altogether epiphenomenal—for example, Cornelia Römer's argument that damage to the dam in the Tuṭūn basin in the fourth century CE caused water to escape through the Maṣraf al-Wādī, thereby decreasing the water supply to the western border canal and desiccating Theadelphia.[118] Römer's argument is tantalizing and, if accepted by archaeologists, will add greatly to our understanding of the functionality of the canal system in the Western Fayyūm. Still, infrastructural problems emerging in the early fourth century CE would only have worsened a problem that had already by this period reached a critical inflection point.

Yet as I have already suggested in the Introduction of this book, even if the remaining Theadelphians had every reason to perceive their situation as a (localized) disaster, it might appear otherwise from a broader perspective. Indeed, Sakaon's own description of the Theadelphian migrants in the nearby Oxyrhynchite and Kynopolite nomes (*P.Sakaon* 44) suggests that, although undeniably disruptive, the disintegration of Theadelphia's irrigation community was not an existential crisis for every one of its former inhabitants. As archaeologist Michael Willcox reminds us, we must think of abandoned inhabitations not as corpses but as shells from which the living organism has moved on to create new life elsewhere.[119] But whether lived upstream in the Fayyūm's wet and fertile central plain or in nearby Nile Valley nomes, the new lives of former Theadelphians almost entirely escape the scholarly gaze. For in departing the tail ends of the Fayyūm's canal system, migrants left behind the arid desert environment in which their papyri—their voices—were preserved, and moved to lands in which the written word fast dissolves in water and returns to the earth.

118. Römer, "Why Did the Villages."
119. Willcox, "Marketing Conquest," 137.

Conclusion

Not Static but Flowing[1]

> Like many of his neighbors, Mostafa's farming days here are coming to an end. Abuxta [Abū Ksā] lies at the end of the Nile's water distribution line that reaches the district. Village residents explain that people up the canal have exceeded their share of water in recent years, such that nothing reaches those at the bottom of the canal in the end.
>
> —Philip Rizk, "Water Scarcity in the Fields: A Glimpse from Fayoum," *Egypt Independent*, 12 July 2010[2]

As historical geographer Peter Thonemann has rightly warned, by yoking a diachronic perspective to a synchronic mode of presentation, historians of premodern landscapes risk "reducing a highly fluid and contingent environmental dialectic to an unchanging web of geological and social constraints." As a result, every past century merges into an undifferentiated unity—"the frozen inertia of 'non-time'"—from which evidence may be plucked at will to illuminate a static and unchanging nature.[3] This risk has been unavoidable here, particularly in the first two chapters. Yet like Thonemann's Maeander Valley, the landscape of the premodern Fayyūm is visible only in fragments, some larger, some smaller, scattered across a considerable timespan. The extent to which my reassembly of these fragments has successfully avoided suggesting a timeless continuity—yet another quaint diorama of Eternal Egypt, here thinly

1. Modifying Keenan, "Deserted Villages," 138, on the Fayyūm's "not static but shifting" village topography.
2. https://egyptindependent.com/water-scarcity-fields-glimpse-fayoum/.
3. Thonemann, *The Maeander Valley*, xv.

obscured by a faddish terminological haze—is for others to judge. The intent, to be certain, was decidedly otherwise. For while flowing water has proved highly effective at eroding disciplinary boundaries, revealing strong continuities, and bringing disparate historiographical and evidentiary streams into communication, its power likewise to disrupt and transform, to reshape landscapes and mark boundaries both in space and in time, has been visible throughout. Moreover, water's distinctly hybrid materiality—its ability to divide and connect, to disperse and collect—did not shape the premodern Fayyūm's rural landscape on its own but rather through the consistent mediation of human agency. This book, in other words, has been a study not only of flowing water but also of choice: the innumerable everyday decisions, occasionally sighted in our evidence but most often passing unseen, that in all periods harnessed water's flow and put its jointly productive and disruptive capacities to work. Above all else, these generative entanglements between flowing water and its human producers cum dependents have been the heart of this study. From this perspective, the intimate links between the sense of wonder and astonishment with which we began and the more sober note of the concluding chapter immediately become clear, for both the luxuriant garden so frequently admired over the centuries and the increasingly barren waste of fourth-century Theadelphia were human landscapes—products of unique local entanglements with water and the routine work of determining where it would flow and, consequently, where it would not.[4]

This attention to the transhistorical, socioenvironmental phenomenology of water flow has borne considerable interpretive fruit. At a bare practical level, it has demonstrated to scholars of the Graeco-Roman period the utility of later comparative evidence for the landscape and environmental history of premodern Egypt. This should encourage future explorations of such sources, Arabic in particular, a wealth heretofore largely untapped by students of earlier periods.[5] This diachronism has also contributed to a more nuanced vision of premodern Egyptian water management by revealing the interpretive poverty of debates pitting state power against local agency and instead illuminating a shifting series of intersections between nature, states, and rural society, each of which bore differently on the flow of water through the Fayyūm's canal system. In the Roman period in particular, this unique canal system was a site of

4. Drawing on the language of Barnes, *Cultivating the Nile*, 2–3.
5. See now, however, the survey in Malleson, *The Fayum Landscape*, 147–87.

intimate encounters between imperial power and rural subjects, the need to continue producing water flow enmeshing both in bonds of mutual obligation and dependence. Here at ground level, following water's flow across the landscape has also revealed significant hydrological and hydraulic disparities between the central and marginal subdivisions of the canal system, thereby resolving the apparent contradiction between the largely single-cropped landscapes visible in the papyri and the perennially cultivated garden environment described by Islamic sources. Far from irreconcilable, each was a manifestation of a productive dialectic between human communities and their local environments. This consistent centering of human agency—the ability even of individual irrigators to alter canal flow and disrupt the lives and landscapes of neighbors downstream—has likewise disclosed the fundamental liquidity of a land- and waterscape that was not static but flowing and thus always in the process of transformation. This essential flux was nonetheless starkly at odds with the aims of Roman administration, which sought to preserve the Fayyūm's shape and extent through the compulsory annual reproduction of extant patterns of water flow. Yet this administrative imperative ultimately could not overcome the need for cultivators to follow the flow of waters whose course had irrevocably changed. Viewed in this light, Valerius Ziper's attempts to resolve Theadelphia's conflicts and keep its few remaining inhabitants in place appear not only inept but also perverse.

Although most apparent in the closing discussion of Theadelphia, the local—local communities, local water flows, local agroenvironments, and so on—has assumed pride of place throughout much of this study. Rightly so, since a bottom-up perspective is essential to any account of irrigation and agricultural practices in small-scale rural communities. Yet this valorization of the local has also come at a cost. By framing irrigation and agricultural practices as the products of an ongoing dialexis between rural communities and the unique environments in which they were embedded, it effectively limits the broader applicability of any conclusions drawn from Fayyūm evidence. In many respects this simply reinforces an emerging consensus that the ancient Fayyūm's marginal villages were unrepresentative of rural Egypt as a whole in the Graeco-Roman period.[6] Indeed, as the second chapter demonstrated, these settlements were atypical even of the wider Fayyūm. This is consequential,

6. For which see Monson, *From the Ptolemies to the Romans*, 50–69; and Monson, "Communal Agriculture." See also the comparisons between the Edfu land survey and Kerkeosiris in Christensen, Thompson, and Vandorpe, *Land and Taxes*, 13–15, 27, and 31–32.

since the Fayyūm's margins are the source of the majority of the surviving papyri of rural provenance. A nuanced understanding of rural landscapes elsewhere in Graeco-Roman Egypt is therefore forestalled by our inability to perceive in detail the certain diversity of local human-water entanglements through which these landscapes were constituted.[7] This frustration notwithstanding, there remains some reason for cheer, namely because estimates of Roman Egypt's total population and aggregate agricultural productivity have been based in large part on extrapolations from data in Fayyūm papyri.[8] As I have argued more thoroughly elsewhere, this likely results in significant underestimates for both figures, a suggestion well supported by later comparative evidence. Paradoxically, then, this study's less-than-optimistic appraisal of the Fayyūm's ancient margins has positive implications for the population and productivity of Roman Egypt as a whole.[9]

In many respects, then, this study has been an attempt to recover, if only in part, the striking peculiarity that transfixed the ancient, medieval, and modern observers quoted in the Introduction—an attempt, in other words, to make the Fayyūm strange again. But beyond the immediate goal of contributing to the emerging environmental historiography of premodern Egypt, the book has broader implications for the environmental history of the ancient Mediterranean. Above all else, it is a programmatic argument for reorienting scholarly attentions away from the stratospheric heights of solar cycles, pandemics, and climate change and toward the everyday lived experience of nature in antiquity. Such an approach necessarily yet productively narrows the field of view. As Horden and Purcell so painstakingly elaborated more than two decades ago, the Mediterranean is a fractured whole: a tightly interconnected tapestry of ecological microregions each with their own unique character.[10] The present work is more microscopic still, embracing even the tightly circumscribed environments of single country villages. Yet this is not microhistory for its own sake; rather, it is an assertion that place, no matter how narrowly construed, matters in environmental history, since it is at ground level that the relationships between human and nonhuman natures are most intimate and meaning-

7. A partial exception is the oblique glimpse of the Mendesian nome in the carbonized papyri from Roman Thmouis. Blouin, *Triangular Landscapes*, esp. 13–169. See also Willems et al., "The Analysis of Historical Maps," for one possible, if still partial, approach to the reconstruction of premodern irrigated landscapes.

8. Rathbone, "Villages, Land and Population," 108–9; and Rathbone, "Roman Egypt," 703–4.

9. Haug, "Agriculture in Roman Egypt," 523–28.

10. Horden and Purcell, *The Corrupting Sea*.

ful. Any environmental history that seeks to hear people speaking must therefore eschew an overreliance on macroscopic generalizations, sterile models, and cliometry and instead locate its subjects in place, embedded in their surrounding nature.[11] To be sure, this is neither to claim that the environmental history of the premodern Mediterranean must always and everywhere be local history, nor is it naïvely to discount the considerable obstacles faced by scholars of regions outside of Egypt, where the granular perspective provided by papyri is altogether lacking. Admittedly, even by Egyptian standards the evidence for the Fayyūm is also singularly robust, microscopic, and continuous, which limits the direct applicability of the approach pursued here. All the same, this study's thoroughgoing demonstration of the centrality of place in human-nature entanglement indicates that investigations of narrow geographical compass yet broad temporal scale are most likely to reveal the generative and productive socioenvironmental relationships through which ancient communities were constituted. Such work—environmental histories of the everyday—would be a far cry from the recent spate of environmental disaster-histories, which conceptually segregate humans from a "cunning and capricious" nature and cast environmental phenomena as actors in what are otherwise traditional declensionist dramas.[12] Of course, even this study concluded on a scene of socioenvironmental dissolution and decline, albeit of a single community. A view from below deeply informed by a sense of place nonetheless enabled us to see both the human complexity and the distinct locality of these events, thereby frustrating any attempt to ascribe causal significance to any single factor or to treat them as representative of contemporary trends.[13] A diachronic perspective, moreover, allowed us to situate the seeming rupture represented by Theadelphia's eventual abandonment within a larger continuity. For when the roughly six centuries of this marginal village's existence are set alongside the continuously uninterrupted lives of more flourishing villages on the central plain, the seeming crisis of its decline comes to look rather more like the early stages of a long transition toward a more locally sustainable socioenvironmental equilibrium.[14]

---

11. Drawing on language in Goldberg, *Sephardi and Middle Eastern Jewries*, 49.
12. See esp. Harper, *The Fate of Rome*, quote on p. 5; and McCormick, "What Climate Science." Cf. the somewhat more nuanced language of Manning et al., "Volcanic Suppression of Nile Summer Flooding."
13. Remembering here that sites at the southern and perhaps even eastern margins would not be abandoned until much later. See chapter 2 above, pp. 78–83.
14. Cf. Fisher, Hill, and Feinman, "Environmental Studies," 8–9.

In sum, although this book has been argued and organized with a view to elaborating the socioenvironmental context of a single dying late-Roman village, it may also be read as an enthusiastic rejoinder to Kristina Sessa's call to move beyond science-driven environmental histories of decline and fall—"the obsession with finding the edges of things"—and instead to study ancient human-nature relationships on their own terms.[15] To this end, this book's most significant contribution is not its discussion of Theadelphia but its exploration of the mundane human-water entanglements that continually remade the liquid landscape of the Fayyūm through the perpetual redirection of water flow. From this inherently diachronic vantage point, notions of a Fayyūm *floruit* and subsequent decline have less explanatory power, appearing not as neutral descriptors but as fallacies of disciplinary specialization and temporal particularism. Indeed, the very concept of the Fayyūm's decline in late antiquity and the early Islamic period effectively establishes the more expansive and productive landscape of the early Roman period—itself largely the production of a revenue-hungry and not infrequently coercive state administration—as an environmental *telos*, thereby necessarily branding all subsequent retrenchment as a regrettable deterioration. If this is indeed so, then the Fayyūm is in decline yet again, since the western extremes of the depression, reclaimed and resettled during the Nasser era in the early 1960s, are once more desiccating, as they did at Theadelphia many centuries before. For want of water, farmers at the tail end of the western border canal are trickling away, their once-verdant fields returning to dust and desert. Yet the waters that once reached these tail end communities have not disappeared; rather, their flow has been redirected for land reclamation projects upstream. Many at the tail end have thus moved up the canal to farm plots on these newly reclaimed lands.[16] Here they will continue to alter the flow of water through the Fayyūm's canal system and, in the process, (re) make the land anew.

---

15. Sessa, "The New Environmental Fall of Rome," 249, citing Bowes and Gutteridge, "Rethinking the Later Roman Landscape," 407.
16. Barnes, *Cultivating the Nile*, 122–25.

Appendix

English translation of al-Maqrīzī (1364–1442) on the canals and villages of the Fayyūm, containing an epitome of Abū Isḥāq's description of its irrigation system (1031 CE).[1]

An Account of What Has Been Said about the Fayyūm, Its Canals, and Its Villages

Al-Yaʿqūbī[2] stated: "In prior days one said 'Egypt and the Fayyūm' because of the splendor of the Fayyūm and its great productivity. Its wheat is renowned and linen-cloth is produced there." Al-Masʿūdī says that the meaning of al-Fayyūm is one thousand days (*alf yawm*).[3]

Al-Quḍāʿī[4] stated: "Al-Fayyūm is a city designed by the prophet Joseph—Peace be upon Him—through divine inspiration (*waḥy*). It had three hundred

1. Translation based upon Ayman Fuʾād Sayyid's 2013 critical edition: Al-Maqrīzī, Khiṭaṭ, 1:669–74. Cf. the French translations by Quatremère, *Mémoires géographiques*, 1:392–402; and Bouriant in *Description topographique*, 2:734–40 (reprinted with minor alterations by Omar Toussoun in *Mémoire sur l'histoire du Nil*, 1:253–58). The manuscript consulted by Quatremère and the edition used by Bouriant—*Kitāb al-mawāʿiẓ wa-al-iʿtibār bi-dhikr al-khiṭaṭ wa-al-āthār*, ed. Muḥammad ibn ʿAbd al-Raḥmān Quṭṭah al-ʿAdawī (Būlāq al-Qāhirah: Dār al-Ṭibāʿah al-Miṣrīyah, 1853)—both contain occasional misspellings of toponyms and canal names, which were subsequently reproduced in translation. These are identified in the notes below.
2. Al-Yaʿqūbī, *Kitāb al-buldān*, 119.
3. Al-Masʿūdī, *Murūj al-dhahab* 2:330. So also Ibn ʿAbd al-Ḥakam, *Futūḥ*, 15.
4. Abū ʿAbdallāh Muḥammad ibn Salāmah ibn Jaʿfar ibn ʿAlī ibn Ḥakmūn al-Quḍāʿī (d. 1062), from his lost work the *Kitāb Khiṭaṭ Miṣr*, otherwise known as al-*Mukhtār fī dhikr al-khiṭaṭ wa-l-āthār*. See Tahere Qutbuddin, ed. and trans., *Al-Qāḍī al-Quḍāʿī: A Treasury of Virtues; Saying, Sermons and Teachings of ʿAlī with the One Hundred Proverbs Attributed to al-Jāḥiẓ*. Library of Arabic Literature (New York: NYU Press, 2013), xx.

and sixty villages, each of which could feed Egypt for a day on their own and could therefore feed Egypt for a year."[5]

"It is cultivated by [a Nile flood of] twelve cubits[6] and is not submerged by any increase over that amount since Joseph—Peace be upon Him—made a stream for [each village] arranged in such a way that their water entered continuously, regulated by stacks of stones. And for this [reason] he built al-Lāhūn."

Ibn Riḍwān[7] stated: "The Fayyūm—the water of the Nile is stored within it and it is sown multiple times during the year. When it is released, one observes that this water changes the color of the Nile and its taste. One perceives this situation most acutely in Buḥayra [i.e., in the western Delta] in the days of summer[8] at Saft and al-Nahīyā[9] and up-country near the Fayyūm. This situation is terrible for the people of the city"—he means Miṣr [i.e., Fusṭāṭ]—"especially when the south wind blows" because the Fayyūm is south of the city of Miṣr at a distance far from its land.

Al-Qāḍī Al-Sa'īd Abū-l-Ḥasan 'Alī ibn al-Qāḍī al-Mu'taman Baqiyyat al-Dawlah Abī 'Amr 'Uthmān ibn Yūsuf al-Qurayshī al-Makhzūmī stated in his book *Minhāj fī 'ilm al-kharāj* [Curriculum for the Understanding of the Land-Tax]: "These districts are among the best with regard to organization. Their lands are very extensive and the country most excellent. But ruin has overcome some of them through the dearth of inhabitants and the encroachment of sand over much of their land. I have consulted a register (*dustūr*) produced by Abū Isḥāq Ibrāhīm ibn Ja'far ibn al-Ḥasan ibn Isḥāq for recording the canals of the abandoned districts and the villages along them. I have presented it here although some of them have already been abandoned and others have changed their names and the locations of others are unknown because of their abandonment. But I have presented it so that it may be learned what of it is presently under cultivation and [so that] whoever desires to restore to cultivation as much of the desolation as he can may be enlightened by it. Citing this has also the benefit of conveying the water quota of each locality. Transcript:

• • •

5. So also Ibn al-Kindī, *Faḍā'il Miṣr*, 33.
6. So also al-Bakrī, *Al-Masālik wa'l-mamālik*, 1:515 (no. 866); and al-Nābulusī, *VF*, 44.
7. Ibn Riḍwān, *Kitāb daf' maḍār al-abdān*, 16. English trans. by Dols *On the Prevention of Bodily Ills*, 110.
8. The text of Ibn Riḍwān edited by Dols reads *fī ayyām al-Nīl* ("in the days of the Nile [flood]") instead of *fī ayyām al-qayẓ* ("in the days of summer") in Maqrīzī's recension of Abū Isḥāq. The two are nonetheless synonymous.
9. Saft al-Laban and al-Nahyā northwest of Giza.

Register of what has been revealed by an investigation into the state of the principal canals in the city [i.e., region] of al-Fayyūm, their locations, what water quota (*shurb*) each village receives from them,[10] the rules for their closing (*al-sadd*), opening (*al-fatḥ*), rectification (*al-ta'dīl*),[11] and cleaning out (*al-taḥrīr*),[12] and the timing of it. Prepared in Jumāda II, Year 422 [March 1031 CE].

We begin, with God's aid and the beauty of his success, by indicating the state of the Grand Canal (*Baḥr al-A'ẓam*),[13] from which this canal [derives], then we will mention the material upon whose soundness its soundness [depends].[14]

Grand Canal of the Fayyūm (*Khalīj al-Fayyūm al-A'ẓam*). Water is conducted to this canal by the small river known as al-Manhā from the Rock of Joseph (*al-Hajar al-Yūsufī*).[15] This river [i.e., the al-Manhā] is above [the canal] in the mountain known as the Sorceress' Seat in upper Ashmūnayn, and several rural areas are watered from it, those of Ashmūn, al-Qays, and al-Ahnās. Along its twin banks are numerous villages irrigated by it, watering from it the vineyards they possess.

The Rock of Joseph. The Rock of Joseph is a wall constructed of brick and the lime known by the ancients as *ṣārūj*,[16] which is lime and oil. Its construction is oriented from north to south and its southern end abuts a wall like it in construction running precisely west to east, and it is enclosed at its end with two

---

10. No such information is preserved in this redaction of the original *dustūr*.
11. Apart from *al-sadd* and *al-fatḥ*, the vocabulary Abū Isḥāq uses to describe canal management is obscure. *Al-ta'dīl* (التَّعْديل) signifies straightening, modification, rectification, regulation, or improvement. Urbain Bouriant regarded the term as a reference to water sharing and translated it as *répartition* or *distribution*, i.e., the distribution of water between irrigators. This is surely incorrect, since in the entries where the timing of *al-ta'dīl* is specified, it occurs only after the annual distribution of water through the canal is complete early in the month of Baramūda/Pharmouthi, i.e., at the beginning of the low-water season. The word should instead be regarded as a reference to the maintenance and repair ("rectification") of public canals after the irrigation season.
12. *Al-taḥrīr* (التَّحْرير) denotes freeing/liberation but also redaction, whence Bouriant's *réglementation*, i.e., the regulations governing water sharing between villages. Once again, this is surely mistaken. Abū Isḥāq elsewhere pairs *al-ta'dīl* with *al-taḥsīn* (التَّحْسين, "improvement") and *taḥyīz* (تحْييز), an obscure term of uncertain meaning (on which see n. 34 below). This once again suggests that the term is a reference to canal maintenance rather than water sharing. I accordingly understand *al-taḥrīr* as the freeing-up, i.e., cleaning out, of public canals.
13. The terminal stretch of the Baḥr Yūsuf within the Fayyūm.
14. فنذكرمادّته التي صلاحه صلاحها. The sense of this passage is unclear. The "material" (*mādda*) in question may refer to the water upon which the "soundness" (*salāḥ*) or more generally the benefit or utility of the canal depends.
15. I.e., the al-Lāhūn dam, on which see below.
16. Possibly from Sahidic Coptic ϭⲱⲣϭ / Bohairic ϫⲱⲣϫ (Crum 831a): "mixed contents, compound, mixture." ϭ frequently interchanges with ϫ (Crum 801), while ϫ can also be rendered as Arabic ص (Crum 745b).

abutments (*maylān*)[17] and its length is two hundred work cubits. After running eighty cubits to the west, the southern tip of the Great Wall reaches this wall. The purpose of the construction of the Great Wall is to return water to Madīnat al-Fayyūm when it has reached the measure of twelve cubits.[18] The length of this wall running from west to east, then rejoining the gap, then descending from the ends of this gap to a gap like it encountering it from the north, is fifty cubits. The portion that is between these two gaps—that is, the lower part—is one hundred and ten cubits. The depth of the lowered part is four cubits. This low area is what is shut by a brushwood-made dam (*jisr*) called a *lamsh*. The width of the part upon which water runs at the time of the Nile [flood]—this being the place of the *lamsh* and that which meets it on the eastern side—is forty cubits. Boats travel over it at the time of the Nile [flood]. The place of the *lamsh* is enclosed by two buttresses whose purpose is to brace the second *lamsh*.

And this gap meets the north face at a length of three hundred seventy-two cubits, then it meets—at the end of its span—a wall laid out in its orientation to the east the structure of the Rock [of Joseph]. And its length along the orientation to the east is one hundred cubits. Then it lowers at the place where it meets this wall at a length of twenty cubits, and the extent of the lowered part is two cubits. And this lowered part is also closed by a dam of brushwood called *lknd*.

And the length of the remainder of which wall until its northern extremity is one hundred and thirty-six cubits, and in front of it along its length is a stone-paved surface (*muballaṭ*), and in it are channels (*qanāṭir*) made of stone, and in ancient times they channeled water to the Fayyūm via an ancient canal in which there are obstructions today. And there were gates in them, and the ancient channels were ten in number. And altogether, the Great Wall from its end is seven hundred and seventy-two work cubits, excluding the opposed east-west wall.

This Great Wall runs complete on all sides until it reaches the mountain, and in the summer its traces are visible, but not in a straight line, and its width varies. And whenever its upper surface is reached its width decreases. The width of its uppermost part with the visible portion of its lower parts is in total

17. ميل (*mayl*), lit. "inclination."
18. I.e., to admit water to the Fayyūm only when it has reached this predetermined height of twelve cubits. On the Fayyūm's need for only a twelve-cubit flood see chapter 1 under "Capturing the Flood."

sixteen cubits. There are holes in it through which water flows, and these are colored glass culverts resembling blue enamel or *sulīmānī*.[19]

And it is among the most beautiful marvels in the greatness of its construction and its perfection. For it is one of the constructions that surpasses the Lighthouse of Alexandria and the building of the Pyramids. For it is miraculous that the Nile has been passing over it since the era of Joseph—Peace be upon Him—until the present yet its stability remains unchanged.

In the present day, the waters come from this canal [i.e., the *Baḥr al-Manhā*] to Madīnat al-Fayyūm via the Grand Canal, which flows between the land of the two villages known as Damūna[20] and al-Lāhūn and irrigates these two villages and others. And their vineyards are irrigated from it by wheels (*dawālīb*) on cattle's necks. When the Nile fails to rise to its cultivated land,[21] it is irrigated by cattle's necks and cultivated.

And the Great Canal comes to the canal known as the Khalīj Al-Awāsī,[22] which has no regulation for closing or opening or rectification (*al-ta'dīl*). And it reaches the village known as Bayāḍ and fills its reservoir (*birka*) and other reservoirs, and on the reservoirs are divisors and every divisor receives a sufficient amount [of water] for the irrigation of the land it is upon. And it [the canal] then reaches the village known as al-Ūsiyya al-Kubrā, which is irrigated from it by two divisors with a gate for their regulation. It irrigates its palms and trees. On its bank is a mill turned by water.

It then arrives at three divisors at its terminus, the village known as Marṭīna.[23] One divisor is for the village, another for multiple tax-farms (*qabālāt*),[24] and the third divisor irrigates one of the groves (*aḥyā'*) of palms.

19. Glass or glazing infused with cobalt to imitate the blue of lapis lazuli. Julian Henderson, *Ancient Glass: An Interdisciplinary Exploration* (Cambridge: Cambridge University Press, 2013), 72–73.

20. Al-Nābulusī's Dumūh al-Lāhūn, perhaps identical with Graeco-Egyptian *chorion* Tmouei (TM Geo 2444), which is linked to this portion of the Fayyūm in two late papyri, *SPP* 20.271 (seventh–eighth century) and *SB* 6.9583 (eighth century).

21. Lit. "to its black[ness]" (*sawād*), referring to the black Nile alluvium that covered the cultivated landscape. Cf. the ancient Egyptian name for the countryside, the "black land" (*km.t*), later Coptic *Kēme* (ⲔⲎⲘⲈ).

22. Possibly identical with the ancient eastern desert canal, later the Bahr Waradān, which was abandoned by al-Nābulusī's day. See chapter 2 under "The Margins."

23. Identification uncertain. Toussoun equated Marṭīna with al-Rubayyāt or al-Rawda but without any justification (*Mémoire sur l'histoire du Nil*, 1:260).

24. Lit. "contract," a *qabāla* is a domain on which the right to collect taxes is ceded to a private person. On tax-farming in the Faṭimid period see Wickham, "The Power of Property."

And on this grove are waterwheels and orchards that have been ruined and sycamore encircles them. And there were in it houses in the palm fields. Then it comes to a second grove like the first.

Then it comes to the village known as al-Khariba[25] and fills its reservoir. Then it comes to three divisors in a row, above which is a nonfunctioning canal, and several villages are irrigated from these divisors. Then the waters from this canal reach al-Baṭs, which is its terminus.

And along the Grand Canal are alluvial deposits, which are irrigated from the water flowing from its mouths. And when the Nile [flood] diminishes, nets are set up on its mouths to regulate the catching of fish.

Then the Grand Canal reaches, on the right of one who heads toward al-Fayyūm [city], the canal known as the Samasṭūs Canal, from which Samasṭūs[26] and other places are irrigated. Considerable alluvial sediment reaches the desert on its east and south and [covers the space] between this canal and the al-Awāsī.

Then the Grand Canal also reaches the Dahāla[27] Canal, from which multiple villages are irrigated and along which rice and other crops are cultivated.

Then the Grand Canal reaches three canals, reaching the Tanbaṭāwa Canal.[28] On this canal are three ancient gates from the time of Joseph. And the width of each of these gates is two work cubits and water passes through them and it comes to two gates of Joseph's day. The regulation of this canal: all of its alternating [canals] are closed when 10 [days] have passed from Hātūr until its end, then opened on the first of Kīyahk until 10 [days] remain in it, then opened until 10 [days] have passed from Ṭūba, then opened on Laylat al-Ghiṭās [i.e., Epiphany] until the end of Ṭūba, then closed on the first of Amshīr until 10 [days] remain in it, then opened when ten [days] remain in [Amshīr] until 10 [days] have passed from Baramhāt, then opened until 10 [days] have passed from Baramūda. Then it is rectified[29] into its proper place. The villages on its

25. بالخربة(*bi-l-Khariba*). Bulāq بالجوبة (*bi-l-Jawba*), rendered *Djoubah* by Quatremère, *Goubah* by Bouriant, and both *Gouba* and *el-Jouba* by Toussoun. *Al-Jawba* ("the Pit") is the name Ibn ʿAbd al-Ḥakam gives to the Fayyūm before its reclamation by Joseph. *Khariba* ("ruins") may be a reference to either ancient Philadelphia (Kawm al-Khariba al-Kabīr) or the nearby site al-Khariba al-Saghīr, although this is far from certain.
26. Al-Nābulusī at *VF*, 47 describes Samasṭūs an abandoned site along the Baḥr Waradān. See chapter 2 under "The Margins."
27. دهالة (*Dahāla*). Bulāq ذهالة (*Dhahāla*). *Dahalah* in Bouriant and *Zahalah* in Toussoun.
28. تنبطاوة (*Tanabṭāwa*). Bulāq بينطاوة. *Beintawah* in Bouriant and *Bintawah* in Toussoun.
29. يُعَدَّل (*yuʿaddal*), i.e., the undertaking of *al-taʿdīl*, here occurring only after the annual waters had been distributed.

northern bank had been abandoned. Multiple villages are irrigated by it. On this canal is built, under the mountain, a vaulted drainage outlet (*mafīḍ bi-qabw*), through which the water flows in the time of its increase.[30]

Then the Grand Canal reaches the Dalah Canal. It is a rotating canal and its closing, opening, rectification (*taʿdīl*), and improvement (*taḥsīn*) are governed as stated above. It is on the left of someone who heads toward the city. It has two gates made of stone from the time of Joseph and their width is two and one quarter cubits. Multiple main villages are watered by it and other places. In its middle is an outlet (*mafīḍ*) opened during the time of the flood (*al-istibḥār*[31]) to channel the water into the Great Lake (*al-Birka al-Uẓmā*, i.e., the Birkat Qārūn). And at the edge of this lake there is another channel with gates that, it is said, were of iron. When [the water] increased, the gates were opened and the water passes toward the west and flows, it is said, toward Santarīya.[32] And along these two canals were orchards and numerous vineyards irrigated by the necks of cattle.

Then the Grand Canal comes to the Majnūna Canal, so-called after the large quantity of water that enters it. Its regulation in opening and the rest is as already stated. It irrigates numerous villages and mills are turned by it. It receives the surplus waters of the southern villages and [carries it] to the lake [i.e., the Birkat Qārūn] at the edge of the Fayyūm adjacent to the mountain known as Abū Qaṭrān.[33] It [i.e., the lake] receives the excess waters that flow from the northern villages. It is the largest lake.

Then the Grand Canal comes to the Talālah Canal. It has two strong stone gates of Joseph's day, the width of each of which is two and two-thirds cubits. There are no regulations for its closing, opening, rectification (*taʿdīl*), or enclosure (*taḥyīz*) except when the Nile is insufficient (*taqṣīr al-Nīl*), at which time it is enclosed by means of brushwood (*yuḥayyiiz bi-ḥashīsh*).[34] Quarters of the

---

30. See above, chapter 2 under "The Margins."
31. Lit. becoming wide or spacious, like the sea (*baḥr*), presumably referring to the waters expanding across the landscape.
32. Arabic geographers referred to Sīwa as Santarīya, likely an Arabicization of its local Egypto-Libyan toponym, perhaps with additional Greek interference. For citations and a proposed etymology see Klaus Peter Kuhlman, "Gleanings from the Texts in the Sanctuary of Amun at Aghūrmī (Siwa Oasis)," *Mitteilungen des Deutschen Archäologischen Instituts, Abteilung Kairo* 57 (2001): 187–203 at 199 ff. In his description of Sīwa, al-Maqrīzī refers to the local language as *al-Siwīyya* (*Khiṭaṭ*, 1:637–38).
33. I.e., the Jabal Qaṭrānī formation to the north of the Birkat Qārūn.
34. تحييز(*taḥyīz*) is Abū Isḥāq's most obscure term. Urbain Bouriant mistook it as a reference to water distribution and translated it as *répartition*. It is derived from حوز (*ḥawaza*), which signifies gaining control, possessing, enclosing, and confining. As a noun, *ḥawz* indicates a bounded or otherwise

[capital] city are irrigated by it as are numerous fields and villages. The mouth of the al-Baṭs canal (*khalīj*) is on it, which receives excess waters. It has gates that are closed until the water rises to a fixed height on the elevated lands. If it happens that the closure is damaged, the expense [for repair] on it comes from the villages irrigated by it, in proportion to their [water] rights in it.

Then the Grand Canal comes to two canals on its southern and northern banks, thereafter reaching the Bamwah[35] Canal, which is on the right of one who approaches Madīnat al-Fayyūm and it is one of the rotating canals. On it are two gates of Joseph's day, each of which is two and one-half cubits in width. It is regulated as stated above. Many quarters are irrigated by it and numerous villages. It arrives at four divisors fitted with gates and a canal in which there is a conduit (*shadhrawān*) into which water flows. And the Grand Canal arrives at numerous canals that irrigate many villages.

The Tabdūd [*l.* Tandūd] Canal.[36] There is a sweet water spring in it that irrigates nearby lands when the canal is closed. This spring appeared when [flood] water was insufficient and the place was dug out to make a well. The spring then emerged from it and was sufficient.

Then the Grand Canal arrives at two canals in which there are ancient conduits and divisors of Joseph's day. They have gates of Joseph's day whose opening and closing are regulated, and numerous villages are irrigated by them. The schedule of these canals: they are both closed from 10 Hātūr until its end. They are opened at the beginning of Kīhak for a period of 20 days and are closed when 10 days remain [i.e., 21 Kīhak] until Epiphany. They are then opened on Epiphany until the end of Ṭūba and closed at the start of Amshīr for a period of 20 days and opened when 10 days remain [i.e., 21 Amshīr] until 20 Baramhāt. They are opened [again] until 10 days pass from Baramūda, then

---

enclosed area. No Form II *maṣdar* is attested in the lexica of Hans Wehr (p. 248); Lane (Book 1:667); or the modern Egyptian Arabic lexicon of Hinds and Badawi (p. 231), although *ḥawwaza* (حوّز) appears with the sense of driving camels to water. Several modern lexica pair the term الهويس (*al-hawīs*, canal lock) with حوز, defining *ḥawz al-hawīs* as a lock chamber, i.e., the enclosed section of the lock whose water level can be raised and lowered. See Hādī al-ʿAlawī, *Qāmūs al-muṣṭalaḥāt al-ṣināʿīyah wa-al-tiknūlūjīyah* (Beirut: Dār al-Kunūz al-Adabīyah, 1999), 420; *Al-Lisān al-ʿArabī* (Rabat: Al-Maktab al-Dāʾim li-Tansīq al-Taʿrīb fī al-waṭan al-ʿArabī, 1973), 9:410; Elias Antoon Elias, *Qāmūs al-ʿaṣrī, ʿArabī-Inkilīzī* (Cairo: al-Maṭbaʿah al-ʿAṣrīyah, 1958), 172. Cf. also حاووز (*ḥāwūz*), meaning reservoir or water-tower. *Taḥyīz* thus seems to designate the enclosure and empondment of the waters of the Talālah Canal by means of one or more unnamed structures constructed from brushwood (*ḥashīsh*). Bouriant's translation here seems nearer the mark, since he renders the following passage, الا في تقصير النيل فانه يُحَيَّز بحشيش, as *quand la crue est insuffisante, on se contente d'établir une digue d'herbes*.

35. بموه (*Bamwah*). Bulāq سموه (*Samwah*). *Samouah* in Bouriant and *Samouh* in Toussoun.

36. تَبْدود (*Tabdūd*) for which read تَنْدود (*Tandūd*). See chapter 2 under "The Center."

they are rectified and cultivation is attended to. And during the rectification (*taʿdīl*) each village receives a share (*qism*) of it [i.e., the work] equal to its water allotment (*shurb*) in accordance with rules well-known among them.

• • •

I have epitomized the names of the villages he mentions since most of them are now ruined. But God knows best.

Bibliography

Papyrological Sources

Papyri are cited following the disciplinary guidelines set out in the *Checklist of Editions of Greek, Latin, Demotic, and Coptic Papyri, Ostraca, and Tablets* (https://papyri.info/docs/checklist). For ease of reference I refer to volume numbers by Arabic rather than Roman numerals. Citations of the online toponymic database Trismegistos Geo refer to each toponym's unique identifier, for example, Karanis, TM Geo 1008.

Medieval Sources

Abū-l-Fidāʾ Ismāʿīl ibn ʿAlī. 1840. *Kitāb taqwīm al-buldān*. Edited by William MacGuckin Slane and Joseph Toussaint Reinaud. Paris: Imprimerie royale.

Al-Bakrī, Abū ʿUbayd. 1992. *Al-Masālik wa'l-mamālik*. Edited by A. P. Van Leeuwen and A. Ferre. Tunis: al-Dār al-ʿArabiyya li'l-Kitāb.

Al-Maqrīzī, Aḥmad ibn ʿAlī. 2002. *Al-Mawāʿiẓ wa-al-iʿtibār bi-dhikr al-khiṭaṭ wa-al-āthār*. Edited by Ayman Fuʾād Sayyid. Vol. 1. London: Muʾassasat al-Furqan li'l-Turāth al-Islāmī.

Al-Masʿūdī, ʿAlī ibn al-Ḥusayn. 1863. *Les prairies d'or* [*Murūj al-dhahab wa-maʿādin al-jawhar*]. Edited and translated by C. Barbier de Meynard and Pavet de Courteille. Vol. 2. Paris: Imprimerie impériale.

Al-Muqaddasī, Muḥammad ibn Aḥmad. 1906. *Kitāb aḥsan al-taqāsīm fī maʿrifat al-āqālīm*. Edited by Michael Jan de Goeje. Leiden: Brill.

Al-Nābulusī, ʿUthmān bin Ibrāhīm. 2018. *The "Villages of the Fayyum": A Thirteenth-Century Register of Rural, Islamic Egypt*. Edited and translated by Yossef Rapoport and Ido Shahar. Turnhout: Brepols.

Al-Yaʿqūbī, Aḥmad ibn Abī Yaʿqūb. 1861. *Kitāb al-buldān*. Edited by Theodor Joynboll. Leiden: Brill.

Ibn al-Ḥakam, Abū-l-Qāsim ʿAbd ar-Raḥman bin ʿAbdullah. 1922. *The History of the Conquest of Egypt, North Africa and Spain, Known as the* Futūḥ Miṣr *of Ibn ʿAbd al-Ḥakam*. Edited by Charles C. Torrey. New Haven: Yale University Press.

Ibn al-Kindī, ʿUmar ibn Muḥammad ibn Yūsuf. 1997. *Faḍāʾil Miṣr al-Maḥrūsa*. Edited by ʿAlī Muḥammad ʿUmar. Cairo: Maktabat al-Khānjī.

Ibn Mammātī, Asʿad ibn al-Khaṭīr. 1943. *Qawānīn al-Dawāwīn*. Edited by ʿAzīz S. ʿAṭiya. Cairo: Royal Agricultural Society.

Ibn Riḍwān al-Maṣrī, Abū-l-Ḥasan ʿAlī. 1984. *Medieval Islamic Medicine: Ibn Riḍwān's Treatise "On the Prevention of Bodily Ills in Egypt."* Edited by Adil S. Gamal. Translated by Michael W. Dols. Berkeley: University of California Press.

Yāqūt ibn ʿAbd Allāh al-Ḥamawī. 1955.*Muʿjam al-buldān*. Vols. 1–5. Beirut: Dar Sader.

Modern Sources

Abd-Elhady, Eman A., Usama Gad, and Cassandre Hartenstein. 2017. "Five 'Penthemeros' Certificates from the Cairo Museum." *Bulletin of the American Society of Papyrologists* 54:59–82.

Alberti, Ben. 2014. "How Does Water Mean?" *Archaeological Dialogues* 21 (2):159–62.

al-Sharqāwī, ʿAbd al-Raḥmān. 1968. *Al-Arḍ*. 3rd ed. Cairo: Dār al-Kātib al-ʿArabī.

Alston, Richard. 2001. "Urban Population in Late Roman Egypt and the End of the Ancient World." In *Debating Roman Demography*, edited by Walter Scheidel, 161–204. Leiden: Brill.

Anagnostou-Canas, Barbara. 1994. "Les différends concernant l'eau dans l'Égypte romaine." In Menu, *Les problèmes institutionnels*, 15–28.

Anagnostou-Canas, Barbara. 2001. "Litiges en rapport avec l'eau dans l'Egypt ptolémaïque." *Atti del XXII congresso internazionale di papirologia, Firenze, 23–29 agosto, 1998*. A cura di I. Andorlini, G. Bastianini, M. Manfredi, e G. Menci. Vol. 1, 41–49. Florence: Istituto G. Vitelli.

Andorlini, Isabella. 2012. "Considerazioni sulla 'peste antonina' in Egitto." In Lo Cascio, *L'impatto della "peste antonina,"* 15–28.

Ast, Rodney. 2001. "Tziper, not Q. Iper." *ZPE* 137:229–30.

Audebeau, Charles, and Commission des domaines de l'État égyptien. 1897. *Carte de la Basse-Egypte et de la province du Fayoum*. Cairo: Commission des domaines d'Egypte.

Bagnall, Roger S. 1982. "The Population of Theadelphia in the Fourth Century." *Bulletin de la Société d'Archéologie Copte* 24:35–57..

Bagnall, Roger S. 1985. "Agricultural Productivity and Taxation in Fourth-Century Egypt." *TAPA* 115:290–308.

Bagnall, Roger S. 1993. *Egypt in Late Antiquity*. Princeton: Princeton University Press.

Bagnall, Roger S. 2000. "*P.Oxy.* 4527 and the Antonine Plague in Egypt: Death or Flight?" *JRA* 13:288–92.

Bagnall, Roger S. 2020. *Reading Papyri, Writing Ancient History*. 2nd ed. London: Routledge.

Bagnall, Roger S., and Dominic Rathbone, eds. 2004. *Egypt from Alexander to the Early Christians: An Archaeological and Historical Guide*. Los Angeles: Getty Museum.

Banaji, Jairus. 2007. *Agrarian Change in Late Antiquity: Gold, Labour, and Aristocratic Dominance*. Oxford: Oxford University Press.

Bannon, Cynthia Jordan. 2017a. "Ancient Roman Water Rights and Commons Theory." In *The Politics of Fresh Water: Access, Conflict and Identity*, edited by Catherine M. Ashcraft and Tamar Mayer, 13–27. London: Routledge.

Bannon, Cynthia Jordan. 2017b. "Rivers, Rights, and 'Romanization.'" In *Ancient Law, Ancient Society*, edited by Dennis P. Kehoe and Thomas A. J. McGinn, 66–83. Ann Arbor: University of Michigan Press.

Barnes, Jessica. 2010. "Governing Flow: The Politics of Water Distribution and Use in Egypt." PhD diss., Columbia University.

Barnes, Jessica. 2013. *Cultivating the Nile: The Everyday Politics of Water in Egypt*. Durham, NC: Duke University Press.

Barois, Eugene J. H. 1887. *Irrigation in Egypt*. Translated by A. M. Miller. Washington, DC: Government Printing Office.

Barois, Eugene J. H. 1904. *Les irrigations en Égypte*. Paris: Libarie Polytechnique.

Bassiouny, Hassan K. 2001. "Bioenvironmental and Meteorological Factors Related to the Persistence of Malaria in Fayoum Governorate: A Retrospective Study." *Eastern Mediterranean Health Journal* 7.6:895–906.

Beccar, Lily, Rutgerd Boelens, and Paul Hoogendam. 2002. "Water Rights and Collective Action in Community Irrigation." In *Water Rights and Empowerment*, edited by R. Boelens and P. Hoogendam, 1–19. Assen: Koninklijke Van Gorcum.

Bell, Harold Idris. 1917. "The Byzantine Servile State in Egypt." *Journal of Egyptian Archaeology* 4.2/3:86–106.

Bender, Matthew V. 2019. *Water Brings No Harm: Management Knowledge and the Struggle for the Waters of Kilimanjaro*. Athens: Ohio University Press.

Berkes, Lajos, and Brendan Haug. 2016. "Villages, Requisitions, and Tax Districts: Two Greek Lists from the Eighth-Century Fayyūm." *BASP* 53:189–222.

Blouin, Katherine. 2014. *Triangular Landscapes: Environment, Society, and the State in the Nile Delta under Roman Rule*. Oxford: Oxford University Press.

Boak, Arthur E. R. 1926. "Irrigation and Population in the Faiyûm, the Garden of Egypt." *Geographical Review* 16.3:353–64.

Boak, Arthur E. R. 1926. "Notes on Canal and Dike Work in Roman Egypt." *Aegyptus* 7.3–4:215–19.

Boak, Arthur E. R. 1955. "The Population of Roman and Byzantine Karanis." *Historia* 2.3:157–62.

Bonneau, Danielle. 1971. *Le fisc et le Nil: incidences des irrégularités de la crue du Nil sur la fiscalité foncière dans l'Égypte grecque et romaine.* Paris: Editions Cujas.

Bonneau, Danielle. 1979a. "Fiscalité et irrigation artificielle en Égypte." In *Points de vue sur la fiscalité antique*, edited by H. van Effenterre, 57–68. Publications de la Sorbonne, Etudes tome 14. Paris: Sorbonnne.

Bonneau, Danielle. 1979b. "Un règlement de l'usage de l'eau au Ve siècle de notre ère." In *Hommages à la mémoire de Serge Sauneron, 1927–1976.* Vol. 2, *Égypte post-pharaonique*, edited by Jean Vercoutter, 3–23. Cairo: Bibliothèque d'étude.

Bonneau, Danielle. 1981. "L'Égypte dans l'histoire de l'irrigation antique de l'époque hellénistique à l'époque arabe." In *Egitto e storia antica dall'ellenismo all'età araba: Bilanco di un confronto. Atti del colloquio internazionale, Bologna 31 agosto–2 settembre, 1987*, edited by Lucia Crisculo and Giovanni Geraci, 301–13. Bologna: CLUEB.

Bonneau, Danielle. 1981. "La haute administration des eaux en Égypte, d'après la documentation papyrologique aux époques grecque, romaine et byzantine." In *Proceedings of the Sixteenth International Congress of Papyrology, New-York, 23–31 July 1980*, edited by Roger S. Bagnall, Gerald M. Browne, Ann E. Hanson, and Ludwig Koenen, 321–28. Chico, CA: Scholars Press.

Bonneau, Danielle. 1982. "Le *Drymos* (δρυμός), marais du Fayoum, d'après la documentation papyrologique." In *L'Égyptologie en 1979: Axes prioritaires de recherches*, 181–90. Paris: Editions du Centre national de la recherche scientifique.

Bonneau, Danielle. 1983. "Loi et coutume en Égypte: Un exemple, les marais du Fayoum appelés *drymoi*." *JESHO* 26.1:1–13.

Bonneau, Danielle. 1989. "Le terre saline (ἁλμυρίς) en Égypte d'après les documents papyrologiques grecs." *Proceedings of the 19th International Congress of Papyrology, Cairo, 2–9 September 1989*, edited by A. H. S. El-Mosallamy, 61–75.

Bonneau, Danielle. 1993. *Le régime administratif de l'eau du Nil dans l'Égypte grecque, romaine et byzantine.* Leiden: Brill.

Borsch, Stuart J. 2000. "Nile Floods and the Irrigation System in Fifteenth-Century Egypt." *Mamluk Studies Review* 4.1:131–45.

Borsch, Stuart J. 2004. "Environment and Population: The Collapse of Large Irrigation Systems Reconsidered." *Comparative Studies in Society and History* 46.3:451–68.

Borsch, Stuart J. 2005. *The Black Death in England and Egypt: A Comparative Study.* Austin: University of Texas Press.

Borsch, Stuart J. 2014. "Plague Depopulation and Irrigation Decay in Medieval Egypt." *Medieval Globe* 1.1:125–56.

Borsch, Stuart J., and Tarek Sabraa. 2017a. "Qānūn al-Riyy: The Water Law of Egypt." *Sophia Journal of Asian, African, and Middle Eastern Studies* 35:87–124.

Borsch, Stuart J., and Tarek Sabraa. 2017b. "Refugees of the Black Death: Quantifying

Rural Migration for Plague and Other Environmental Disasters." *Annales de démographie historique* 2:63–93.

Bouriant, Urbain. 1895. *Description historique et topographique de l'Egypte*. Paris: Mission archéologique française du Caire and E. Leroux.

Bowes, Kim, and Adam Gutteridge. 2005. "Rethinking the Later Roman Landscape." *Journal of Roman Archaeology* 18:405–18.

Bowman, Alan K. 2005. "Egypt from Septimius Severus to Diocletian." In *The Cambridge Ancient History*. Vol. 12, *The Crisis of Empire, AD 193–337*, edited by Alan K. Bowman, Peter Garnsey, and Averil Cameron, 313–26. Cambridge: Cambridge University Press.

Bowman, Alan K. 2013. "Agricultural Production in Egypt." In Bowman and Wilson, 219–53.

Bowman, Alan K., and Eugene Rogan, eds. 1999. *Agriculture in Egypt from Pharaonic to Modern Times*. Proceedings of the British Academy 96. Oxford: Oxford University Press.

Bowman, Alan K., and Andrew N. Wilson, eds. 2013. *The Roman Agricultural Economy: Organization, Investment, and Production*. Oxford: Oxford University Press.

Brashear, William. 1979. "Before the Penthemeros: Government Funds for the Canals." *Bulletin of the American Society of Papyrologists* 16.1–2:25–29.

Braun, Bruce. 2002. *The Intemperate Rainforest: Nature, Culture, and Power on Canada's West Coast*. Minneapolis: University of Minnesota Press.

Bravard, Jean-Paul. 2019. "Water Resources and Irrigation in Two Oases of the Western Desert of Egypt: Kharga and Dakhla." In *The Great Oasis of Egypt: The Kharga and Dakhla Oases in Antiquity*, edited by Roger S. Bagnall and Gaëlle Tallet, 17–29. Oxford: Oxford University Press.

Briant, Pierre, ed. 2001. *Irrigation et drainage dans l'antiquité: Qanāts et canalisations souterraines en Iran, en Égypte et en Grèce*. Paris: Thotm Editions.

Brown, Robert Hanbury. 1887. "The Bahr Yusuf, Roughly Describing Its Present State and Uses." *Proceedings of the Royal Geographical Society and Monthly Record of Geography* 9.10:614–17.

Brown, Robert Hanbury. 1892. *The Fayûm and Lake Moeris*. London: Edward Stanford.

Bruun, Christer. 2003. "The Antonine Plague in Rome and Ostia." *Journal of Roman Archaeology* 16:426–34.

Bruun, Christer. 2007. "The Antonine Plague and the 'Third-Century Crisis.'" In *Crises and the Roman Empire: Proceedings of the Seventh Workshop of the International Network Impact of Empire (Nijmegen, June 20–24, 2006)*, edited by Olivier Hekster, Gerde de Kleijn, and Daniëlle Slootjes, 201–17. Leiden: Brill.

Bryen, Ari. 2013. *Violence in Roman Egypt: A Study in Legal Interpretation*. Philadelphia: University of Pennsylvania Press.

Bryen, Ari. 2021. "Citizenship and Its Alternatives: A View from the East." In *Imperial and Local Citizenship in the Long Second Century AD*, edited by Clifford Ando and Myles Lavan. Oxford: Oxford University Press, 41–68.

Buckley, Robert Burton. 1893. *Irrigation Works in India and Egypt*. London: E. and F. N. Spon.

Bülow-Jacobsen, Adam, and Sten Ebbesen. 1971. "Five Copenhagen Papyri." *Cahiers de l'institut du moyen-âge grec et latin* 6:181–221.

Butler, Alfred J. 1902. *The Arab conquest of Egypt and the last thirty years of the Roman dominion*. Oxford: Clarendon Press.

Butzer, Karl W. 1976. *Early Hydraulic Civilization in Egypt*. Chicago: Oriental Institute.

Cappers, René, Emily Cole, Daniel Jones, Simon Holdaway, and Willike Wendrich. 2013. "The Fayyûm Desert as an Agricultural Landscape." In *Das Fayyûm in Hellenismus und Kaiserzeit: Fallstudien zu multikulturellem Leben in der Antike*, edited by Carolin Arlt and Martin A. Stadler, 37–50. Wiesbaden: Harrassowitz.

Capponi, Livia. 2005. *Augustan Egypt: The Creation of a Roman Province*. Studies in Classics 13. London: Routledge.

Castiglione, Dario. 2016. "Social Learning and the Bonds of Self-Governing Communities." In *Institutional Diversity in Self-Governing Societies: The Bloomington School and Beyond*, edited by Filipo Sabetti and Dario Castiglione, 129–56. New York: Rowman and Littlefield.

Chauveau, Michel. 2001. "Les *qanāts* dans les ostraca de Manāwīr." In Briant, 137–42.

Chélu, Alfred Jacques. 1891. *De l'équateur à la Méditerranée: Le Nil, le Soudan, l'Égypte*. Paris: Librairie Chaix.

Christensen, Thorolf, Dorothy J. Thompson, and Katelijn Vandorpe, eds. 2017. *Land and Taxes in Ptolemaic Egypt: An Edition, Translation and Commentary for the Edfu Land Survey (P. Haun. IV 70)*. Cambridge: Cambridge University Press.

Clarke, Alison, and Rosalind Malcolm. 2016. "The Role of Property in Water Regulation: Locating Communal and Regulatory Property Rights on the Property Rights Spectrum." In *Regulatory Property Rights: The Transforming Notion of Property in Transnational Business Regulation*, edited by Christine Godt, 121–40. Leiden: Brill.

Clarysse, Willy. 1988. "A New Fragment of a Zenon Papyrus from Athens." In *Proceedings of the 18th International Congress of Papyrology*, edited by Basil G. Mandilaras, 77–81. Athens: Greek Papyrological Society.

Clarysse, Willy. 2005. "Tebtynis and Soknopaiou Nesos: The Papyrological Documentation through the Centuries." In Lippert and Schentuleit, 19–28.

Clarysse, Willy. 2007. "Toponymy of Fayyum Villages in the Ptolemaic Period." In *New Archeological and Papyrological Researches on the Fayyum*, edited by M. Capasso and P. Davoli, 67–81. Galatina: Congedo.

Clarysse, Willy. 2008. "Graeco-Roman Oxyrhyncha, a Village in the Arsinoite Nome." In Lippert and Schentuleit, 55–73.

Clarysse, Willy. 2016. "Village Names in Greco-Roman Egypt and in the Fayum." In *Décrire, imaginer, construire l'espace. Toponymie égyptienne de l'Antiquité au Moyen-Âge*, edited by Sylvain Dhennin and Somaglino Claire, 155–66. Cairo: L'institut français d'archéologie orientale.

Clarysse, Willy, and Dorothy Thompson. 2006. *Counting the People in Hellenistic Egypt*. Vol. 2. Historical Studies. Cambridge: Cambridge University Press.

Clarysse, Willy, and Bart Van Beek. 2002. "Philagris, Perkethaut and Hermopolis: Three Villages or One?" *ZPE* 140:195–200.

Claytor, Graham. 2013. "*Penthemeros* Certificates from the Granary C123, Karanis." *Bulletin of the American Society of Papyrologists* 50:49–75.

Claytor, Graham. 2017. "More *Penthemeros* Certificates from Karanis." *BASP* 54:83–102.

Conermann, Stephan, and Gül Şen, eds. 2017. *The Mamluk-Ottoman Transition: Continuity and Change in Egypt and Bilad al-Sham in the Sixteenth Century*. Göttingen: V und R Unipress.

Cook, James Ronald. 2011. "Landscapes of Irrigation in the Ptolemaic and Roman Fayum: Interdisciplinary Archaeological Survey and Excavation Near Kom Aushim (Ancient Karanis), Egypt." PhD diss., University of Michigan.

Cooper, John. 2014. *The Medieval Nile: Route, Navigation, and Landscape in Islamic Egypt*. Cairo: American University in Cairo Press.

Cooper, Richard Stefan. 1974. "Ibn Mammātī's Rules for the Ministries: Translation with Commentary of the Qawānīn al-Dawāwīn." PhD diss., University of California, Berkeley.

Corburn, Jason. 2005. *Street Science: Community Knowledge and Environmental Health Justice*. Cambridge, MA: MIT Press.

Crawford, Dorothy J. 1971. *Kerkeosiris: An Egyptian Village in the Ptolemaic Period*. Cambridge: Cambridge University Press.

Crawford, Stanley G. 1988. *Mayordomo: Chronicle of an Acequia in Northern New Mexico*. Albuquerque: University of New Mexico Press.

Dankoff, Robert, Nuran Tezcan, and Michael D. Sheridan. 2018. *Ottoman Explorations of the Nile: Evliya Çelebi's "Matchless Pearl These Reports of the Nile" Map and His Accounts of the Nile and the Horn of Africa in The Book of Travels*. Chicago: University of Chicago Press.

Daris, Sergio. 1984. "Toponimi della Meris di Polemone." *Aegyptus* 64:101–20.

Davoli, Paola. 1998. *L'archeologia urbana nel Fayyum di età ellenistica e romana*. Naples: G. Procaccini.

Davoli, Paola. 2010. "Archaeological Research in Roman Soknopaiou Nesos: Results and Perspectives." In *Tradition and Transformation: Egypt under Roman Rule; Proceedings of the International Conference, Hildesheim, Roemer- and Pelizaeus-Museum, 3–6 July 2008*, edited by Katja Lembke, Martina Minas-Nerpel, and Stefan Pfeiffer, 53–77. Leiden: Brill.

Davoli, Paola. 2012. "The Archaeology of the Fayum. In, *The Oxford Handbook of Roman Egypt*, edited by Christina Riggs, 152–70. Oxford: Oxford University Press.

Dee, Sylvia G., Kim. M. Cobb, Julien Emile-Geay, Toby R. Ault, R. Lawrence Edwards, Hai Cheng, Christopher D. Charles. 2020. "No Consistent ENSO Response to Volcanic Forcing over the Last Millennium." *Science* 367.6485:1477–81.

Derda, Tomasz. 2006. ΑΡΣΙΝΟΙΤΗΣ ΝΟΜΟΣ*: Administration of the Fayum under Roman Rule.* Supplement 7, *JJP*. Warsaw: Warsaw University Press.

Derr, Jennifer L. 2019. *The Lived Nile: Environment, Disease, and Material Colonial Economy in Egypt*. Stanford: Stanford University Press.

Di Giminiani, Piergiorgio. 2018. *Sentient Lands: Indigeneity, Property, and Political Imagination in Neoliberal Chile.* Tuscon: University of Arizona Press.

Duncan-Jones, Richard. 1996. "The Impact of the Antonine Plague." *Journal of Roman Archaeology* 9:108–36.

Duncan-Jones, Richard. 2018. "The Antonine Plague Revisited." *Arctos* 52:41–72.

Echols, Stuart, and Hala F. Nassar. 2006. "Canals and Lakes of Cairo: Influence of Traditional Water System on the Development of Urban Form." *Urban Design International* 11:203–12.

Edgeworth, Matt. 2011. *Fluid Pasts: The Archaeology of Flow*. London: Bristol Classical Press.

Egypt, Government of, Ministry of the Interior. Sanitary Department. 1897. *Report on the Epidemic of Cholera in Egypt during the Years 1895 & 1896*. Cairo: National Printing Office.

Egypt, Government of, with Danish International Development Assistance. 2008. "Environmental Action Plan for Fayoum Governate." [online] https://www.eeaa.gov.eg/Uploads/Reports/Files/20221128131502548.pdf

Egypt. Maṣlaḥat al-Misāḥah. 1914. *Atlas of Egypt Compiled at the Offices of the Survey Department.* Vol. 2, *Upper Egypt Comprising Maps of the Cultivated Area between Cairo and the Sudan Boundary*. Cairo: Government Press.

Elliott, Colin P. 2016. "The Antonine Plague, Climate Change and Local Violence in Roman Egypt." *Past and Present* 231:3–31.

Elliott, Colin P. 2020. "Disease Proxies and the Diagnosis of the Late Antonine Economy." In *Complexity Economics: Building a New Approach to Ancient Economic History*, edited by Koenraad Verboven, 297–326. Cham: Palgrave Macmillan.

Eman, Abd-Elhady, Usama Gad, and Cassandre Hartenstein. 2017. "Five *Penthemeros* Certificates from the Cairo Museum." *BASP* 54:59–82.

Erdkamp, Paul. 2019. "War, Food, Climate Change and the Decline of the Roman Empire." *Journal of Late Antiquity* 12.2:422–65.

Ermatinger, James W. 1997. "The Economic Death of Theadelphia during the Early Fourth Century A.D." *Münstersche Beiträge zu antiken Handelsgeschichte* 16:1–10.

Ertsen, Maurits Willem. 2010. "Structuring Properties of Irrigation Systems: Understanding Relations between Humans and Hydraulics through Modeling." *Water History* 2:165–83.

Eyre, Christopher J. 1994. "The Water Regime for Orchards and Plantations in Pharaonic Egypt." *JEA* 80:57–80.

Fisher, Christopher T., J. Brett Hill, and Gary T. Feinman. 2009. "Environmental Studies for Twenty-First-Century Conservation." In *The Archaeology of Environmental Change: Socionatural Legacies of Degradation and Resilience*, edited by Christopher T. Fisher, J. Brett Hill, and Gary T. Feinman, 1–12. Tuscon: University of Arizona Press.

Flemming, Rebecca. 2019. "Galen and the Plague." In *Galen's Treatise* Περὶ Ἀλυπίας *(De indolentia) in Context: A Tale of Resilience*, edited by Caroline Petit, 219–44. Leiden: Brill.

Gad, Abd-Allah, and Ahmed El Zeiny. 2016. "Spatial Analysis for Sustainable Development of El Fayoum and Wadi El Natrun Desert Depressions, Egypt, with the Aid of Remote Sensing and GIS." *Journal of Geography, Environment and Earth Science International* 8.3:1–18.

Garbrecht, Gunther. 1996. "Historical Water Storage for Irrigation in the Fayum Depression (Egypt)." *Irrigation and Drainage Systems* 10:47–76.

Garbrecht, Günther, and Horst Jaritz. 1990. *Untersuchung antiker Anlagen zur Wasserspeicherung im Fayum/Ägypten.* Leichtweiss-Institut für Wasserbau der Technischen Universität. Cairo: Braunschweig.

Gaubert, Christian, and Jean-Michel Mouton. 2014. *Hommes et villages du Fayyoum dans la documentation papyrologique arabe (Xe–XIe siècles).* Hautes études orientales. Moyen et Proche-Orient 6. Geneva: Librarie Droz.

Geertz, Clifford. 1983. *Local Knowledge: Further Essays in Interpretive Anthropology.* New York: Basic Books.

Gilliam, James Frank. 1961. "The Plague under Marcus Aurelius." *American Journal of Philology* 82.3:225–51.

Girard, Pierre Simon. 1801. "Mémoire sur les irrigations, l'agriculture, et le commerce de la province du Fayoum." In *Mémoires sur l'Égypte, publiés dans les années VII, VIII et IX: Tome troisieme (an X)*, 329–56. Paris: L'Imprimerie de P. Didot l'Ainé.

Girard, Pierre Simon. 1824. "Mémoire sur l'agriculture, l'industrie, et le commerce de l'Égypte." In *Description de l'Égypte.* 2nd ed. Vol. 17, *État Moderne*, 1–436. Paris: C. L. F. Panckoucke.

Glick, Thomas F. 1970. *Irrigation and Society in Medieval Valencia.* Cambridge, MA: Harvard University Press.

Goitein, Shelomo Dov. 1983. *A Mediterranean Society: The Jewish Communities of the Arab World as Portrayed in the Documents of the Cairo Geniza.* Vol. 4, *Daily Life.* Berkeley: University of California Press.

Goldberg, Harvey E. 1996. *Sephardi and Middle Eastern Jewries: History and Culture in the Modern Era.* Bloomington: Indiana University Press.

Gouda, Dalia M. 2016. *Social Capital and Local Water Management in Egypt.* Cairo: American University in Cairo Press.

Greenberg, James. 2003. "Plagued by Doubt: Reconsidering the Impact of a Mortality Crisis in the 2nd C. A.D." *Journal of Roman Archaeology* 16:413–25.

Grey, Cam. 2011. *Constructing Communities in the Late Roman Countryside.* Oxford: Oxford University Press.

Habermann, Wolfgang. 1998. "Aspekte des Bewässerungswesens im kaiserzeitlichen Ägypten I: Die 'Erklärungen für nicht-überflutetes Land' (Abrochia-Deklarationen)." In *Miscellanea oeconomica: Studien zur antiken Wirtschaftsgeschichte H. Winkel zum 65. Geburtstag*, edited by Kai Ruffing and Bernhard Tenger, 213–83. St. Katharinen: Scripta Mercatura Verlag.

Habermann, Wolfgang. 2000. *Zur Wasserversorgung einer Metropole im kaiserzeitlichen Ägypten: Neuedition von P. Lond. III 1177; Text, Übersetzung, Kommentar.* Munich: C. H. Beck.

Haldon, John, Hugh Elton, Sabine R. Huebner, Adam Izdebski, Lee Mordechai, and Timothy P. Newfield. 2018. "Plagues, Climate Change, and the End of an Empire: A Response to Kyle Harper's *The Fate of Rome*, 1–3." *History Compass* 16, issue 12. Online, no pagination. https://doi.org/10.1111/hic3.12506.

Haldon, John, Lee Mordechai, Timothy P. Newfield, Arlen F. Chase, Adam Izdebski, Piotr Guzowski, Inga Labuhn, and Neil Roberts. 2017. "History Meets Palaeoscience: Consilience and Collaboration in Studying Past Societal Responses to Environmental Change." *PNAS* 115.13:3210–18.

Harper, Kyle. 2016. "People, Plagues, and Prices in the Roman World: The Evidence from Egypt." *Journal of Economic History* 76.3:803–39.

Harper, Kyle. 2017. *The Fate of Rome: Climate, Disease, and the End of an Empire.* Princeton: Princeton University Press.

Harris, William. 2012. "The Great Pestilence and the Complexities of the Antonine-Severan Economy." In Lo Cascio, 331–38.

Harris, William. 2013. "What Kind of Environmental History for Antiquity?" In *The Ancient Mediterranean Environment between Science and History*, edited by William Harris, 1–10. Leiden: Brill.

Hassan, Fekri, and Geoff Tassie. 2006. "Modelling Environmental and Settlement Change in the Fayum." *Egyptian Archaeology* 29:37–40.

Haug, Brendan. 2021a. "Agriculture in Roman Egypt." In *A Companion to Ancient Agriculture*, edited by David Hollander and Timothy Howe, 517–38. London: Wiley Blackwell.

Haug, Brendan. 2021b. "Civilizing the Past: Egyptian Irrigation in the Colonial Imagination." *Journal of Egyptian History* 14.1:1–44.

Hengstl, Joachim. 1992. "Wasser in den Urkunden des griechisch-römischen Ägypten." In Menu, 213–28.

Hill, J. Brett, Christopher T. Fisher, and Gary M. Feinman. 2009. "The Socionatural Connection." In Fisher, Hill, and Feinman, 249–58.

Hobson, Deborah. 1984. "Agricultural Land and Economic Life in Soknopaiou Nesos." *BASP* 21:89–109.

Hobson, Deborah. 1985. "The Village of Heraklia in the Arsinoite Nome." *BASP* 22:101–15.

Hodgson, Stephen. 2006. *Modern Water Rights: Theory and Practice.* FAO Legislative Study 92. Rome: Food and Agricultural Organization of the United Nations.

Horden, Peregrine, and Nicholas Purcell. 2000. *The Corrupting Sea: A Study of Mediterranean History.* Oxford: Blackwell.

Huebner, Sabine. 2020. "Climate Change in the Breadbasket of the Roman Empire—Explaining the Decline of Fayum Villages in the Third Century CE." *Studies in Late Antiquity* 4.4:486–518.

Ibrahim, Fouad N., and Barbara Ibrahim. 2003. *Egypt: An Economic Geography*. London: I.B. Tauris.

Jacotin, Pierre. 1818. *Carte topographique de l'Égypte et de plusieurs parties des pays limitrophes*. Paris: Dépôt de la guerre.

Jomard, Edme-François. 1821. "Description des antiquités du nome Arsinoïte, aujourd'hui le Fayoum." In *Description de l'Égypte.* 2nd ed. Vol. 4, *Antiquités-Descriptions*, 437–516. Paris: C. L. F. Panckoucke.

Jomard, Edme-François. 1822. "Mémoire sur le lac de Moeris compare au lac du Fayoum." In *Description de l'Égypte.* 2nd ed. Vol. 6, *Antiquités-Mémoires*, 155–226. Paris: C. L. F. Panckoucke.

Jördens, Andrea. 2011. "Communicating with Tablets and Papyri." In *The Oxford Handbook of Social Relations in the Roman World*, edited by Michael Peachin, 227–47. Oxford: Oxford University Press.

Karpouzoglou, Timothy, and Sumit Vij. 2017. "Waterscape: A Perspective for Understanding the Contested Geography of Water." WIREs Water 4. Online, no pagination. https://doi.org/10.1002/wat2.1210

Kater, E. A., T. S. Abd-El-Aal, A. A. Abou-Arab, and A. A. Awadalla. 1997. "Mineralogical and Chemical Composition of the Main Soil Types in Fayoum Area." *Egyptian Journal of Soil Science* 37.2:153–74.

Keenan, James G. 2003. "Deserted Villages: From the Ancient to the Medieval Fayyum." *BASP* 40:119–39.

Keenan, James G. 2005. "Egyptian Villages in the 13th Century: Al-Nabulsi's Tarikh al-Fayyum." In Lefort, Morrison, and Sodini, 567–76.

Keenan, James G. 2007a. "Byzantine Egyptian Villages." In *Egypt in the Byzantine World*, edited by Roger S. Bagnall, 226–43. Cambridge: Cambridge University Press.

Keenan, James G. 2007b. "Fayyum Villages in *SPP* XX 229." In *Proceedings of the Twenty-Fourth International Congress of Papyrology, Helsinki, 1–7 August 2004*, edited by Jaakko Frösén, Tiina Purola, and Erja Salmenkivi, 487–96.

Kehoe, Dennis. 2008. "Economics and the Law of Water Rights in the Roman Empire." In *Vers une gestion intégrée de l'eau dans l'Empire romain: Actes dul Colloque*

International, Université Laval, Octobre 2006, edited by Ella Hermon, 243–52. Rome: "L'Erma" di Bretschneider.

Kehoe, Dennis. 2013. "Property Rights over Land and Economic Growth in the Roman Empire." In *Ownership and Exploitation of Land and Natural Resources in the Roman World*, edited by Paul Erdkamp, Koenraad Verboven, and Arjan Zuiderhoek, 88–106. Oxford: Oxford University Press.

Kelly, Benjamin. 2011. *Petitions, Litigation, and Social Control in Roman Egypt*. Oxford: Oxford University Press.

König, Ingeborg. 1996. "Die Oase al-Fayyûm nach 'Uthmân ibn Ibrahîm an-Nâbulsî': Ein Beitrag zur Wirtschaftsgeschichte Ägyptens um die Mitte des 13. Jahrhunderts n. Chr." *Zeitschrift für Geschichte der arabisch-islamischen Wissenschaften* 10:189–253.

Kopp, Peter. 2020. "Water Management in Philoteris." in Römer, *News from Texts and Archaeology*, 123–35.

Kotb, Tarek H. S., Tsugihiro Watanabe, Yoshihiko Ogino, and Kenneth K. Tanji. 2000. "Soil Salinity in the Nile Delta and Related Policy Issues in Egypt." *Agricultural Water Management* 43.2:239–61.

Kraemer, Bryan. 2010. "The Meandering Identity of a Fayum Canal: The Henet of Moeris / Dioryx Kleonos / Bahr Wardan / Abdul Wahbi." In *Proceedings of the Twenty-Fifth International Congress of Papyrology, Ann Arbor 2007*, edited byTraianos Gagos, Adam Hyatt, Arthur Verhoogt, and Terry Wilfong, 365–76. Ann Arbor: University of Michigan Press.

Krikē, Eirēnē. 2013. "Μορφοσυντακτικοί νεωτερισμοί στη γλώσσα των μη φιλολογικών παπύρων: οι αναφορικές δομές." PhD diss., Aristotle University of Thessaloniki.

Kumakura, Wakako. 2017. "The Early Ottoman Rural Government System and Its Development in View of Water Administration." In Conermann and Şen, 87–114.

Lane, Edward William. 2000. *Description of Egypt: Notes and Views in Egypt and Nubia, Made during the Years 1825, -26, -27, and -28*. Edited and with an introduction by Jason Thompson. Cairo: American University in Cairo Press.

Langellotti, Micaela. 2020. *Village Life in Roman Egypt: Tebtunis in the First Century AD*. Oxford: Oxford University Press.

Lefort, Jacques, Cécile Morrisson, and Jean-Pierre Sodini, eds. 2005. *Les villages dans l'empire byzantin, IVe–XVe siècle*. Paris: Lethielleux.

Leone, Anna. 2012. "Water Management in Late Antique North Africa: Agricultural Irrigation." *Water History* 4.1:119–33.

Lewis, Naphtali. 1983. *Life in Egypt under Roman Rule*. Oxford: Clarendon Press.

Linant de Bellefonds, Louis Maurice Adolphe. 1843. *Mémoire sur le lac Moeris présenté et lu à la Société égyptienne le 5 juillet 1842 par Linant de Bellefonds*. Alexandria: Imprimerie de A.E. Ozanne..

Linant de Bellefonds, Louis Maurice Adolphe. 1854. *Carte hydrographique de la moyenne Égypte: Où sont indiqués les travaux exécutés ou à exécuter d'après les or-*

dres de son altesse Méhémet-Ali, vice-roi d'Égypte; avec le projet de communication directe des deux mers au travers de l'Isthme. Paris: Dépot de la Guerre, Imprimerie de Kaeppelin.

Linant de Bellefonds, Louis Maurice Adolphe. 1855. *Carte hydrographique de la partie septentrionale de la haute Égypte où sont indiqués les travaux d'arrosage exécutés et à exécuter d'après les ordres de son altesse Méhémet-Ali, vice-roi d'Égypte*. Paris: Dépot de la Guerre, Imprimerie de Kaeppelin.

Linant de Bellefonds, Louis Maurice Adolphe. 1873. *Mémoires sur les principaux travaux d'utilité publique, exécutés en Egypte depuis la plus haute antiquité jusqu'à nos jours*. Paris: Arthus Bertrand.

Linton, Jamie. 2010. *What Is Water? The History of a Modern Abstraction*. Vancouver, BC: The UBC Press.

Linton, Jamie, and Jessica Budds. 2014. "The Hydrosocial Cycle: Defining and Mobilizing a Relational-Dialectical Approach to Water." *Geoforum* 57:170–80.

Lippert, Sandra, and Maren Schentuleit, eds. 2008. *Graeco-Roman Fayum—Texts and Archaeology: Proceedings of the Third International Fayum Symposion, Freudenstadt, May 29–June 1, 2007*. Wiesbaden: Harassowitz.

Litinas, Nikos. 2007. "Habent sua fata fragmenta: 'Donum Borgianum.'" In *Akten des 23. Internationalen Papyrologenkongresses, Wien, 22.–28. Juli 2001*, edited by Bernhard Palme, 399–405. Vienna: Verlag der Österreichischen Akademie der Wissenschaften.

Littman, Robert J., and M. L. Littman. 1973. "Galen and the Antonine Plague." *American Journal of Philology* 94.3:243–55.

Livi Bacci, Massimo. 2012. "Note demografiche ed epidemiologiche a margine della 'Peste antonina.'" In Lo Cascio, 339–45.

Lloris, Francisco Beltrán. 2006. "An Irrigation Decree from Roman Spain: 'The *Lex Rivi Hiberiensis*.'" *JRS* 96:147–97.

Lloris, Francisco Beltrán. 2014. "Irrigation Infrastructures in the Roman West: Typology, Financing, Management." In *Infrastruktur und Herrschaftsorganisation im Imperium Romanum: Herrschaftsstrukturen und Herrschaftspraxis III. Akten der Tagung in Zürich 19.-20.10.2012*, edited by Anne Kolb, 121–36. Berlin: De Gruyter.

Lo Cascio, Elio, ed. 2012. *L'impatto della "peste antonina."* Bari: Edipuglia.

Lucas, Alfred. 1902. *A Preliminary Investigation of the Soil and Water of the Fayum Province*. Cairo: National Printing Department.

Lucas, Paul. 1712. *Voyage du sieur Paul Lucas*. Vols. 1–2. Paris: Nicolas Simart.

Luyendijk, Joris. 2009. *People Like Us: Misrepresenting the Middle East*. Berkeley: Soft Skull Press.

Mabry, Jonathan B., ed. 1996a. *Canals and Communities: Small-Scale Irrigation Systems*. Tucson: University of Arizona Press.

Mabry, Jonathan B. 1996b. "The Ethnology of Local Irrigation." In Mabry, 3–30.

Mahaffy, John Pentland. 1895. *The Empire of the Ptolemies*. London: Macmillan.

Malleson, Claire J. 2019. *The Fayum Landscape: Ten Thousand Years of Archaeology, Texts, and Traditions in Egypt*. Cairo: American University in Cairo Press.

Malouta, Myrto, and Andrew Wilson. 2013. "Mechanical Irrigation: Water-Lifting Devices in the Archaeological Evidence and in the Egyptian Papyri." In Bowman and Wilson, 273–305.

Manning, Joseph G. 2002. "Irrigation et État en Égypte antique." *Annales: Histoire, Sciences Sociales* 57.3:611–23.

Manning, Joseph G. 2003. *Land and Power in Ptolemaic Egypt*. Cambridge: Cambridge University Press.

Manning, Joseph G. 2010. *The Last Pharaohs: Egypt under the Ptolemies, 305–30 BC*. Princeton: Princeton University Press.

Manning, Joseph G. 2018. *The Open Sea: The Economic Life of the Ancient Mediterranean World from the Iron Age to the Rise of Rome*. Princeton: Princeton University Press.

Manning, Joseph G., Francis Ludlow, Alexander R. Stine, William R. Boos, Michael Sigl, and Jennifer R. Marlon. 2017. "Volcanic Suppression of Nile Summer Flooding Triggers Revolt and Constrains Interstate Conflict in Ancient Egypt." *Nature Communications* 8, article no. 900.

Marks, Leszek, Alaa Salem, Fabian Welc, Jerzy Nitychoruk, Zhongyuan Chen, Maarten Blaauw, Abdelfattah Zalat, Aleksandra Majecka, Marcin Szymanek, Marta Chodyka, Anna Tołoczko-Pasek, Qianli Sun, Xiaoshuang Zhao, and Jun Jiang. 2018. "Holocene Lake Sediments from the Faiyum Oasis in Egypt: A Record of Environmental and Climate Change." *Boreas* 47.1:62–79.

Martin, Pierre-Dominique. 1826. "Description hydrographique des provinces de Beny-Soueyf et du Fayoum." In *Description de l'Égypte*. 2nd ed. Vol. 16, *État Moderne*, 1–72. Paris: C. L. F. Panckoucke.

Marx, Werner, Robin Haunschild, and Lutz Bornmann. 2018. "Climate and the Decline and Fall of the Western Roman Empire: A Bibliometric View on an Interdisciplinary Approach to Answer a Most Classic Historical Question." *Climate* 6.4:90.

Maspero, Jean, and Gaston Wiet. 1919. *Matériaux pour servir à la géographie de l'Égypte*. Cairo: L'institut français d'archéologie orientale.

Maury, Patrick. 1987. "Irrigation et agriculture en Égypte à la fin du XVIIIe siècle." In *L'homme et l'eau en Méditerranée et au Proche-Orient, IV*, edited by Pierre Louis, François Métral, and Jean Métral, 77–93. Paris: Ed. du Centre national de la recherche scientifique.

McCormick, Michael. 2013. "What Climate Science, Ausonius, Nile Floods, Rye, and Thatch Tell Us about the Environmental History of the Roman Empire." In Harris, 61–88.

McCormick, Michael, Ulf Büntgen, Mark A. Cane, Edward R. Cook, Kyle Harper, Peter Huybers, Thomas Litt, Sturt W. Manning, Paul Andrew Mayewski, Alexander F. M. More, Kurt Nicolussi, and Willy Tegel. 2012. "Climate Change during and after the Roman Empire: Reconstructing the Past from Scientific and Historical Evidence." *Journal of Interdisciplinary History* 43.2:169–220.

McHugh, Kevin E. 2000. "Inside, Outside, Upside Down, Backward, Forward, Round and Round: A Case for Ethnographic Studies in Migration." *Progress in Human Geography* 24.1:71–89.

McNeill, John. 2017. Review of *Liquid Power: Contested Hydro-Modernities in Twentieth-Century Spain*, by Erik Swyngedouw. *Environmental History* 22:160–61.

Mehanna, Sohair, Richard Huntington, and Rachad Antonius. 1984. *Irrigation and Society in Rural Egypt*. Cairo Papers in Social Science 7, Monograph 4. Cairo: American University in Cairo Press.

Mehari, Abraham, Frank van Steenbergen, and Bart Schultz. 2007. "Water Rights and Rules, and Management in Spate Irrigation Systems in Eritrea, Yemen and Pakistan." In Van Koppen, Giordano, and Butterworth, 114–29.

Meinzen-Dick, Ruth, and Leticia Nkonya. 2007. "Understanding Legal Pluralism in Water and Land Rights: Lessons from Africa and Asia." In Van Koppen, Giordano, and Butterworth, 12–27.

Menu, Bernadette, ed. 1992. *Les problèmes institutionnels de l'eau en Egypte ancienne et dans l'antiquité méditerranéenne*. Paris: IFAO.

Michel, Nicolas. 2005. "Travaux aux digues dans la vallée du Nil aux époques papyrologique et ottomane: Une comparaison." *Cahiers de Recherches de l'Institut de Papyrologie et d'Égyptologie de Lille* 25:253–76.

Micucci, Federica. 2019. "The Egerton Collection of Papyri and a Prefect's Disappointment over a Failing Irrigation System." Paper read at the *29th International Congress of Papyrology, Lecce, Italy, 28 July–3 August 2019*.

Mikhail, Alan. 2010. "An Irrigated Empire: The View from Ottoman Fayyum." *International Journal of Middle East Studies* 42:569–90.

Mikhail, Alan. 2011. *Nature and Empire in Ottoman Egypt: An Environmental History*. Cambridge: Cambridge University Press.

Mikhail, Alan. 2013. "Unleashing the Beast: Animals, Energy, and the Economy of Labor in Ottoman Egypt." *American Historical Review* 118.2:317–48.

Mikhail, Alan. 2017. *Under Osman's Tree: The Ottoman Empire, Egypt, and Environmental History*. Chicago: University of Chicago Press.

Monson, Andrew. 2008. "Communal Agriculture in the Ptolemaic and Roman Fayyum." In Lippert and Schentuleit, 173–86.

Monson, Andrew. 2012. *From the Ptolemies to the Romans: Political and Economic Change in Egypt*. Cambridge: Cambridge University Press.

Monson, Andrew. 2013. "Salinization and Agricultural Productivity in the Fayyûm." In *Das Fayyum in Hellenismus und Kaiserzeit*, edited by Caroline Arlt and Martin A. Stadler, 123–40. Wiesbaden: Harrassowitz.

Morelli, Federico. 2004. "I χωρία in α dell'Arsinoite." *ZPE* 149:125–37.

Moreno García, Juan Carlos. 2014. "Ancient States and Pharaonic Egypt: An Agenda for Future Research." *Journal of Egyptian History* 7.2:203–40.

Morini, Anna. 2007. "La nuova gestione indrica del Fayyum Tolemaico." In *La terra, gli uomini e gli dei: Il paesaggio agricolo nell'antico Egitto*, edited by S. Pernigotti and M. Zecchi, 111–20. Imola: La Mandragora.

Moustafa, F. B., F. M. Ibrahim, and S. A. Fikry. 1990. "Drainage Efficiency and Soil Productivity in Fayoum Governate." *Agricultural Research Review* 68.2:253–61.

Muhs, Brian P. 2005. *Taxes, Taxpayers, and Tax Receipts in Early Ptolemaic Thebes: Demotic and Greek Ostraca from the Oriental Institute Museum, Chicago*. Oriental Institute Publications 126. Chicago: Oriental Institute.

Newfield, Timothy P., and Inga Labuhn. 2017. "Realizing Consilience in Studies of Pre-Instrumental Climate and Pre-Laboratory Disease." *Journal of Interdisciplinary History* 48.2:211–40.

Omar, Hussein. 2013. "'The Crinkly-Haired People of the Black Earth': Examining Egyptian Identities in Ibn ʿAbd al-Ḥakam's *Futūḥ*." In *History and Identity in the Late Antique Near East*, edited by Philip Wood, 149–67. Oxford: Oxford University Press.

Omran, El-Sayed Ewis. 2019. "Land and Groundwater Resources in the Egypt's [*sic*] Nile Valley, Delta, and Its Fringes." In *Groundwater in the Nile Delta*. Handbook of Environmental Chemistry 73, edited by Abdelazim M. Negm, 45–105. Cham: Springer International.

Ostrom, Elinor. 1990. *Governing the Commons: The Evolution of Institutions for Collective Actions*. Cambridge: Cambridge University Press.

Ostrom, Elinor. 1998. "A Behavioral Approach to the Rational Choice Theory of Collective Action." Presidential Address, American Political Science Association, 1997. *American Political Science Review* 92.1:1–22.

Ostrom, Elinor. 1999. "Coping with Tragedies of the Commons." *Annual Review of Political Science* 2:493–535.

Ostrom, Elinor. 2000. "Collective Action and the Evolution of Social Norms." *Journal of Economic Perspectives* 14.3:137–58.

Ostrom, Elinor. 2002. "How Farmer Managed Irrigation Systems Build Social Capital to Outperform Agency Managed Systems That Rely Primarily on Physical Capital." In *Farmer Managed Irrigation Systems in the Changed Context*, edited by Prachanda Pradhan and Upendra Gautam, 15–65. Kathmandu: Farmer Managed Irrigation Systems Trust.

Ostrom, Elinor, and Roy Gardner. 1993. "Coping with Asymmetries in the Commons: Self-Governing Irrigation Systems Can Work." *Journal of Economic Perspectives* 7.4:93–112.

Owen, Charles A., and Charles C. Torrey. 1955. "Scandal in the Egyptian Treasury: A Portion of the Lumaʿ al-Qawānīn of ʿUthaman ibn Ibrāhīm al-Nābulusī." *Journal of Near Eastern Studies* 14: 70–96.

Pearl, Orsamus M. 1951. "ΕΞΑΘΥΡΟΣ: Irrigation Works and Canals in the Arsinoite Nome." *Aegyptus* 31:223–30.

Pearl, Orsamus. 1954. "ΑΡΓΑΙΤΙΣ and ΜΟΗΡΙΣ." *Aegyptus* 34:27–34.

Perreault, Tom. 2015. "Introduction to Part II." In *Negotiating Water Governance: Why the Politics of Scale Matter*, edited by Emma S. Norman, Christina Cook, and Alice Cohen, 117–23. London: Routledge.

Peust, Carsten. 2010. *Die Toponyme vorarabischen Ursprungs im modernen Ägypten: Ein Katalog*. Göttinger Miszellen, Beihefte Nr. 8. Göttingen: Alfa-Druck.

Pierre, Jacotin. 1818. *Carte topographique de l'Egypte et de plusieurs parties des pays limitrophes large pendant l'expédition de l'armée française*. Paris: Dépot de la guerre. Pococke, Richard. 1745. *A Description of the East and Some Other Countries*. London: W. Bowyer.

Pollard, Nigel. 1998. "The Chronology and Economic Condition of Late Roman Karanis: An Archaeological Reassessment." *JARCE* 35:147–62.

Price, David H. 1993. "The Evolution of Irrigation in Egypt's Fayoum Oasis: State, Village and Conveyance Loss." PhD diss., University of Florida, Gainesville.

Price, David H. 1995a. "The Cultural Effects of Conveyance Loss in Gravity-Fed Irrigation Systems." *Ethnology* 34.4:273–91.

Price, David H. 1995b. "Water Theft in Egypt's Fayoum Oasis: Emics, Etics, and the Illegal." In *Science, Materialism and the Study of Culture*, edited by Martin F. Murphy and Maxine Margolis, 96–110. Gainesville: University of Florida Press.

Quatremère de Quincy, Antoine-Chrysostome. 1811. *Mémoires géographiques et historiques sur l'Égypte*. Paris: F. Schoell.

Rapoport, Yossef. 2018. *Rural Economy and Tribal Society in Islamic Egypt: A Study of al-Nābulusī's "Villages of the Fayyum."* The Medieval Countryside 19. Leuven: Peeters.

Rapoport, Yossef, and Ido Shahar. 2012. "Irrigation in the Medieval Islamic Fayyum: Local Control in a Large-Scale Hydraulic System." *JESHO* 55:1–31.

Rathbone, Dominic. 1990. "Villages, Land and Population in Graeco-Roman Egypt." *Proceedings of the Cambridge Philological Society* 36:103–42.

Rathbone, Dominic. 1991. *Economic Rationalism and Rural Society in Third-Century AD Egypt: The Heroninos Archive and the Appianus Estate*. Cambridge: Cambridge University Press.

Rathbone, Dominic. 1997. "Surface Survey and the Settlement History of the Ancient Fayum." In *Archeologia e papiri nel Fayyum: Storia della ricerca, problemi e prospettivi; Atti del convegno internazionale, Siracusa, 24–25 Maggio 1996*, edited by Corrado Basile and Anna di Natale, 7–19. Syracuse: Instituto internazionle del papiro.

Rathbone, Dominic. 2001. "Mapping the South-West Fayyum: Sites and Texts." *Atti del XXII congresso internazionale di papirologia, Firenze, 23–29 agosto, 1998*. A cura di I. Andorlini, G. Bastianini, M. Manfredi, e G. Menci, 1109–17. Florence: Istituto G. Vitelli.

Rathbone, Dominic. 2005. "Economic Rationalism and the Heroninos Archive." *Topoi* 12–13:261–69.

Rathbone, Dominic. 2007a "Roman Egypt." In Scheidel, Morris, and Saller, 698–719.

Rathbone, Dominic. 2007b. "*Mēchanai* (Waterwheels) in the Roman Fayyum." In *New Archaeological and Papyrological Researches on the Fayyum: Proceedings of the International Meeting of Egyptology and Papyrology, Lecce, June 8th–10th, 2005*, edited by Mario Capasso and Paola Davoli, 251–62. Pap.Lup. XIV. Galatina: Congedo.

Rathbone, Dominic. 2008. "Villages and Patronage in Fourth-Century Egypt: The Case of *P.Ross.Georg.* 3.8." *Bulletin of the American Society of Papyrologists* 45:189–207.

Rathwell, Kaitlyn Joanne, Derek Armitage, and Fikret Berkes. 2015. "Bridging Knowledge Systems to Enhance Governance of the Environmental Commons: A Typology of Settings." *International Journal of the Commons* 9.2:851–80.

Raymond, Andre. 2002. *Cairo*. Cambridge, MA: Harvard University Press.

Rea, John. 1993. "P. Haun. III 58: Caranis in the Fifth Century." *ZPE* 99:89–95.

Rebentisch, Matthias. 2020. "Notes on the Maps 'Fayyum Blue Moeris.'" In Römer, *News from Texts and Archaeology*, 187–93, with the map inserts 'Fayyum Blue Moeris."

Rivera, José A. 1998. *Acequia Culture: Water, Land, and Community in the Southwest.* Albuquerque: University of New Mexico Press.

Römer, Cornelia. 2004. "Philoteris in the Themistou Meris: Report on the Archaeological Survey Carried Out as Part of the Fayum Survey Project." *ZPE* 147:281–305.

Römer, Cornelia. 2013. "Why Did the Villages in the Themistou Meris Die in the 4th Century AD? New Ideas about an Old Problem." In *Das Fayyûm in Hellenismus und Kaiserzeit: Fallstudien zu multikulturellem Leben in der Antike*, edited by Carolyn Arlt and Martin Stadler, 169–79. Wiesbaden: Harrassowitz.

Römer, Cornelia. 2017. "The Nile in the Fayum: Strategies of Dominating and Using the Water Resources of the River in the Oasis in the Middle Kingdom and the Graeco-Roman Period." In Harco Willems and Jan-Michael Dahms, 171–92.

Römer, Cornelia. 2019. *The Fayoum Survey Project: The Themistou Meris.* Vol. A, *The Archaeological and Papyrological Survey.* Collectanea Hellenistica—KVAB 8. Leuven: Peeters.

Römer, Cornelia E., ed. 2020. *News from Texts and Archaeology: Acts of the 7th International Fayoum Symposium, 29 October–3 November 2018 in Cairo and the Fayoum*. Wiesbaden: Harrassowitz.

Ross, Justin C. 1893. "Irrigation and Agriculture in Egypt." *Scottish Geographical Magazine* 9.4:169–93.

Rowlandson, Jane. 1996. *Landowners and Tenants in Roman Egypt: The Social Relations of Agriculture in the Oxyrhynchite Nome*. Oxford: Clarendon Press.

Rowlandson, Jane. 2005. "The Organization of Public Land in Roman Egypt." *Cahiers de Recherches de l'Institut de Papyrologie et d'Égyptologie de Lille* 25:173–96.

Said, Rushdi. 1993. *The River Nile: Geology, Hydrology and Utilization*. Oxford: Pergamon Press.

Sallares, Robert. 2002. *Malaria and Rome: A History of Malaria in Ancient Italy*. Oxford: Oxford University Press, 2002.

Salmon, Georges. 1900. "Le nom de lieu Babîdj dans la géographie égyptienne." *BIFAO* 1:235–39.

Salmon, Georges. 1901. "Répertoire géographique de la province du Fayyoûm d'après le *Kitâb Târîkh al-Fayyoûm* d'Al-Nâboulsî." *BIFAO* 1:29–77.

Sampsell, Bonnie M. 2014. *The Geology of Egypt: A Traveler's Handbook*. Cairo: American University in Cairo Press.

Sato, Tsugitaka. 1972. "Irrigation in Rural Egypt from the Twelfth to the Fourteenth Centuries." *Orient: Report of the Society for Near Eastern Studies in Japan* 8:81–92.

Sato, Tsugitaka. 1997. *State and Rural Society in Medieval Islam: Sultans, Muqta's and Fellahun*. Leiden: Brill.

Savary, Claude Étienne. 1786. *Lettres sur l'Égypte, où l'on ofre le parallèle des mœurs anciennes & modernes de ses habitans, où l'on décrit l'état, le commerce, l'agriculture, le gouvernement du pays, & la descente de S. Louis à Damiette*. Amsterdam: Chez les libraires associeés.

Scheidel, Walter. 2001. *Death on the Nile: Disease and the Demography of Roman Egypt*. Leiden: Brill.

Scheidel, Walter. 2002. "A Model of Demographic and Economic Change in Roman Egypt after the Antonine Plague." *Journal of Roman Archaeology* 15:97–114.

Scheidel, Walter. 2007. "Demography." In Scheidel, Morris, and Saller, 38–86.

Scheidel, Walter. 2012. "Roman Wellbeing and the Economic Consequences of the Antonine Plague." In Lo Cascio, 265–95.

Scheidel, Walter, Ian Morris, and Richard P. Saller, eds. 2007. *The Cambridge Economic History of the Greco-Roman World*. Cambridge: Cambridge University Press.

Schubert, Paul. 2007. *Philadelphie: Un village égyptien en mutation entre le II*e *et le III*e *siècle ap. J.-C.* Basel: Schwabe.

Scott, Anthony. 2008. *The Evolution of Resource Property Rights*. Oxford: Oxford University Press.

Scott, James C. 1998. *Seeing Like a State: How Certain Schemes to Improve the Human Condition Have Failed*. New Haven: Yale University Press.

Scott-Moncrieff, Colin. 1888. *Irrigation Report for the Year 1887*. Cairo: National Printing Office.

Scott-Moncrieff, Colin. 1892. Prefatory note to *The Fayûm and Lake Moeris*, by R. H. Brown.

Sessa, Kristina. 2019. "The New Environmental Fall of Rome: A Methodological Consideration." *Journal of Late Antiquity* 12.1:211–55.

Shafei, Ali. 1939. "Fayoum Irrigation as Described by Nabulsi in 1245 A.D. with a Description of the Present System of Irrigation and a Note on Lake Moeris." *Bulletin de la Société Royale de Geographie d'Egypte* 20:283–327.

Sharp, Michael. 1999. "The Village of Theadelphia in the Fayyum: Land and Population in the Second Century." In Bowman and Rogan, 159–92.

Shaw, Brent D. 1982. "Lamasba: An Ancient Irrigation Community." *Antiquités Africaines* 18:61–102.

Shaw, Brent D. 1984. "Water and Society in the Ancient Maghrib: Technology, Property and Development." *Antiquités Africaines* 20:121–73.

Shendi, Mahmoud M., Mahmoud A. Abdelfattah, and Ahmed Harbi. 2013. "Spatial Monitoring of Soil Salinity and Prospective Conservation Study for Sinnuris District Soils, Fayoum, Egypt." In *Developments in Soil Salinity Assessment and Reclamation*, edited by Shabbir A. Shahid, Mahmoud A. Abdelfattah, and Faisal K. Taha, 199–217. Dordrecht: Springer.

Sidebottom, Harry. 2020. "Fire from Innermost Depths: How Ancients Experienced Plague." *Times Literary Supplement*, issue 6116, 19 June 2020, 13.

Sijpesteijn, Petra. 2014. *Shaping a Muslim State: The World of a Mid-Eighth-Century Egyptian Official*. Oxford: Oxford University Press.

Sijpesteijn, Pieter J. 1964. *Penthemeros-Certificates in Graeco-Roman Egypt*. Papyrologica Lugduno-Batava 12. Leiden: Brill.

Silvestre de Sacy, Antoine Isaac. 1826. *Chrestomathie arabe*. Vol. 2. Paris: Imprimerie Royale.

St. John, James Augustus. 1845. *Egypt and Nubia*. London: Chapman and Hall.

Strang, Veronica. 2004. *The Meaning of Water*. London: Bloomsbury.

Subias, Eva, I. Fiz, and R. Cuesta. 2013. "The Middle Nile Valley: Elements in an Approach to the Structuring of the Landscape from the Greco-Roman Era to the Nineteenth Century." *Quaternary International* 312:27–44.

Swyngedouw, Erik. 2004. *Social Power and the Urbanization of Water: Flows of Power.* Oxford Geographical and Environmental Studies Series. Oxford: Oxford University Press.

Swyngedouw, Erik. 2015. *Liquid Power: Contested Hydro-Modernities in Twentieth-Century Spain*. Cambridge, MA: MIT Press.

Thommen, Lukas. 2012. *An Environmental History of Ancient Greece and Rome*. Cambridge: Cambridge University Press.

Thompson, Dorothy J. 1996. "Fayum Irrigation in the Third Century BC: Evidence from the Greek Papyri." In *16th Congress on Irrigation and Drainage, Cairo 1996*. Vol. 1G, History Seminar, 43–59.

Thompson, Dorothy J. 1999. "Irrigation and Drainage in the Early Ptolemaic Fayyum." In Bowman and Rogan, 107–22.

Thompson, Dorothy J. 2014. "The Fayum in the *Description de l'Égypte*." In *Von der Pharaonenzeit bis zur Spätantike: Kulturelle Vielfalt im Fayum; Akten der 5. Internationalen Fayum-Konferenz, 29. Mai bis 1. Juni 2013, Leipzig*, edited by Nadine Quenouille, 163–71. Wiesbaden: Harrasowitz.

Thonemann, Peter. 2011. *The Maeander Valley: A Historical Geography from Antiquity to Byzantium*. Cambridge: Cambridge University Press.

Tillier, Mathieu, and Naïm Vanthieghem. 2020. "Un voyageur témoin de manœuvres militaires dans le Fayoum (IIIe–IVe/IXe–Xe siècle)." *Chronique d'Égypte* 95:347–67.

Timm, Stefan. 1984–92. *Das christlich-koptische Ägypten in arabischer Zeit.* Vols. 1–6. Wiesbaden: Ludwig Reichert.

Toussoun, Omar. 1926. *Mémoire sur l'histoire du Nil.* Mémoires de l'Institut de l'Égypte 8–10. Cairo: L'Institut français d'archéologie orientale.

Trawick, Paul. 2001a. "The Moral Economy of Water: Equity and Antiquity in the Andean Commons." *American Anthropologist*, n.s., 103.2:361–79.

Trawick, Paul. 2001b. "Successfully Governing the Commons: Principles of Social Organization in an Andean Irrigation System." *Human Ecology* 29.1:1–25.

Tvedt, Terje. 2010. "'Water Systems,' Environmental History and the Destruction of Nature." *Environment and History* 16:143–66.

Tvedt, Terje, and Richard Coopey. 2011. "A 'Water Systems' Perspective on History." In *A History of Water.* Series II. Voume 2, *Rivers and Society: From Early Civilizations to Modern Times*, edited by Terje Tvedt and Richard Coopey, 3–26. London: I.B. Tauris.

Uytterhoeven, Inge, and Ingrid Blom-Böer. 2002. "New Light on the Egyptian Labyrinth: Evidence from a Survey at Hawara." *Journal of Egyptian Archaeology* 88.1:1111–20.

Van Beek, Bart. 2018. *The Archive of the Architektones Kleon and Theodoros (P. Petrie Kleon).* Collectanea Hellenistica—KVAB 7. Leuven: Peeters.

Vandorpe, Katelijn. 2004. "The Henet of Moeris and the Ancient Administrative Division of the Fayum in Two Parts." *Archiv für Papyrusforschung* 50:61–78.

Vandorpe, Katelijn, ed. 2019. *A Companion to Greco-Roman and Late Antique Egypt.* Hoboken, NJ: John Wiley and Sons.

Van Koppen, Barbara, Mark Giordano, and John. Butterworth. 2008. *Community-Based Water Law and Water Resource Management Reform in Developing Countries*. Wallingford: CAB International.

Van Minnen, Peter. 1995. "Deserted Villages: Two Late Antique Town Sites in Egypt." *BASP* 32:41–56.

Van Minnen, Peter. 2001. "*P.Oxy.* LXVI 4527 and the Antonine Plague in Egypt." *ZPE* 135:175–77.

Van Minnen, Peter. 2006. "The Changing World of the Cities of Later Roman Egypt." In *Die Stadt in der Spätantike, Niedergang oder Wandel? Akten des internationalen Kolloquiums in München am 30. und 31. Mai 2003*, edited by Christian Witschel and Jens-Uwe Krause, 153–80. Stuttgart: Franz Steiner.

Van Minnen, Peter. 2019. "Economic Growth and the Exploitation of Land." In K. Vandorpe, 251–65.

Van Steenbergen, Frank. 1997. "Understanding the Sociology of Spate Irrigation: Cases from Balochistan." *Journal of Arid Environments* 35.2:349–65.

Varisco, Daniel Martin. 1983. "*Sayl* and *Ghayl*: The Ecology of Water Allocation in Yemen." *Human Ecology* 11.4:365–83.

Verhoogt, Arthur M. F. W. 1998. *Menches, Komogrammateus of Kerkeosiris: The Doings and Dealings of a Village Scribe in the Late Ptolemaic Period (120–110 B.C.)*. Papyrologica Lugduno-Batava 29. Leiden: Brill.

Volney, Constantin François de Chassebœuf. 1787. *Voyage en Syrie et en Egypte, pendant les annees 1783, 1784, et 1785*. 2nd ed. Vol. 1. Paris: Royal Library.

Wansleben, Johann Michael. 1678. *The Present State of Egypt; or, A New Relation of a Late Voyage into That Kingdom: Performed in the Years 1672, and 1673. By F. Vansleb.* Englished by M. D. B. D. London: J. Starkey.

Ward, John. 1900. *Pyramids and Progress: Sketches from Egypt*. London: Eyre and Spottiswoode.

Ward, John. 1902. "The Oasis of Roses." *Wide World Magazine* 9 (April–September): 89–93.

Westermann, W. L. 1917. "Aelius Gallus and the Reorganization of the Irrigation System of Egypt under Augustus." *Classical Philology* 12.3:237–43.

Westermann, W. L. 1919. "The Development of the Irrigation System of Egypt." *CP* 14.2:158–64.

Westermann, W. L. 1927. "Egyptian Agricultural Labor under Ptolemy Philadelphus." *Agricultural History* 1.2:34–47.

White, Richard. 1995. *The Organic Machine: The Remaking of the Columbia River*. New York: Hill and Wang.

Whitehouse, Frederick Cope. 1887. "Le Bahr Youssouf d'après les traditions musulmanes." *Bulletin de l'Institut Égyptien*, 2nd ser., no. 7:102–15.

Wickham, Chris. 2019. "The Power of Property: Land Tenure in Fāṭimid Egypt." *Journal of the Economic and Social History of the Orient* 62.1:67–107.

Willcocks, William W. 1899. *Egyptian Irrigation*. 2nd ed. London: E. and F. N. Spon.

Willcocks, William, and James Ireland Craig. 1913. *Egyptian Irrigation*. 3rd ed. London: E. and F. N. Spon.

Willcox, Michael. 2010. "Marketing Conquest and the Vanishing Indian: An Indigenous Response to Jared Diamond's Archaeology of the American Southwest." In *Questioning Collapse: Human Resilience, Ecological Vulnerability, and the Aftermath of Empire*, edited by Patricia A. McAnany and Norman Yoffee, 113–41. Cambridge: Cambridge University Press:

Willems, Harco. 2013. "Nomarchs and Local Potentates: The Provincial Administration in the Middle Kingdom." In *Ancient Egyptian Administration*. Handbuch der Orientalistik, sec. 1, Ancient Near East 104, edited by Juan Carlos Moreno García, 341–92. Leiden: Brill.

Willems, Harco, Hanne Creylman, Véronique de Laet, and Gert Verstraeten. 2017. "The Analysis of Historical Maps as an Avenue to the Interpretation of Pre-Industrial Irrigation Practices in Egypt." In Willems and Dahms, 255–343.

Willems, Harco, and Jan-Michael Dahms, eds. 2017. *The Nile: Natural and Cultural Landscape in Egypt; Proceedings of the International Symposium held at the Johannes Gutenberg-Universitat Mainz, 22–23 February 2013*. Bielefeld: Transcript Verlag.

Winkler, Andreas. 2018. "Mouchis and Its Crocodiles: Topography, Toponymy, and Theonymy." *Bulletin of the American Society of Papyrologists* 55:229–51.

Wolters, W., Nadi Selim Ghobrial, and M.G. Bos. 1987. "Division of Irrigation Water in the Fayoum, Egypt." *Irrigation and Drainage Systems* 1,159–72.

Wolters, W., Nadi Selim Ghobrial, H. M. Van Leeuwen, and M. G. Bos. 1989. "Managing the Water Balance of the Fayoum Depression, Egypt." *Irrigation and Drainage Systems* 3:103–23.

Wuttmann, Michel. 2001. "Les *qanāts* de ʿAyn-Manāwīr (oasis de Kharga, Égypte)." In Briant, 109–36.

Xoplaki, Elena, Jürg Luterbacher, Sebastian Wagner, Eduardo Zorita, Dominik Fleitmann, Johannes Preiser-Kapeller, Abigail M. Sargent, Sam White, Andrea Toreti, John F. Haldon, Lee Mordechai, Deniz Bozkurt, Sena Akçer-Ön, and Adam Izdebski. 2018. "Modelling Climate and Societal Resilience in the Eastern Mediterranean in the Last Millennium." *Human Ecology* 46:363–79.

Zelener, Yan. 2012. "Genetic Evidence, Density Dependence and Epidemiological Models of the Antonine Plague." In Lo Cascio, 167–77.

Index Toporum

Note: Alternative toponyms referring to the same place are given in parentheses after the main entry. These toponyms also appear as entries with cross references to the main entries. An asterix indicates that the toponym referred to can be found in the Index Verborum.

Index Verborum

Printed and bound by CPI Group (UK) Ltd, Croydon, CR0 4YY
16/09/2026
14963490-0001